IET TELECOMMUNICATIONS SERIES 103

Terahertz Dielectric Resonator Antennas for High Speed Communication and Sensing

Other volumes in this series:

Volume 9 **Phase Noise in Signal Sources** W.P. Robins
Volume 12 **Spread Spectrum in Communications** R. Skaug and J.F. Hjelmstad
Volume 13 **Advanced Signal Processing** D.J. Creasey (Editor)
Volume 19 **Telecommunications Traffic, Tariffs and Costs** R.E. Farr
Volume 20 **An Introduction to Satellite Communications** D.I. Dalgleish
Volume 26 **Common-Channel Signalling** R.J. Manterfield
Volume 28 **Very Small Aperture Terminals (VSATs)** J.L. Everett (Editor)
Volume 29 **ATM: The broadband telecommunications solution** L.G. Cuthbert and J.C. Sapanel
Volume 31 **Data Communications and Networks, 3rd Edition** R.L. Brewster (Editor)
Volume 32 **Analogue Optical Fibre Communications** B. Wilson, Z. Ghassemlooy and I.Z. Darwazeh (Editors)
Volume 33 **Modern Personal Radio Systems** R.C.V. Macario (Editor)
Volume 34 **Digital Broadcasting** P. Dambacher
Volume 35 **Principles of Performance Engineering for Telecommunication and Information Systems** M. Ghanbari, C.J. Hughes, M.C. Sinclair and J.P. Eade
Volume 36 **Telecommunication Networks, 2nd Edition** J.E. Flood (Editor)
Volume 37 **Optical Communication Receiver Design** S.B. Alexander
Volume 38 **Satellite Communication Systems, 3rd Edition** B.G. Evans (Editor)
Volume 40 **Spread Spectrum in Mobile Communication** O. Berg, T. Berg, J.F. Hjelmstad, S. Haavik and R. Skaug
Volume 41 **World Telecommunications Economics** J.J. Wheatley
Volume 43 **Telecommunications Signalling** R.J. Manterfield
Volume 44 **Digital Signal Filtering, Analysis and Restoration** J. Jan
Volume 45 **Radio Spectrum Management, 2nd Edition** D.J. Withers
Volume 46 **Intelligent Networks: Principles and applications** J.R. Anderson
Volume 47 **Local Access Network Technologies** P. France
Volume 48 **Telecommunications Quality of Service Management** A.P. Oodan (Editor)
Volume 49 **Standard Codecs: Image compression to advanced video coding** M. Ghanbari
Volume 50 **Telecommunications Regulation** J. Buckley
Volume 51 **Security for Mobility** C. Mitchell (Editor)
Volume 52 **Understanding Telecommunications Networks** A. Valdar
Volume 53 **Video Compression Systems: From first principles to concatenated codecs** A. Bock
Volume 54 **Standard Codecs: Image compression to advanced video coding, 3rd Edition** M. Ghanbari
Volume 59 **Dynamic Ad Hoc Networks** H. Rashvand and H. Chao (Editors)
Volume 60 **Understanding Telecommunications Business** A Valdar and I Morfett
Volume 65 **Advances in Body-Centric Wireless Communication: Applications and state-of-the- art** Q.H. Abbasi, M.U. Rehman, K. Qaraqe and A. Alomainy (Editors)
Volume 67 **Managing the Internet of Things: Architectures, theories and applications** J. Huang and K. Hua (Editors)
Volume 68 **Advanced Relay Technologies in Next Generation Wireless Communications** I. Krikidis and G. Zheng
Volume 69 **5G Wireless Technologies** A. Alexiou (Editor)
Volume 70 **Cloud and Fog Computing in 5G Mobile Networks** E. Markakis, G. Mastorakis, C.X. Mavromoustakis and E. Pallis (Editors)
Volume 71 **Understanding Telecommunications Networks, 2nd Edition** A. Valdar
Volume 72 **Introduction to Digital Wireless Communications** Hong-Chuan Yang
Volume 73 **Network as a Service for Next Generation Internet** Q. Duan and S. Wang (Editors)
Volume 74 **Access, Fronthaul and Backhaul Networks for 5G & Beyond** M.A. Imran, S.A.R. Zaidi and M.Z. Shakir (Editors)
Volume 76 **Trusted Communications with Physical Layer Security for 5G and Beyond** T.Q. Duong, X. Zhou and H.V. Poor (Editors)

Volume 77 **Network Design, Modelling and Performance Evaluation** Q. Vien
Volume 78 **Principles and Applications of Free Space Optical Communications** A.K. Majumdar, Z. Ghassemlooy, A.A.B. Raj (Editors)
Volume 79 **Satellite Communications in the 5G Era** S.K. Sharma, S. Chatzinotas and D. Arapoglou
Volume 80 **Transceiver and System Design for Digital Communications, 5th Edition** Scott R. Bullock
Volume 81 **Applications of Machine Learning in Wireless Communications** R. He and Z. Ding (Editors)
Volume 83 **Microstrip and Printed Antenna Design, 3rd Edition** R. Bancroft
Volume 84 **Low Electromagnetic Emission Wireless Network Technologies: 5G and beyond** M.A. Imran, F. Héliot and Y.A. Sambo (Editors)
Volume 86 **Advances in Communications Satellite Systems Proceedings of the 36th International Communications Satellite Systems Conference (ICSSC-2018)** I. Otung, T. Butash and P. Garland (Editors)
Volume 87 **Real Time Convex Optimisation for 5G Networks and Beyond** T.Q. Duong, L.D. Nguyen and H.D. Tuan
Volume 89 **Information and Communication Technologies for Humanitarian Services** M.N. Islam (Editor)
Volume 90 **Communication Technologies for Networked Smart Cities** S.K. Sharma, N. Jayakody, S. Chatzinotas and A. Anpalagan (Editors)
Volume 91 **Green Communications for Energy-Efficient Wireless Systems and Networks** Himal Asanga Suraweera, Jing Yang, Alessio Zappone and John S. Thompson (Editors)
Volume 92 **Flexible and Cognitive Radio Access Technologies for 5G and Beyond** H. Arslan and E. Başar (Editors)
Volume 93 **Antennas and Propagation for 5G and Beyond** Q. Abbasi, S.F. Jilani, A. Alomainy and M.A. Imran (Editors)
Volume 94 **Intelligent Wireless Communications** G. Mastorakis, C.X. Mavromoustakis, J.M. Batalla and E. Pallis (Editors)
Volume 95 **ISDN Applications in Education and Training** R. Mason and P.D. Bacsich
Volume 96 **Edge Caching for Mobile Networks** H. Vincent Poor and Wei Chen (Editors)
Volume 98 **Artificial Intelligence Applied to Satellite-based Remote Sensing Data for Earth Observation** M.P. Del Rosso, A. Sebastianelli and S.L. Ullo (Editors)
Volume 99 **Metrology for 5G and Emerging Wireless Technologies** T.H. Loh (Editor)

Terahertz Dielectric Resonator Antennas for High Speed Communication and Sensing

From theory to design and implementation

Rajveer S. Yaduvanshi

The Institution of Engineering and Technology

Published by The Institution of Engineering and Technology, London, United Kingdom

The Institution of Engineering and Technology is registered as a Charity in England & Wales (no. 211014) and Scotland (no. SC038698).

First published 2021

The Institution of Engineering and Technology
Michael Faraday House
Six Hills Way, Stevenage
Herts, SG1 2AY, United Kingdom

www.theiet.org

British Library Cataloguing in Publication Data
A catalogue record for this product is available from the British Library

ISBN 978-1-83953-355-6 (hardback)
ISBN 978-1-83953-356-3 (PDF)

Typeset in India by MPS Limited

Contents

About the author

Rajveer S. Yaduvanshi is a professor in the Department of Electronics & Communication Engineering at Netaji Subhas University of Technology, Sec-3, Dwarka, New Delhi, India. He has more than 35 years of experience in the fields of teaching and research in radio frequency and microwave engineering. He has written 3 books on the topic of antennas and published more than 107 research papers. He has organized the international conference ICPR-2017. His research interests involve quantum antennas, terahertz dielectric resonator antennas, nano dielectric resonator antenna design, optical antennas, radio frequency sensors, microwave devices, millimeters wave (mm-wave) antennas, vehicular antennas and retinal quantum antennas. He has written three books on the topic of antennas and organized the international conference ICPR-2017 in Delhi. He holds a PhD from Delhi University in Electronics and Communication Engineering. He has served as Dean R&D at East Campus. He was selected as Director EQDC (Electronics & Quality Development Centre) by Govt. of Gujarat.

Prof. Rajveer Yaduvanshi received Academic Excellence awards and Excellence in Management awards by Delhi Government, India. He has supervised seven PhD theses and thirty-three PG theses. Seven more PhD scholars are currently registered under him. His current research interests are nano DRA, sensors, retinal photoreceptors and photonics.

Preface

- This book covers microwave and terahertz dielectric resonator antenna (DRA) technologies and their applications to high-speed communications and sensing. Microwave, terahertz and optical DRAs use ceramic/silicon/TiO_2/teflon and graphene materials. These DRAs are efficient and used for high-speed communication due to their wide bandwidth and less lossy property. They can perfectly work at microwave, terahertz and optical spectrum regime. These are compact in size that makes them very well suited for advanced applications in sensing, scanning, imaging and biomedical applications. They have futuristic capability to design optical DRAs in visible frequencies, so that DRAs can be used as implant antennas for retinal photoreceptors. The authors introduced new DRA geometries such as conical optical DRAs, cylindrical optical DRAs and rectangular optical DRAs, which can provide ultrafast data-transfer rate due to large bandwidth and multimode operations.
- Optical DRAs are useful for devices such as sensors (LiDAR) for scanning and imaging.
- Spherical optical DRAs have features of super directivity that can be used in quantum radars.
- Cylindrical optical DRAs with photodiodes can be used for wireless energy harvesting, i.e. rectennas.
- Conical optical DRAs can be used at basic R, G, B color wavelengths.
- Optical DRAs of rectangular shape, conical shape, cylindrical shape and spherical shape and structures if embedded with photodiodes can be used for eye retinal prosthesis applications.

The authors explained thoroughly the differences among microwave DRAs, terahertz DRAs and other optical DRAs.

- A mathematical study of rectangular, cylindrical, conical and spherical microwave, terahertz and optical DRAs has been developed. The concepts on Maxwell's and Schrödinger equations and Dirac equations have been worked out for far-field equations. The resonant frequency and plasmon frequency, light–matter interaction solutions have been presented with computational capabilities.
- A theoretical study of directivity (super directivity), quality factor, radiated quantum fields, radiation pattern and quantum entanglement features of quantum DRAs has also been included.

- The miniaturization techniques of optical DRAs using higher order modes and metal or graphene crafting are included.
- A mathematical description of optical DRAs-equivalent circuits and quality factor and impedance equations has been well defined by using dynamic impedance terms.
- Design and simulations of RF absorbers are to be used for renewable energy harvesting and absorption of unwanted RF signals in chip-to-chip high-speed communication and sensing applications. Absorbers have special applications in MIMO designs to improve isolation.
- The mechanisms of quantum eye gage optics for estimating exact focal point of the images have been derived.

This book will help one to solve the following problems:

- Design and implementation of microwave/terahertz/optical DRAs useful for biomedical applications. Special geometries DRAs to improve directivity and other radiation parameters tuning. The high-speed communication compact RF devices design techniques along with bandwidth and polarization tuning.
- Terahertz designs of DRAs will have large bandwidth for 5G and beyond applications to fetch high data rates. LiDAR, radar, satellite communication and wireless personal communication antennas using various dielectric, semiconductor, graphene PDMS and polymer materials.
- Design and implementation of terahertz/optical/microwave DRAs for self-driving autonomous cars, sensor and scanners for noninvasive biomedical applications.
- Design and implementations of small size terahertz/optical DRAs for chip-to-chip communications in SOC (system on chip).
- DRAs for use as photodetectors.
- Analysis of equivalent circuits at microwave, terahertz and quantum devices.

Case studies with prototype models have been included at microwave and mm-wave frequencies. This book covers new geometry DRAs at a terahertz and optical spectrum. This book will be very useful to Industries, academia, research scholars, engineers, scientists, PG and UG students working in the fields of electronics, microwave, communication, sensing, radars, satellites and remote sensing, navigation sensing, healthcare sensors, smart building sensing and renewable energy harvesting, etc.

Prof. Rajveer S. Yaduvanshi, PhD
NSUT, Dwarka, New Delhi 110078

Chapter 1
Dielectric resonator antennas (DRAs) and its synthesis

Abstract

Dielectric resonator antenna (DRA) has high efficiency and low losses at high frequencies. Terahertz (THz) and optical DRA are new trend topics of research in millimeter wave (mm wave), optics and photonics. They are simple in design, highly efficient, built with high dielectric constant ceramics and polymers materials (ε_r = 10–1,600), have design flexibility, easy to fabricate and stable operations at high temperatures. DRAs have excellent features of working throughout the frequency spectrum, right from microwave to the optical regime. Light (photons) are bosons and metal (electron–positrons) at terahertz frequency, metal acts as gaseous (plasma) at terahertz frequencies; hence it is termed fermions. The plasmon frequency gets coupled to terahertz DRA (TDRA) in proximity feed thus creates radiations in TDRA at its resonant frequency. Bosons and fermions scattering and transportation mechanism can be solved by giving quantum-mechanical treatment to these TDRAs. Dirac and Maxwell's equations have used to get desired solution using creation and annihilation operators theory. Electromagnetic (e.m.) far fields are solved as desired outcome. The radiation pattern is based on input to feed, **current density fluctuations** and retardation potentials. **The fluctuations in quantum antennas must be controlled to minimum threshold value. The radiated field is a state of jointly coherent for the bosons (i.e. photons) and for the fermions (positrons–electrons).** These are basic concept of radiations into DRAs at terahertz and optical frequencies. Hence, interaction between these **bosonic and fermionic fields causes the e.m. field to change** and induce surface currents that radiate out in space. **This change in e.m. fields must be optimized to a threshold value. Light-imaging detection and ranging operates at 200-THz frequency. Polymer, ceramics and polymer composites can be used at mm wave or terahertz and optical frequencies.**

TDRA can be built with low-quality factor materials, and absorbers are designed using materials having high-quality factor materials. In TDRAs, input is LASER, it interacts with noble metals such as gold or silver, light–matter interaction takes place resulting into surface plasmon resonance. **The silver nano waveguides** have been used to provide **feed mechanism to TDRAs** of different shapes and geometries, i.e. cylindrical, rectangular, spherical and conical shape DRAs. **LASER input interacts with silver metal and dielectric SiO_2** substrate. Hence, in TDRA, **surface plasmon polytron waves are generated** due to light–matter interaction, these will give rise to plasmon frequency generation. **This plasmon frequency is always kept lesser than LASER input frequency**, so as to enable it to propagate in forward direction. Here, we propose a TDRA fully integrated with a photonic crystal waveguide for broadside radiation pattern.

1.1 Introduction

Microwave dielectric resonator antennas (DRAs) have been in use due to its high-performance parameters, i-phone 12 mobile phone device is built by millimeter-wave (mm-wave) technology for high-speed communication. DRAs are efficient and stable against temperature variations. DRA has design flexibility for choosing different aspect ratios. Hence, DRAs have capability to excite application-specific resonant modes. These modes can be operated in a controlled manner such as merged modes, multimodes (harmonics) and single operating mode as dominant mode. Terahertz devices will be bridging the gap between electronics and photonics. The noninvasive THz imaging has become an effective tool of application in many industrial and research sectors. Gas sensors are new devices at optical frequencies for monitoring environmental parameters such as sulfur, carbon mono oxide, oxygen and nitrogen contents in the environment. These parameters can be monitored on real-time basis. Microfluidic and biosensing are new applications for terahertz DRAs (TDRAs). Wireless sensors and Internet of Things (IoT) are two main areas of applications for TDRAs. Input is given by LASER to nano waveguide that becomes proximity feed to DRA [1–9]. The surface plasmon polytron (SPP) phenomenon takes place at metal dielectric interface, when laser is connected at the edge of nano waveguide. In terahertz antennas, we have an expression for the are interaction Hamiltonian between the electromagnetic (e.m.) fields of photons and the Dirac fields of electrons and positrons comparing it with bosonic and fermionic operators and these canonical communication rules (CCR) of the bosonic operators and the canonical anticommutation relations (CAR) of the fermionic operators can be used to calculate all the matrix elements of interaction Hamiltonian and all the quantum-statistical properties of the field formed. This interaction Hamiltonian may also be used to calculate radiation pattern correction to the Dirac wave function. Thus, the Dirac current field in turn manifests itself as correction to the far-field e.m. field radiation pattern, thus, correction to the mean-square fluctuation-radiated quantum e.m. fields in any state [10–19]. It defines a trilinear form in the electron–positron–photon creation–annihilation operators based on the theory of the Dyson series. Feynman diagrams are applied to compute or estimate radiated quantum e. m. fields. The light is in fact a quantum e.m. field, when it falls on material cavity, it starts immediately interacting with the matter within the cavity [19–37]. A coarse approximation of this matter is to regard as a sea of electrons and positrons and then apply the formalism of quantum electrodynamics to describe its interaction with the incident photons, both before the light hits the cavity matter directed based on boundary conditions. Bosonic creation and annihilation operators satisfy the CCR [38–42]. If the field is fermionic, they will satisfy CARs. Finally, we shall be able to compute the mean and covariance of the e.m. field in a state. Hence, due to Planck's constant in quantum antennas, radiation energy comes in quanta. The mean-square fluctuations are at least of the order of magnitude of Planck's constant. The frequency bands in the range of 275–3,000 GHz, which are known as terahertz communication bands, are useful for high-speed communication [43–47]. Hence, it can be used for wireless communication in 5G and beyond or IoT for high data rate transmission. Nano-photonics is a branch of science that deals with transmission and reception of signals at terahertz, optical frequencies and travel in nanometer dimensions devices such as silver nano waveguide. DRAs have been developed at nanotechnology due to high efficiency, stability and less loss. A nano antenna has two important features, such as high radiation efficiency and super directivity [48–59].

Basic working principle of a quantum or TDRAs (cavity resonator antenna) of different shapes enclosing a terahertz or quantum **e.m. field comprising photons interacting with electrons and positrons using the canonical quantization method of Feynman, Schwinger, Tomonaga and Dyson. The quantum e.m. field is regarded as an ensemble of quantum harmonic oscillators,** whose Hamiltonian is treated as a quadratic form in the **creation and annihilation operator fields** and the matter field within the cavity in which it interacts is regarded as an ensemble of electrons and positrons modeled using the second quantized Dirac Hamiltonian. The interaction between these bosonic and fermionic fields causes the e.m. field to change and induce surface currents which radiate out in space. The statistics of this radiated field, namely that of quantum fluctuations in this radiated field in a given by state of the electron–positron–photon field is analyzed [58,59]. This quantum field theoretical analyses of the cone, rectangular, spherical and cylindrical cavities have been solved in this book. An example of TDRA design and modeling is given in Figure 1.1.

TDRA dimension and volume is selected based on operating frequencies. Basic physics of working of DRA differs in microwave and optical region as metal acts as gaseous state in optical or quantum frequencies. The feeding mechanism of the TDRA consists of the silver (Ag) nanostrip feed in the place of microstrip feed. The Gaussian pulse excitation is given to nano-feed line for proper phase matching in the place of wave port signal. As the nano waveguide is inserted into substrate in place surface, an interface is created by metal and dielectric substrate. This interface acts as channel to carry SPPs waves. This is plasmon frequency, which depends on dielectric properties of materials for forward propagations. Hence, its configuration is wavelength dependent. Dispersion characteristics are given by Drude's theory [1–7].

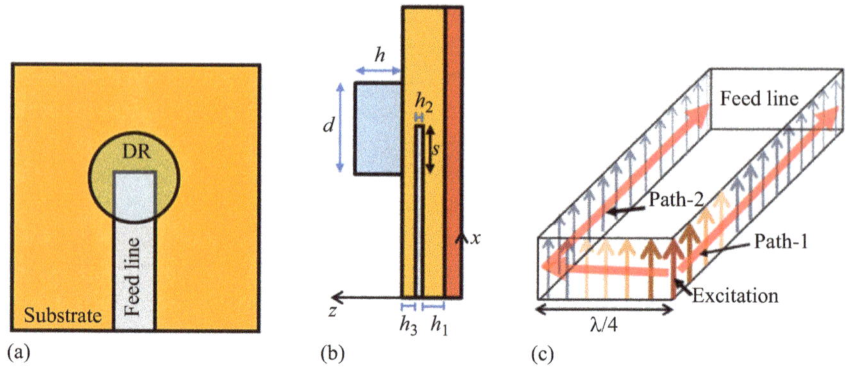

Figure 1.1 *CDRA design with input excitation of Gaussian laser beam to silver nano waveguide of TDRA: (a) DR element with feed and substrate—top view, (b) DR with inserted nano wave—side view guide, (c) nano waveguide fields—inner view*

1.2 CDRA (cylindrical DRA): design and modeling using silicon-radiating element

The TDRA, cylindrical DRA (CDRA) of the silicon (Si) with relative permittivity $\epsilon_r = 11.56$ is placed at the top of the substrate, dimensions diameter $(d) = 0.510\,\mu$m, height $(h) = 0.325\,\mu$m. The proposed CDRA is excited using a rectangular nano waveguide of width w with the stub of lengths. The value of the stub length is $s = 0.375\,\mu$m for impedance match, 3-dB AR bandwidth is fine here. The dispersive properties of silver nano waveguide can be described based on **Drude's model** as $\epsilon_{Ag} = \epsilon_o\left[\epsilon_\infty - \left(f_p^2/\{f(f+i\gamma)\}\right)\right]$. Here, $\epsilon_\infty = 5$, $f_p = 2{,}175$ THz and $\gamma/2\pi = 4.35$ THz, f is the operating frequency, ϵ_{Ag} is the permittivity of silver, f_p is the plasma frequency, γ is the collision frequency and it is inverse of the relaxation time, ϵ_∞ is the offset of real part of the dielectric constant and $\epsilon_o = 8.85 \times 10^{-12}$ F/m. These E-fields tell us about excited states of TDRA, i.e. resonant modes. The numbers of half-wave variation in x, y, z directions tell us that fields are propagating in particular direction. These are state of **short magnetic dipoles or short electric dipoles**. The E-fields and H-fields provide physical **insight** of TDRA. These are shown in Figures 1.2–1.4.

Nanotechnology is multidisciplinary area of science, merges physics, chemistry, material, biology, medicine and engineering branches for innovations on current trends of optical DRAs at nano regime. The physics of optical antennas involves a dominance behavior of skin effects, SPPs and light–matter interaction (creation and annihilation). It is well known that light can penetrate through metallic walls at optical frequencies. Light–matter interaction at optical domain is studied to design quantum devices [11,55–57].

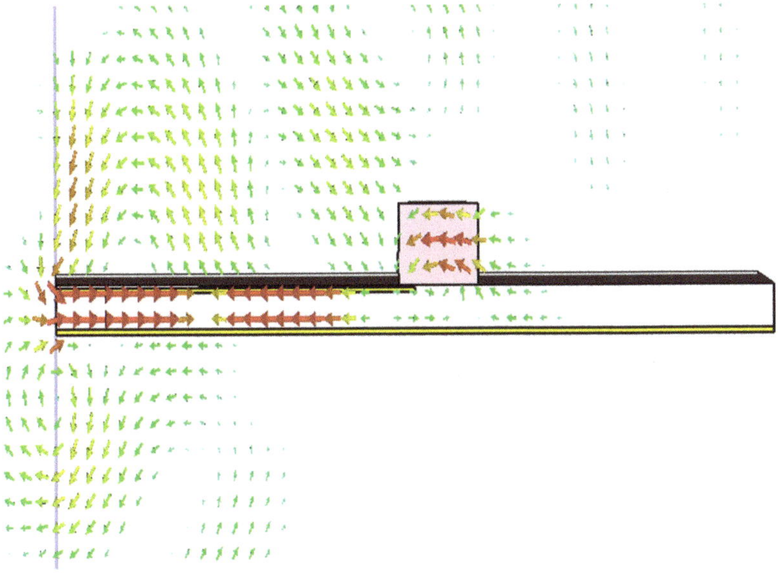

Figure 1.2 TDRA feed to generate SPP waves and E-fields

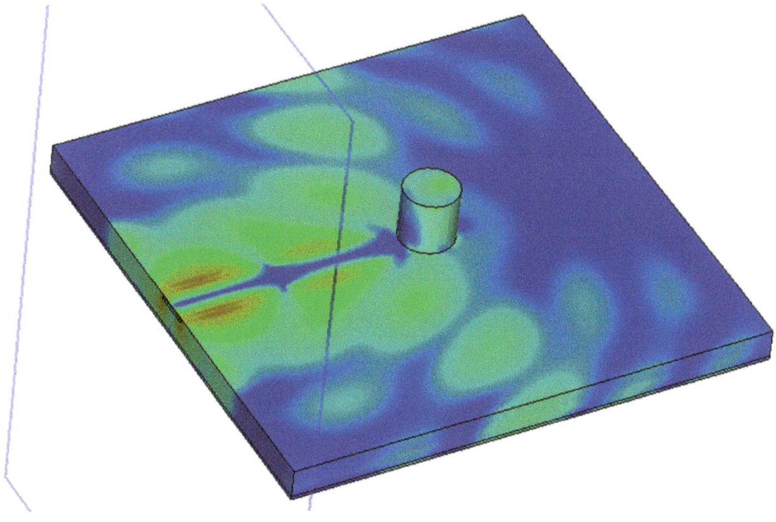

Figure 1.3 Surface currents in TDRA

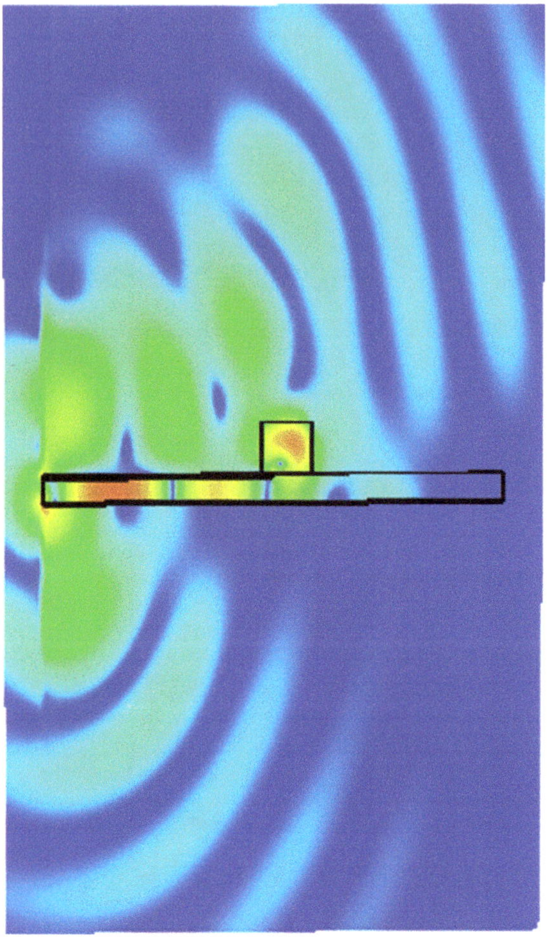

Figure 1.4 Radiated fields in TDRA

1.3 Terahertz or quantum devices characteristics

1.3.1 Theory of TDRA

In TDRA, there can be zero, one or more than one particle having a specified momentum and helicity. In the case of fermionic fields like the second quantized Dirac field, we cannot have more than one particle having a definitive value of momentum and spin. Complete frequency spectrum is shown in frequency in Figure 1.5.

If, $a(k)^*, a(k), k = 1, 2, \ldots$ are canonical bosonic creation and annihilation operators, they satisfy the CCR $[a(m)^*, a(k)] = \delta[k - m], [a(k), a(m)] = [a(k)^*, a(m)^*] = 0$, while if $c(k)^*, c(k),\ k = 1, 2, \ldots$ are canonical fermionic

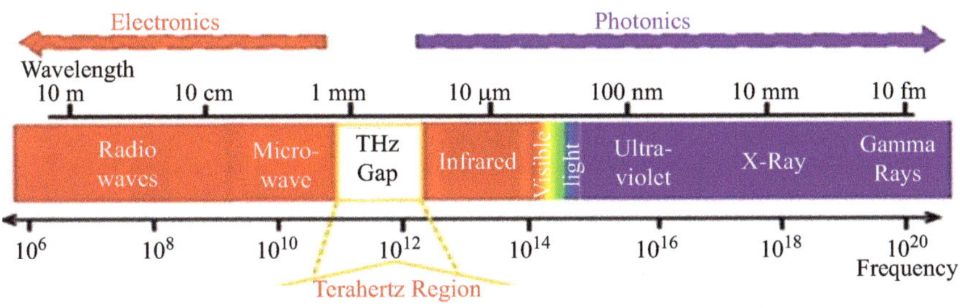

Figure 1.5 Terahertz spectrum is vacant for high-speed communications

creation and annihilation operators, they satisfy the CAR $[c(k), c(m)^*]_+ = \delta[k-m], [c(k), c(m)]_+ = [c(k)^*, c(m)^*]_+ = 0$. Here, $[A, B] = AB - BA, [A, B]_+ = AB + BA$. It is clear therefore that $c(k)^2 = 0 = c(k)^{*2}$. Here, k represents a definite value of momentum and spin/helicity. If $|0>$ is a vacuum boson state, then $a(k_1)^{*n_1} \cdots a(k_r)^{*n_r}/\sqrt{n_1! \cdots n_r!}|0 \geq |(k_1, n_1), \ldots, (k_r, n_r)>$ represents a normalized state of the bosonic field in which there are n_j bosons having momentum and helicity index k_j for each $j = 1,2, \ldots, r$. It is clear then from the CCR that

$$a(k_j)|(k_1, n_1), \ldots, (k_r, n_r) \geq \sqrt{n_j}|(k_1, n_1), \ldots, (j_j, n_j - 1), \ldots, (k_r, n_r)>, \quad 1 \leq j \leq r \tag{1.1}$$

While if $k \notin \{k_1, \ldots, k_r\}$, then

$$a(k)|(k_1, n_1), \ldots, (k_r, n_r) \geq 0$$

In short, $a(k)$ annihilation, a boson, having momentum–helicity k and if there is no such boson with momentum–helicity k in the state, then $a(k)$ annihilates the entire states giving zero. Likewise, $a(k)^*$ creates a boson having momentum–helicity k:

$$a(k_j)^*|(k_1, n_1), \ldots, (k_r, n_r) \geq \sqrt{n_j}|(k_1, n_1), \ldots, (k_j, n_j + 1), \ldots, (k_r, n_r) >, 1 \leq k \leq r,$$
$$a(k)^*|(k_1, n_1), \ldots, (k_r, n_r) \geq |(k_1, n_1), \ldots, (k_r, n_r), \ldots, (k, 1) >, k \notin \{k_1, \ldots, k_r\}$$

It can be verified that these two rules are in agreement with the CCR. From these CCR, it is evident that $N(k) = a(k)a(k)^*$ is the number operator, i.e. when acting on a state, it gives the number of bosons in that state:

$$N(k)\Big|(k_1, n_1), \ldots, (k_r, n_r) \geq \left(\sum_{j=1}^{r} \delta[k - k_j] n_j\right)\Big|(k_1, n_1), \ldots, (k_r, n_r) >$$

The state described before are called the occupation number states or simply the number sates. They are Eigen states of the number operators but not of the creation and annihilation operators. The second quantized bosonic field is a superposition of the annihilation and creation operators:

$$\phi(x) = \sum_k [a(k)\chi_k(x) + a(x)^* \tilde{\chi}(x)]$$

where satisfy the classical bosonic field equation

$$L\chi_k(x) = 0, x = (t, r)$$

with, for example

$$L = \left(\frac{1}{2}\right)\partial_\mu \partial^{jmu} + m^2$$

In the Klein–Gordon case, or in e.m. field case where the bosons are photon having zero mass, L is the wave operator:

$$L = \left(\frac{1}{2}\right)\partial_\mu \partial^{jmu}$$

(Klein–Gordon particles have zero spin while the photon has spin 1 with, helicities ± 1, the zero helicity not being allowed. This corresponds to the fact that any state of photon polarization can be expressed as a superposition of left and right circularly polarized states.) It follows that to get a state definite field amplitude for positive frequencies, the state should be an Eigen state of the annihilation operators $a(k)$ and such a state called a coherent state can be obtained (in the discrete momentum–helicity setting) as

$$\begin{aligned}\varnothing(u) &= \exp\left(-\left(\frac{1}{2}\right)\Sigma\{n_k, k \in I\}|u(k)|^2\right)\Sigma\prod\nolimits_{k\in I}\left\{\frac{u(k)^{n_k} a(k)^{*n_k}}{n_k!}\right\}|0> \\ &= \exp\left(-\left(\frac{1}{2}\right)\sum_{k\epsilon I}|u(k)|^2\right)\sum_k \prod\nolimits_k u(k)^{n_k}/\sqrt{n_k!}|\{n_j, j \in I\}>\end{aligned}$$

The second quantized Dirac fermionic field can be expressed on the other hand in terms of fermionic creation and annihilation operators. In the photon case, the antiparticle of a photon is again a photon and so its creation operator at a given momentum–helicity is the adjoint of the corresponding annihilation operator. On the other hand, the antiparticle of the electron is another particle, a positron, and hence the Dirac field should be expressed as a superposition of electron annihilation operators and positron creation operators. The creation of a positron of the positive energy is according to Dirac equivalent to annihilation an electron of negative energy. Thus, the **Dirac field** is expressed as the superposition

$$\psi(x) = \sum_k [b(k)\chi_k(x) + c(x)^* \eta_k(x)]$$

where $b(k)$ annihilation, an electron, with momentum–spin k and $c(k)$ annihilations, a positron, with momentum–spin k. Equivalently, $c(x)^*$ creates a positron with momentum–spin k. $\chi_k(x)$ and $\eta_k(x)$ are solutions to the free Dirac equation:

$$\begin{aligned}&\left(i\gamma^{\mu}\partial_{\mu} - m\right)\chi_k(x) = 0,\\&\left(i\gamma^{\mu}\partial_{\mu} - m\right)\eta_k(x) = 0\end{aligned}$$

Given a quantum matter field $\eta(x)$, either bosonic or fermionic, it satisfies a classical wave equation of the Klein–Gordon type, or the 3D wave type or the Dirac type with certain boundary conditions. This second quantized field can therefore be expressed as superposition of Eigen functions corresponding to the boundary condition with coefficients being particle creation and annihilation operators. **If the field is bosonic, these creation and annihilation operators will satisfy CCR**s while if the field **is fermionic, they will satisfy CARs.** If H corresponds to the Hamiltonian of the first quantized theory, then we can write the **second quantized Humiliation** of field as

$$\int \eta(x)^* H\eta(x)d^3x$$

This is a function of an infinite number of particle operators acting in **a bosonic or fermionic Fock space.**

In optical DRA or TDRA, SPP phenomenon occurs due to strong light–matter interactions. A simplified analysis would be to model the matter as just the second quantized electron–positron field using Dirac's relativistic wave equation. In this process, wave function is an operator field, and then to include the photon interaction term in the usual way. In this way, the total **Hamiltonian of the photon and matter field splits into three** terms as follows:

1. One, the Hamiltonian of the e.m. field described as a quadratic form in the photon creation and annihilation operators within the cavity.
2. Two, the Hamiltonian of the Dirac field with cylindrical cavity (in CDRA) boundary conditions on the wave function described as a quadratic form in the electron–positron creation and annihilation operators.
3. Three, the Hamiltonian of the interaction between the photon and electron–positron field described as a quadratic form in the electron–positron creation–annihilation operators multiplied with a linear form in the photon creation–annihilation operators.

This **interaction Hamiltonian can be used to compute the amplitudes for scattering**, but since we are primarily interested in the statistics of the e.m. wave field pattern, we shall describe this interaction using the Dirac current density expressed as a quadratic form in the electron–positron creation–annihilation operator fields that drive the photon field using the wave equation for the e.m. field in the presence of a current density.

The previous quantum theory can be applied to understand the concept of quantum physics as stated later: the cavity can be modeled as cylindrical cavity or conical shape cavity or spherical cavity.

Since light is in fact a quantum e.m. field, when it falls on cavity, it starts immediately interacting with the matter within the cavity. A coarse approximation of this matter is to regard as a sea of electrons and positrons and then apply the formalism of quantum electrodynamics to describe its interaction with the incident photons, both before the light hits the cavity matter directed based on boundary conditions. The field in the cavity is actually a planar quantum e.m. field in certain state and we wish to calculate the probability that such a field will be formed.

1.3.2 Terahertz DRA or quantum DRA near fields/far fields

To perform cavity quantum e.m. field computation, we must solve the joint Maxwell–Dirac equations within the cavity with the appropriate boundary conditions corresponding to what shape of the cavity we are using. The cavity shape can be cone, cylindrical or spherical. This means that we expand the Maxwell field in terms of basis functions that vanish on the boundary and then the coefficients in this expansion will be the photon creation and annihilation operator fields. Likewise, we must expand the Dirac field within the cavity using the same orthogonal basis functions whose coefficients will now be the electron–positron creation and annihilation operator fields.

The resulting Hamiltonian of these fields will then comprise three components: one, the Hamiltonian of the free photon field expressed as a quadratic form in the photon creation and annihilation operators, two the Hamiltonian of the free electron–positron field expressed as a quadratic form in the electron–positron creation and annihilation operators and finally three, the interaction Hamiltonian between the electron–positron field and the photon field.

It is this last interaction term that is of importance in scattering theory of light and the formation of fields. Using the interaction picture of Dyson, we can expand the solution to the Schrödinger evolution operator for this quantum system as a Dyson series in powers of the interaction Hamiltonian. Then we can, using the technology of Feynman diagrams, calculate various terms of the scattering matrix to determine the probability amplitude of obtaining a certain image field on the retinal screen. The basic physics behind this computation does not depend on what shape we choose for the cavity. No matter what shape we choose, we can in principle determine a complete orthonormal basis of test functions that satisfy the appropriate boundary conditions for the given cavity and then expand the Maxwell and the Dirac field in terms of these test functions. Before the interaction between light and matter, the system was in an initial state and after this interaction, including the interaction of light with the matter, the state of the whole system changes with time and this change can be determined by the action of the unitary evolution operator on the initial state with the unitary evolution operator constructed from the total Hamiltonian of the interacting electrons, positrons and photons using the Dyson series. The final state of this system is then equivalently determined once we know the matrix elements of the evolution operator between any given initial state and a final state of the particles. From this final state, the probability of getting a given field is computed from the basic rules for quantum probability.

Thus, we shall by perturbation theory be able to calculate the change in the e. m. fields pattern caused by the Dirac current in term of operators and then by assuming a definitive state of the photon–electron–positron field, we shall be able to compute the mean and covariance of the e.m. field in this state. Their operating wavelengths are very small, mostly in the nanometer size. Optical antennas work on Drude's theory and have nature of nonlinearity. Optical antenna have far field with finite values of Planck's constant.

1.3.3 Radiation parameters

The signal-to-noise ratio (SNR), mean-square fluctuation in S_{11}, S_{12}, S_{21} and S_{22} in a given state, fluctuation variance from state to state, second quantized current density, mean of current density dependence of state of system, status of photons, electrons and positron builds current density in system at any state, average current with fluctuations develop field pattern, i.e. **radiation pattern is defined as theta vs mean square of e.m. fields**, as well as theta vs mean value of current, frequency of minimum fluctuations value will be considered important, and state will change if large fluctuation occurs. Hence, optimization by minimizing the mean-square value of fluctuation is a Lagrangian problem**. The mean-square fluctuation of power along with SNR is an important term for power in radiated fields**. Thus, SNR is an average value of the field square divided by mean-square value of the fluctuating component in terms of Planck's constant. Quantum field radiations occur due to light–matter interactions.

In fact, as Feynman had observed in his path integral approach to quantum mechanics, the quantum amplitude for a particle to go from one space–time point to another is given by a sum over paths of the phase function that is action integral over the given path. The probability of going from the first to the second is then the modulus square of this sum and it contains interference terms between any two paths. On the other hand, in the limit when Planck's constant h goes to zero, the rapid phase variations between neighboring paths will cancel out except in the vicinity of the classical path where the action integral has zero variation. Thus, in the limit as Planck's constant goes to zero, the particle will move only along the classical path. Letting Planck's constant converge to zero is equivalent to making S/h very large, which can equivalently be accomplished by increasing the dimensions of our system so that S becomes large. This means that for large sized objects, only classical mechanics will matter.

1.3.4 Drude's model theory

$$\varepsilon_{Ag} = \varepsilon_0 \left[\varepsilon_a + \frac{f_p^2}{f(f + i\gamma)} \right]$$

where γ is the collision frequency, f_p is the plasmon frequency, e is the electron charge, T is the temperature, ω is the radian frequency, μ_c is the chemical potential, Γ is the scattering rate.

In order of Planck's constant in quantum antennas, radiation energy comes in quanta. Thus, discrete particle distributions of electrons, positrons and photons do matter and this discreetness is also responsible for producing quantum fluctuation in the radiation field pattern, which are emerging technology.

$$\varepsilon_{Ag} = \varepsilon_0 \left[\varepsilon_\alpha + \frac{f_p^2}{f(f + i\gamma)} \right]$$

$$\varepsilon_0 = 8.85 \times 10^{-12} \frac{F}{m}$$

$$\varepsilon_\alpha = 5$$

Drude's model for dispersion in dielectrics, i.e. plasma frequency:

$$\varepsilon_p = \sqrt{\frac{ne^2}{\varepsilon_0 m}}$$

1.4 Terahertz MIMO DRA parameters

Any deviation in the excitation position at the input port can disturb the field components resulting in linear polarization. A wideband circular polarization (CP) response can be obtained due to the merging of the fundamental and higher order orthogonal degenerate mode pairs. The antenna can provide high peak gain in the operating impedance pass band of the antenna. The MIMO diversity performance parameters, including the envelope correlation coefficient (ECC), diversity gain (DG), channel capacity loss (CCL), mean effective gain (MEG) and total active reflection coefficient (TARC), have to be within the acceptable limits. The MIMO measurements using far-field parameters are usually calculated as given in the following:

$$\rho_{ei\,j} = \left| \frac{\iint \left(E_{\theta i} \cdot E_{\theta j}^* + E_{\phi i} \cdot E_{\phi j}^* \right) d\Omega}{\left| \iint \iint \left(E_{\theta i} \cdot E_{\theta i}^* + E_{\phi i} \cdot E_{\phi i}^* \right) d\Omega \iint \left(E_{\theta j} \cdot E_{\theta j}^* + E_{\phi j} \cdot E_{\phi j}^* \right) d\Omega \right|} \right|^2$$

The DG of the antenna is calculated using mathematical calculations as explained in the following equation:

$$G_d = 10 e_p \text{ where } e_p = \sqrt{(1 - |0.99\,\rho_e|^2)}$$

Similarly, the CCL of the proposed antenna remains as follows:

$$CCL = -\log_2 \det(\psi^R)$$

$$\psi^R = \begin{pmatrix} \rho_{e11} & \rho_{e12} \\ \rho_{e21} & \rho_{e22} \end{pmatrix}$$

The MEG details are given as follows:

$$MEG_i = 0.5\eta_{i,rad} = 0.5\left[1 - \sum_{j=1}^{M} |S_{ij}|^2\right]$$

where M is the number of ports in MIMO antenna and is the radiation efficiency of the antenna. Furthermore, the TARC is the ratio of square root of total reflected power to the square root of total incident power.

TARC based on S-parameters is defined as follows:

$$\Gamma_a^t = \sqrt{\frac{|S_{11} + S_{12}e^{j\theta}|^2 + |S_{21} + S_{22}e^{j\theta}|^2}{2}}$$

Theta is phase of input signal. Here are few parameters listed in the following, which differentiates classical antennas from optical.

1.4.1 Microwave DRAs vs optical DRA parameters

The basic difference between a classical and a quantum antenna is the size. Classical antennas have surface currents. The former has length dimensions of order ranging from a few centimeters to several meters, while the latter has length dimension of the order nanometer, which is the typical order of magnitude of the radius of an atom. The physical quantities measured by a classical antenna are actually quantum averages w.r.t. a given pure or mixed state of electrons, positrons and photons. The classical case, the SNR is infinite, which is equivalent to setting Planck's constant to zero after doing a quantum-mechanical analysis.

1.4.2 Optical DRAs

The optical antennas shall have SPP or surface charge density due to the involvement of Planck's constant [1–4]. The physical quantities measured by a quantum antenna are not only quantum averages of the field amplitude but also quantum averages of the higher moments of the e.m. field in a given state of the electrons, positrons and photons. Design of spherical optical DRA is given in Figure 1.6.

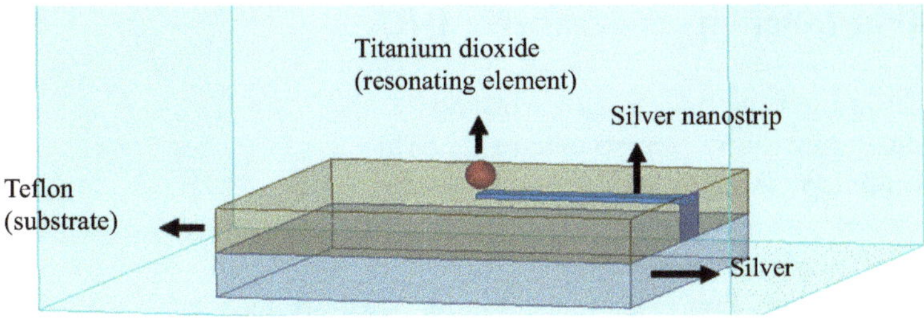

Figure 1.6 Spherical optical DRA

These higher order moments are taken about its mean value.

In particular, apart from the average field amplitude, the mean-square fluctuations of the e.m. field around its mean value become significant in a quantum antenna.

The quantum fluctuations of the field around a given mean value is to be interpreted as quantum noise in the system and the ratio of the square of the mean field amplitude to the mean-square amplitude fluctuation is called the SNR of the antenna.

The mean-square fluctuations in the field arise due to the Heisenberg uncertainty principle or equivalently due to the noncommutative of the operator field.

These mean-square fluctuations are at least of the order or magnitude of Planck's constant and hence the SNR of a quantum antenna is usually bounded above by Planck's constant.

1.4.3 Radiated fields

A typical example of how the **analysis of the radiation field** is carried out in quantum field theory is as follows:

1. We interpret the e.m. four potential as a canonical position vector and its time derivative as a canonical momentum vector.
2. In quantum mechanics, we have the Heisenberg commutation relations.
3. The mean field energy in the Gibb's state is defined by quantum-mechanical process as given by the following:

Define the annihilation and creation operators, derive Heisenberg commutation relations and express the field energy as radiated far fields.

The reason for this discrepancy lies in the fact first observed by Max Planck when he proposed his quantum-mechanical law of black-body radiations as an improvement over the earlier Wien's displacement law: radiation energy comes in quanta and not as a continuum as the Maxwell wave theory of radiation predicts. Thus, in an optical or quantum DRA, the discrete particle distribution of electrons, positrons and photons do matter and this discreteness is also responsible for producing quantum fluctuations in the radiation field pattern, which are of the order of Planck's constant.

1.5 Main functions of terahertz DRA

1. Light in light out principle of working.
2. Input excitation by semiconductor LASER.
3. Plasmonic resonance takes place into substrate.
4. Electromagnetic coupling (proximity feed) takes place to quantum DRA through plasmonic resonance.
5. Capacitive coupling takes place due to proximity feed.

6. Screening of electron–hole takes place near band gap due to accelerated charge carriers.
7. Initially very small band gap energy was observed.
8. LASER provided higher photon energy then initial band gap energy.
9. LASER accelerated photons due to excitation.
10. LASER created clouds of charge carrier (screening of fields).
11. Quick change in radiated field takes place (due to fluctuations).
12. Quantum states.

1.5.1 Some important parameters of microwave and terahertz DRA

1. Career multiplication ratio.
2. Signal-to-power ratio.
3. Receiving gain.
4. Transmitting gain.
5. Dispersion.
6. Spectral response.
7. Beam forming.
8. Bandwidth.
9. Directivity.
10. Localized energy.
11. Coupled energy.
12. Light emitting.
13. Spectroscopy.
14. Electromagnetic states.
15. Scattering for microwave imaging.
16. Creation and annihilation.
17. Absorption.
18. Coherent control.
19. Field enhancement for scanning.
20. Wave fronts.
21. Top down fabrication.
22. Control and utilize material at nanoscale.
23. Electron–hole pair combines to emit a photon.
24. Plasmons—electron oscillations.
25. Electromagnetic response.
26. Device that efficiently converts localized energy into propagating optical radiations, and vice versa.
27. Diffraction limit.
28. Metals do not behave as perfect conductors at optical frequency.
29. Capacitive driven by proximity feed.
30. Surface plasmon resonances are material and shape dependent.
31. Quantum absorber or emitter.
32. Control light–matter interaction.

33. Local density of electromagnetic states.
34. Energy dissipation by dipoles.
35. Quantum-mechanical treatment.
36. Power dissipated.
37. Antenna efficiency.
38. MIMO radar for scanning.
39. Graphene absorber due to strong light–matter interaction.
40. The field enhancement, field localization, absorption, high-resolution microscopy and spectroscopy, photovoltaic (rectenna), light emission and coherent control.
41. Optical antennas are still infancy stage study.
42. Classical antenna downsizing does not work due to radiation penetrates into metals, which results into plasma oscillations.
43. Optical antenna increases interaction with absorber or emitter radiation, enhances light–matter interaction.
44. Bottom up designs by colloidal solution.
45. Resonant light scattering.
46. Fundamental and higher order modes.
47. Electron energy loss spectroscopy.
48. Near-field microscopy for scanning.
49. Field intensity distribution.
50. Impedance matches for maximum energy transfer from source to antenna.

1.6 THz DRA model design parameters

The dimensions of the proposed rectangular DRA (RDRA) and feeding mechanism parameters are determined by considering the following dimension guidelines: the DRA length and width is selected as and optimized height is taken in order to maintain the aspect ratio equivalent to while splitting the DRA into two equal halves using the single-port configuration. This high value of aspect ratio allows the excitation of the fundamental and higher order design dimensions, let $a = 2b$, aspect ratio $a/h = 2$.

The isolation between the ports is maintained by gap, which is optimized using the parametric analysis. It is interesting to note that the good isolation is maintained at gap, which separates feed line by the distance equivalent to $\lambda 0/2$. For CP generation, the excitation is given at the corner of the feed lines which allows the traveling wave field to travel with quarter-wavelength path difference along the edges of the respective feed lines. In this motif, the width of the feed lines is taken equivalent to $\approx\lambda 0/4$, where $\lambda 0$ is the operating wavelength of the DRA. This path difference in the traveling field at the edges of the feed line provides the 90° phase difference. The MIMO diversity performance parameters include ECC, DG, CCL, MEG and TARC. The specific feeding technique excites the orthogonal degenerate mode pairs inside the DR resulting in CP.

1.7 Rectangular nano-DRA design parameters

The design step of nano-DRA is given in Figure 1.7.

The formulations used to compute resonant frequency for terahertz RDRA is

$$k_z \tan\left(\frac{k_z d}{2}\right) = \sqrt{(\epsilon_r - 1)k_0^2 - k_z^2}$$

1.7.1 Design steps

To design nano-RDRA, for ground plane, assign the material silver (Ag) of minimum height. The substrate of same size is placed on ground plane. Assign this substrate material as silicon dioxide (SiO_2). Use of low loss tangent materials for substrate can provide high gain, greater efficiency and lower dielectric losses. The nanostrip waveguide of silver material (Ag), having length half the length of ground plane, desired width and desired height based on frequency of excitation is used. The impedance matching of feed is obtained for 50 Ω. For input excitation, we have to define the 50-Ω lumped port. The port height is equal to the height of substrate and width greater than width of nanostrip. The nano-RDRA of silicon of required dimensions, based on resonant frequency, is placed on SiO_2 (permittivity = 3.9) substrate with proximity feed arrangements. The materials are Si, SiO_2 from material library. The Gaussian pulse excitation is given through the nanostrip feed line of TDRA. The excitation position is chosen at the edge of the feed line having width equivalent to the quarter wavelength. SPP generates due to edge feeding into silver waveguide. The RDRA allows the generation of specific resonant mode based on aspect ratio. The substrate used for TDRA is SiO_2 having relative permittivity, (3.9) and equal to dimensions of the ground plane. Nano rectangular RDRA design

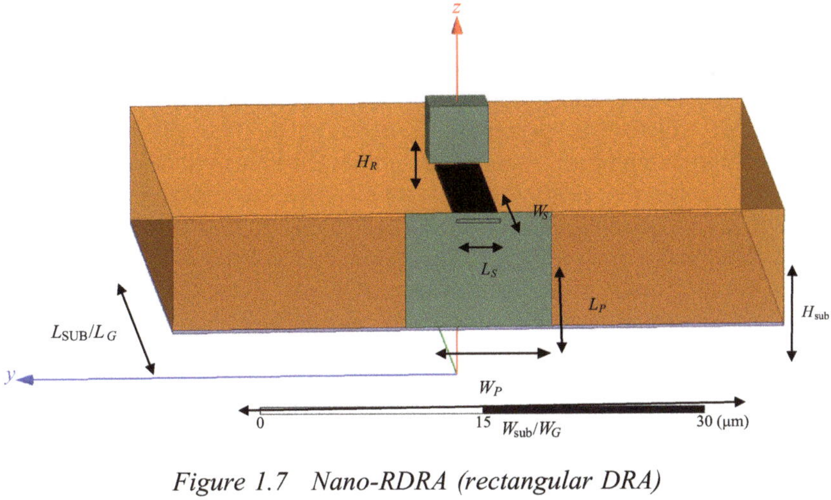

Figure 1.7 Nano-RDRA (rectangular DRA)

Table 1.1 Design dimensions

Parameters	Material	Dimensions (μ_r, m)
L_{SUB}	Silicon dioxide ($\varepsilon_r = 4$, $\mu_r = 1$ and mass density = 2,220)	45
W_{SUB}		45
H_{SUB}		5
L_G	Silver ($\varepsilon_r = 1$, $\mu_r = 0.99998$ and mass density = 10,500)	45
W_G		45
H_G		5
L_{N1}	Silver ($\varepsilon_r = 1$, $\mu_r = 0.99998$ and mass density = 10,500)	21
W_{N1}		3
L_{N2}		21
W_{N2}		3
W_S		0.2
L_{R1}	Silicon ($\varepsilon_r = 11.9$, $\mu_r = 1$ loss tangent = 0.001 and mass density = 2,330)	6
B_{R1}		6
H_{R1}		6
L_{R2}		6
B_{R2}		6
H_{R2}		6

specifications are given in Table 1.1. The nanostrip feed line is made of silver of dimensions and stub length, substrate layers of height h.

1.8 Conclusion

Microwave DRAs have been in use in mobile communication such as Wi-Fi, Wi-max, WLAN and for mm-wave applications. TDRA has excellent applications for sensing and high-speed communication applications. These TDRAs can be developed as light-imaging detection and ranging (LiDAR) for self-driving cars. The terahertz antennas radiation pattern controls on the wave function operator of the Dirac field of electrons and positrons. The quantum cavity photon and electron–positron fields will then be expressible in terms of the free quantum fields plus additional perturbation terms involving the classical current and field sources. Once this is done, we can in principle calculate the far-field antenna pattern produced by the cavity surface currents induced by the tangential components of the quantum magnetic field operators as well as that produced by the Dirac field of electrons and positrons and then design these classical control fields so that the far-field quantum Poynting radiation pattern has a mean value and correlations in a given quantum coherent state of the photons and electrons–positrons within the cavity as close as possible to specified values. The applications of terahertz frequency in communication are intra-chip<length 1-mm IC CMOS/monolithic, link-type line of sight, nm size, terahertz, speed, 10 Tbps. LiDAR is an application of terahertz radars for imaging and ranging in automotive vehicles.

DRAs at terahertz frequency have been characterized based on importance. Important parameters of TDRAs have been described with few new geometry models. They are useful for practical applications such as high-speed communication and other medical applications such scanning, imaging, etc. Planck's constant has an important role at terahertz frequencies. Controlling current density fluctuations to threshold value is an important parameter in quantum antennas. For fabrication, the .gds II design file is compatible with commercial foundry to embed silicon CDRA on silver surface or SiO_2 substrate. Terahertz radars can be used to detect stealth aircraft. Terahertz radar transmits photons. These photons have short wavelengths of between 0.3 and 3 THz. These photons can penetrate nonmetallic objects and see inside for imaging or scanning. A terahertz imaging camera can be used at airports for security.

References

[1] P. Bharadwaj and L. Novotny, "Spectral dependence of single molecule fluorescence enhancement," *Opt. Express*, vol. 15, pp. 14266–14274, 2007.

[2] R. Loudon, *The Quantum Theory of Light*, 2nd ed., Oxford Science Publications, Clarendon, 1983.

[3] L. Tang, S. K. Kocabas, S. Latif *et al.*, "Nanometre-scale germanium photodetector enhanced by a near-infrared dipole antenna," *Nat. Photonics*, vol. 2, pp. 226–229, 2008.

[4] P. Bharadwaj, P. Anger, and L. Novotny, "Nanoplasmonic enhancement of single-molecule fluorescence," *Nanotechnology*, vol. 18, p. 044017, 2007.

[5] L. Novotny, "Optical antennas: A new technology that can enhance light-matter interactions," *Front. Eng.*, vol. 39, no. 4, pp. 100–120, 2012.

[6] S. Abadal, E. Alarcón, A. Cabellos-Aparicio, M. Lemme, and M. Nemirovsky, "Graphene-enabled wireless communication for massive multicore architectures," *IEEE Commun. Mag.*, vol. 51, no. 11, pp. 137–143, 2013.

[7] D. W Pohl, "Near-field optics seen as an antenna problem," in *Near-field Optics, Principles and Applications*, X. Zhu and M. Ohtsu (eds.), pp. 9–21, World Scientific, Singapore, 2000.

[8] P. Mühlschlegel, H. J. Eisler, O. J. F. Martin, B. Hecht, and D. W. Pohl, "Applied physics: Resonant optical antennas," *Science*, vol. 308, no. 5728, pp. 1607–1609, 2005.

[9] J. Wen, S. Romanov, and U. Peschel, "Excitation of plasmonic gap waveguides by nanoantennas," *Opt. Express*, vol. 17, no. 8, p. 5925, 2009.

[10] M. Schnell, P. Alonso-González, L. Arzubiaga, *et al.*, "Nanofocusing of mid-infrared energy with tapered transmission lines," *Nat. Photonics*, vol. 5, no. 5, pp. 283–287, 2011.

[11] W. T. Sethi, H. Vettikalladi, H. Fathallah, and M. Himdi, "Nantenna for standard 1550 nm optical communication systems," *Int. J. Antennas Propag.*, vol. 2016, pp. 1–9, 2016.

[12] L. Novotny and N. Van Hulst, "Antennas for light," *Nat. Photonics*, vol. 5, no. 2, pp. 83–90, 2011.

[13] H. Iizuka, N. Engheta, H. Fujikawa, and K. Sato, "Arm-edge conditions in plasmonic folded dipole nanoantennas," *Opt. Express*, vol. 19, no. 13, p. 12325, 2011.

[14] L. Tang, S. Kocabas, S. Latif, *et al.*, "Nanometre-scale germanium photodetector enhanced by a near-infrared dipole antenna," *Nat. Photonics*, vol. 2, no. 4, pp. 226–229, 2008.

[15] N. Pfullmann, C. Waltermann, M. Noack, *et al.*, "Bow-tie nano-antenna assisted generation of extreme ultraviolet radiation," *New J. Phys.*, vol. 15, 2013.

[16] T. Kosako, Y. Kadoya, and H. F. Hofmann, "Directional control of light by a nano-optical Yagi-Uda antenna," *Nat. Photonics*, vol. 4, no. 5, pp. 312–315, 2010.

[17] I. S. Maksymov, I. Staude, A. E. Miroshnichenko, and Y. S. Kivshar, "Optical Yagi-Uda nanoantennas," *Nanophotonics*, vol. 1, no. 1, pp. 65–81, 2012.

[18] Y. Yang, D. Zhao, H. Gong, Q. Li, and M. Qiu, "Plasmonic sectoral horn nanoantennas," *Opt. Lett.*, vol. 39, no. 11, p. 3204, 2014.

[19] B. Lahiri, S. G. McMeekin, R. M. De La Rue, and N. P. Johnson, "Resonance hybridization in nanoantenna arrays based on asymmetric splitring resonators," *Appl. Phys. Lett.*, vol. 98, no. 15, pp. 2012–2015, 2011.

[20] M. Walther, D. G. Cooke, C. Sherstan, M. Hajar, M. R. Freeman, and F. A. Hegmann, "Terahertz conductivity of thin gold films at the metal-insulator percolation transition," *Phys. Rev. B – Condens. Matter Mater. Phys.*, vol. 76, no. 12, pp. 1–9, 2007.

[21] R. K. Mongia and P. Bhartia, "Dielectric resonator antennas—a review and general design relations for resonant frequency and bandwidth," *Int. J. Microwave Millimeter-Wave Comput. Eng.*, vol. 4, no. 3, pp. 230–247, 1994.

[22] A. Petosa and A. Ittipiboon, "Dielectric resonator antennas: A historical review and the current state of the art," *IEEE Antennas Propag. Mag.*, vol. 52, no. 5, pp. 91–116, 2010.

[23] Y. Pan, K. W. Leung and K. K. Luk, "Design of the millimeter-wave rectangular dielectric resonator antenna using a higher-order mode," *IEEE Trans. Antennas Propag.*, vol. 59, no. 8, pp. 2780–2788, 2011.

[24] D. Kajfez, A. Elsherbeni, and A. Mokaddem, "Higher order modes in dielectric resonators," *IEEE Antennas Propag. Soc. Symp. Dig.*, vol. 1, pp. 306–309, 1996.

[25] D. Kajfez, A. W. Glisson, and J. James, "Computed modal field distributions for isolated dielectric resonators," *IEEE Trans. Microwave Theory Tech.*, vol. 32, no. 12, pp. 1609–1616, 1984.

[26] D. Hou, Y. Z. Xiong, W. Hong, W. L. Goh, and J. Chen, "Silicon-based onchip antenna design for millimeterwave/THz applications," 2011 IEEE Electr. Des. Adv. Packag. Syst. Symp. EDAPS 2011, pp. 1–4, 2011.

[27] H. Yi, S. W. Qu, and C. H. Chan, "Wideband dielectric resonator terahertz reflectarray," ICCEM 2015 – 2015 IEEE Int. Conf. Comput. Electromagn., pp. 273–274, 2015.

[28] P. H. Bolivar, M. Brucherseifer, J. G. Rivas, *et al.*, "Measurement of the dielectric constant and loss tangent of high dielectric-constant materials at terahertz frequencies," *IEEE Trans. Microwave Theory Tech.*, vol. 51, no. 4, pp. 1062–1066, 2003.

[29] G. Varshney, S. Gotra, J. Kaur, V. S. Pandey, and R. S. Yaduvanshi, "Obtaining the circular polarization in a nanodielectric resonator antenna for photonics applications," *Semicond. Sci. Technol.*, vol. 34, no. 7, pp. 07L–T01, 2019.

[30] T. P. H. Sidiropoulos, M. P. Nielsen, T. R. Roschuk, A. V. Zayats, S. A. Maier, and R. F. Oulton, "Compact optical antenna coupler for silicon photonics characterized by third-harmonic generation," *ACS Photonics*, vol. 1, no. 10, pp. 912–916, 2014.

[31] R. Maram, S. Kaushal, J. Azaña, and L. R. Chen, "Recent trends and advances of silicon-based integrated microwave photonics," *Photonics*, vol. 6, no. 1, 2019.

[32] G. N. Malheiros-Silveira, G. S. Wiederhecker, and H. E. Hernández-Figueroa, "Dielectric resonator antenna for applications in nanophotonics," *Opt. Express*, vol. 21, no. 1, p. 1234, 2013.

[33] W. T. Sethi, H. Vettikalladi, and H. Fathallah, "Dielectric resonator nanoantenna at optical frequencies," 2015 Int. Conf. Inf. Commun. Technol. Res. ICTRC 2015, pp. 132–135, 2015.

[34] H. Zhou, X. Chen, D. S. Espinoza, A. Mickelson, and D. S. Filipovic, "Nanoscale optical dielectric rod antenna for on-chip interconnecting networks," *IEEE Trans. Microwave Theory Tech.*, vol. 59, no. 10 PART 2, pp. 2624–2632, 2011.

[35] B. Y. Toh, R. Cahill, and V. F. Fusco, "Understanding and measuring circular polarization," *IEEE Trans. Educ.*, vol. 46, no. 3, pp. 313–318, 2003.

[36] G. Rui, R. L. Nelson, and Q. Zhan, "Circularly polarized unidirectional emission via a coupled plasmonic spiral antenna," *Opt. Lett.*, vol. 36, no. 23, p. 4533, 2011.

[37] S. Perhirin and Y. Auffret, "Circularly polarized nanoring antenna for uniform overheating applications," *Microwave Opt. Technol. Lett.*, vol. 55, no. 11, pp. 2562–2568, 2013.

[38] S. Gotra, G. Varshney, R. S. Yaduvanshi, and V. S. Pandey, "Dual-band circular polarisation generation technique with the miniaturisation of a rectangular dielectric resonator antenna," *IET Microwaves Antennas Propag.*, vol. 13, no. 10, pp. 1742–1748, 2019.

[39] M. S. Sharawi, "Current misuses and future prospects for printed multiple-input, multiple-output antenna systems [wireless corner]," *IEEE Antennas Propag. Mag.*, vol. 59, no. 2, pp. 162–170, 2017.

[40] I. F. Akyildiz, J. M. Jornet, and C. Han, "Terahertz band: Next frontier for wireless communications," *Phys. Commun.*, vol. 12, pp. 16–32, 2014.

[41] I. F. Akyildiz and J. M. Jornet, "Realizing ultra-massive MIMO (1024×1024) communication in the (0.06-10) terahertz band," *Nano Commun. Netw.*, vol. 8, pp. 46–54, 2016.

[42] Z. Xu, X. Dong, and J. Bornemann, “Design of a reconfigurable MIMO system for THz communications based on graphene antennas,” *IEEE Trans. Terahertz Sci. Technol.*, vol. 4, no. 5, pp. 609–617, 2014.

[43] G. Varshney, S. Gotra, V. S. Pandey, and R. S. Yaduvanshi, “Proximity-coupled two-port multi-input-multi-output graphene antenna with pattern diversity for THz applications,” *Nano Commun. Netw.*, vol. 21, 2019.

[44] H. Yoon, C. Forsythe, L. Wang, *et al.*, “Measurement of collective dynamical mass of Dirac fermions in graphene,” *Nat. Nanotechnol.*, vol. 9, no. 8, pp. 594–599, 2014.

[45] D. Takase and T. Ohtsuki, “Optical wireless MIMO communications (OMIMO),” *GLOBECOM – IEEE Glob. Telecommun. Conf.*, vol. 2, no. 5, pp. 928–932, 2004.

[46] Y. Chen, B. Weng, and J. Liu, “A novel photonic-based MIMO radar architecture with all channels sharing a single transceiver,” *IEEE Access*, vol. 7, pp. 165093–165102, 2019.

[47] G. Ding, C. Clavero, D. Schweigert, and M. Le, “Thickness and microstructure effects in the optical and electrical properties of silver thin films,” *AIP Adv.*, vol. 5, no. 11, 2015.

[48] G. Varshney, A. Verma, V. S. Pandey *et al.*, “A proximity coupled wideband graphene antenna with the generation of higher order TM modes for THz applications,” *Opt. Mater.*, vol. 85, pp. 456–463, 2018.

[49] R. Garg, P. Bhartia, I. Bahl, and A. Ittipiboon, *Microstrip Antenna Design Handbook*, Artech House, London, UK, 2001.

[50] A. Sharma and A. Biswas, “Wideband multiple-input–multiple-output dielectric resonator antenna,” *IET Microwaves Antennas Propag.*, vol. 11, no. 4, pp. 496–502, 2016.

[51] S. Blanch, J. Romeu, and I. Corbella, “Exact representation of antenna system diversity performance from input parameter description,” *Electron. Lett.*, vol. 39, no. 9, p. 705, 2003.

[52] S. F. Roslan, M. R. Kamarudin, M. Khalily, and M. H. Jamaluddin, “An MIMO rectangular dielectric resonator antenna for 4G applications,” *IEEE Antennas Wirel. Propag. Lett.*, vol. 13, pp. 321–324, 2014.

[53] S. W. Su, C. T. Lee, and F. S. Chang, “Printed MIMO-antenna system using neutralization-line technique for wireless USB-dongle applications,” *IEEE Trans. Antennas Propag.*, vol. 60, no. 2, pp. 456–463, 2012.

[54] J. Nasir, M. H. Jamaluddin, M. Khalily, M. R. Kamarudin, I. Ullah, and R. Selvaraju, “A reduced size dual port MIMO DRA with high isolation for 4G applications,” *Int. J. RF Microwave Comput. Eng.*, vol. 25, no. 6, pp. 495–501, 2015.

[55] M. S. Sharawi and E. Rajo-iglesias, “Printed multi-band MIMO antenna systems and their performance metrics,” *IEEE Antennas Propag. Mag.*, vol. 55, no. 5, pp. 218–232, 2013.

[56] G. N. Malheiros-Silveira, G. S. Wiederhecker, and H. E. Hernández Figueroa, “Dielectric resonator antenna for applications in nanophotonics,” *Opt. Express*, vol. 21, no. 1, p. 1234, 2013.

[57] https://youtu.be/Tb_5t6SbEWY.

[58] A. Tang, T. Reck, and G. Chattopadhyay, "CMOS system-on-chip techniques in millimeter-wave/THz instruments and communications for planetary exploration," *IEEE Commun. Mag.*, vol. 54, no 10, 2016.

[59] C. Fumeaux, C. Zou, D. Headland, *et al.*, "Terahertz and optical dielectric resonator antennas: Potential and challenges for efficient designs," 2016 10th European Conference on Antennas and Propagation (EuCAP), Davos, 2016, pp. 1–4, doi: 10.1109/EuCAP.2016.7481118.

Chapter 2
Dielectric resonator antennas—a comprehensive review

Abstract

Design and implementation of a dielectric resonator antenna (DRA) using ceramic materials having permittivity greater than ten have been used for multiple applications such as mobile communication, radars and satellite communication. Various techniques have been suggested for circular polarization. Trends of getting higher gain by exciting higher order modes have been used in microwave DRAs. Wide bad DRAs have been developed by mode merging techniques. Aspect ratios of 0.5–2.5 have been used in rectangular designs of DRA for exciting multimodes. Rectangular DRA reported as a favorable choice of many designers for simplicity. Terahertz DRA can be developed as miniature DRAs, they can offer high speed and high data rate for communication applications at terahertz or optical frequencies. DRAs at photonic wavelength can be used as artificial retinal photoreceptors and energy harvesting. Also, terahertz frequency can result into wide bandwidth and shall be able to provide terabits per second data rate. In this chapter, a comprehensive review on microwave, terahertz and optical DRAs has been surveyed. Terahertz is new technology, and developing this hardware is complex and requires high-resolution fabrication. Its fabrication is costly due to sophisticated fabrication infrastructure, i.e. an LPKF laser machine can be used to design antennas up to 50 μm. The manufacturing cost can be minimized once technology is matured and large-scale industrial production is taken up.

2.1 Introduction

A dielectric resonator antenna (DRA) is an efficient radiator that has been proved by Richtmyer in 1939, and later SA long experimentally used a DRA for practical implementations. Since then RK Mongia, Kishk, YMM Antaar, Guha, Van Bladel, Pan, Petosa, Leung, Gangawar, Rashidan, Tammor, KM Luk, Withwatt W, Yaduvanshi and many other researchers have done lots of work on DRAs [1–11]. There are few books, recently published by Yaduvanshi in Springer Publishing, 2016 on rectangular DRA theory and design and nano-DRA for 5G applications in Taylor and Francis CRC Press 2020. Terahertz is a new field of applications for

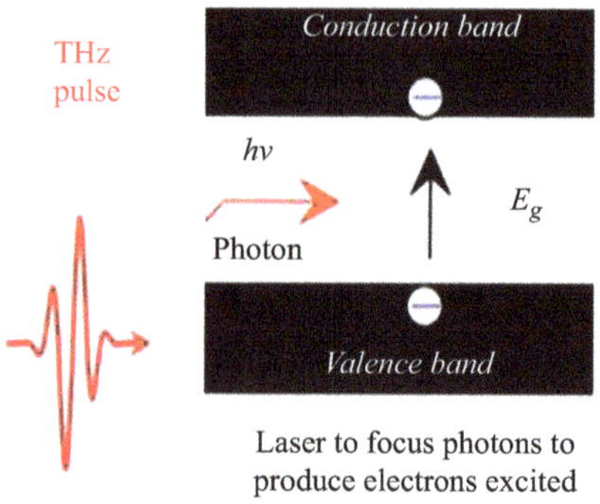

Figure 2.1 Photon energy responsible for electron to jump valence band to conduction band (E = h·v)

Table 2.1 Design dimensions

Parameters	Material	Dimensions (μm)
L_{SUB}	Silicon dioxide (ε_r = 4, μ_r = 1 and mass density = 2,220)	42
W_{SUB}		42
H_{SUB}		8
L_G	Silver (ε_r = 1, μ_r = 0.99998 and mass density = 10,500)	42
W_G		42
H_G		8
L_N	Silver (ε_r = 1, μ_r = 0.99998 and mass density = 10,500)	21
W_N		3
L_R	Silicon (ε_r = 11.9, μ_r = 1, loss tangent = 0.001 and mass density = 2,330)	3.5
B_R		3.5
H_R		3

high-speed communications as this 300-GHz to 10-THz band is still lying vacant for application [1–5]. Terahertz technology has miniaturized the size of DRA. They consume less power. Hence, terahertz DRAs (TDRAs) are in much demand nowadays. The design of TRDRA is shown in Figure 2.1 and dimensions are shown in Table 2.1. Figures 2.2–2.5 are simulated results. The frequency spectrum of microwave, terahertz and photonics is given in Figure 2.1. Here, most of our work shall be in the range of 1,011–1,015 Hz (THz, infrared and optical spectrum). The DR materials used were Al_2O_3, TMM, sapphire, (Ba,Sr)TiO_2 at microwave frequencies and silicon, graphene, TiO_2, Arlon and Teflon at terahertz and optical spectrum. DRAs are efficiently radiating elements throughout the entire frequency spectrum. The permittivities of these materials are available from 10 to 1,600. Hence, due to high permittivity, the materials help to reduce the device size as it is inversely proportional to size, i.e. high permittivity corresponds to a small size. They have design flexibility as DRA command two different aspect ratios height/base length or height/base width. They have capability to generate higher order

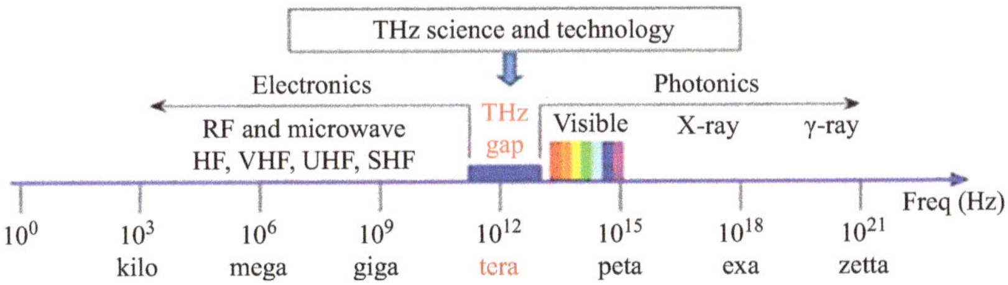

Figure 2.2 Microwave, terahertz and photonic science

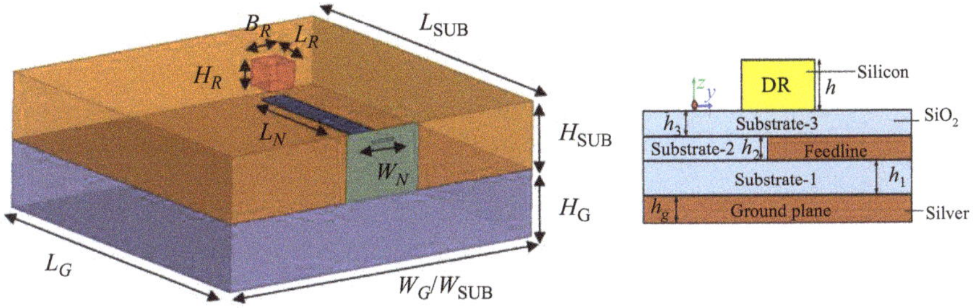

Figure 2.3 Rectangular terahertz DRA (TDRA) with feed and laser source for excitation

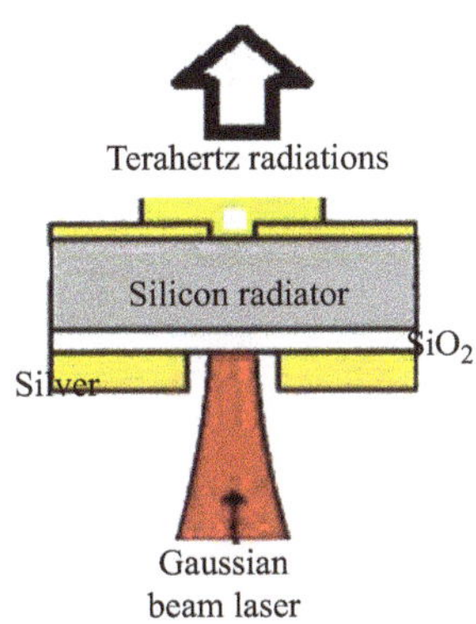

Figure 2.4 Terahertz DRA radiations with a Gaussian beam laser

modes along with fundamental modes depending upon aspect ratios. It is generally chosen between 0.5 and 2.5 for generating resonant modes of our choice. Higher order modes can have higher gain and better directivity as compared to fundamental mode. A terahertz regime can offer higher bandwidth. Hence, terahertz is suitable for higher speed communication. This terahertz band is lying vacant thus can be used for high-speed communication. Due to skin effect, metal behaves like

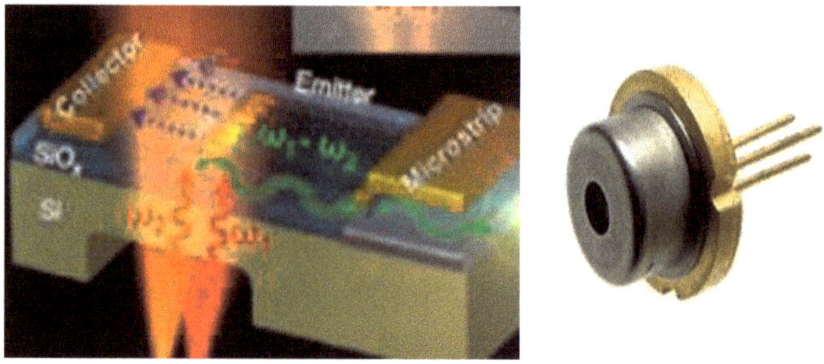

Figure 2.5 TDRA with laser input (quantum cascade semiconductor laser diode)

plasma and Drude's analysis will be required to assess scattering properties. Wave port analysis of a terahertz device may not give accurate results; however, Gaussian pulse along with sliver nano-wave guide shall give accurate results. The Gaussian beam at terahertz is used for phase matching, and proximity feed is used to accommodate electrical effects of surface plasmon polaritons (SPP) length. Plasma frequency of each metal is different, and for propagation it is always kept lower than laser input excitation frequency. The Dirac equation is used to compute radiated fields at optical frequencies. The development of quantum antennas has been a slow process, and operators are used for the mathematical analysis of quantum antennas. There is a great contribution of P Bhardwaj, Bradely, Filonov, S Sriram, L Zou, Malherios-Silveria and L Novelity in the development of light antennas or optical antennas [12–17]. Propagation loss is main disadvantage at higher frequencies. These losses can be minimized by making the use of DRAs at terahertz and optical frequencies. The quantum cascade lasers 100 fs can be used as 1–5 THz, 90 mW, narrow width line and single-mode operation source to feed. Noble metal silver is used as waveguide at higher frequency in the place of micro-strip line feed. With this antenna, Tbps of data rate can be achieved. This high data rate is good for 5G and beyond communication systems. Microwave antennas have limited capability to deliver data speed only up to few Gbps. These terahertz antennas can be placed in chip (CMOS). Of course, they will be costly to develop, but cost can be reduced if they are to be produced at large industrial scale. Initially light antennas were to be operated only near field. By making the use of DR, their gain can be enhanced and they are able to produce far fields and most suitable for long-distance wireless communications at THz and optical frequencies. Terahertz DRAs of cylindrical, conical and spherical shape at optical frequency and dimension can be very useful for the development of artificial retinal photoreceptors as these DRA photonic antennas are efficient [18,19]. Terahertz measurements can be held using vector network analyzers (VNAs) and free-space methods. Compact test range facility for terahertz measurements can be used for echo-free environments also known as CATR (compact antenna test range). It has probe 1.1-THz probe station for test and measurements. For developing THz hardware, the tools required are direct laser writing, THz spectrometers below 90-fs pulse width, THz source/signal generators, THz VNA, electron beam lithography 10-mm resolution, wet and dry etching, metal and dielectric depositions and clean room are essential. High-resolution lithography system is used to create nano-structures to introduce SPP at interface of metal dielectric. Plasmonics is a study, excitation and control of SPP. Design, fabricate, measure and investigate plasmonics devices at terahertz to optical regime is a new science of present and future era. Nano-wires have been used in optoelectronics to develop optical and terahertz devices such as a waveguide. Photon microwave oscillations phenomenon is used in radio over fiber.

2.2 Propagation of light

The understanding of light since Sir Isaac Newton started with first assuming light to be **corpuscles**, namely **small particles** of different color corresponding to the different frequencies. This idea was first propounded by Newton and came to be known as the corpuscular theory of light. Several years later, based on experiments involving interference of light, Huygens came to understand that light is to be regarded as being composed not of particles but rather is a **wave** and that any light field can be regarded as a superposition of plane waves of different frequencies and wave numbers. This meant that during the process of interference involving the computation of the total intensity of light coming from different sources, one should not add the respective intensities, but rather the amplitudes, then square the resultant amplitude and then form its time average. This meant that in the resultant total intensity, there would appear cross terms that are oscillatory and lead us to the phenomenon of interference. Later on, Fermat propounded the ray theory of light in order to explain Snell's theorems on reflection and refraction of light, according to which a light ray always follows the path of minimum time. The wave nature of light as propounded by Christian Huygens gained more support when **Maxwell unified electromagnetism** with light shows that the basic equations of electromagnetism imply that the electric and magnetic fields in space satisfy the wave equation with the velocity of propagation being equal to that of light. This fundamental discovery gave firm support to the wave theory of light. However, with the birth of the quantum theory in the early part of the nineteenth century, it became clear that light comes in discrete packets called quanta and each quantum of radiation carries an energy proportional to its frequency and that the intensity of light is proportional to the number of quanta. These quanta came to be known as photons and it appeared therefore that with the advent of the quantum theory, physicists had reverted back to Newton's corpuscular theory of light! However, while **Planck was busy creating his quantum theory of light**, simultaneously Einstein wrote some beautiful papers proving that light can behave sometimes as a particle and sometimes as a wave, specifically showing that the variance of fluctuations in the energy of black-body radiation has two components: a particle component in agreement with **Planck's quantum hypothesis** and a wave component in agreement with **Rayleigh's theory on the number** of "modes" of a wave field within a given energy shell. The particle theory of light got further support from Einstein's special theory of relativity giving the relationship between the energy and momentum of a particle of any given mass and in particular for light particles having zero rest mass. However, the crucial breakthrough into the wave–particle duality matter and in particular of light came with de Broglie's discovery of the fundamental relationship between the wavelength of a wave and the momentum of the associated particle. The de Broglie theory was applicable to all of matter not just light. It gave conclusive evidence that even particles like electrons exhibit wave–particle duality, i.e. they sometimes behave like particles and sometimes like waves. The final crunch describing the wave–particle duality came with the wave equation for particles discovered by Schrödinger using which he was able to calculate the energy levels of an electron bound to a nucleus in terms of the Eigen values of a partial differential equation. The corresponding Eigen functions were complex wave fields that Schrödinger could not immediately interpret. These Eigen functions approximately corresponded to the kind of "matter-waves" predicted by Louis de Broglie having frequency determined by Planck's quantum hypothesis, relating frequency of waves to the energy of the quanta and having wavelength determined by de Broglie's theory relating the wavelength of waves to the momentum of the particles. Schrödinger's wave equation was a nonrelativistic wave equation in that it could be "derived" by assuming the nonrelativistic relationship between energy and momentum and then assigning operators to energy and momentum. The correct interpretation of the Eigen functions of Schrödinger's equations was provided about a year later by Max Born with the suggestion that the modulus square of the wave function gave the probability density of the particle to be present at a given point in space at a given time. Equivalently, this modulus square could be interpreted as the number density of particles or the intensity of the wave at that point in space at that time. In fact, Schrödinger's equation could also be generalized to determine the wave functions and energy levels for any particle with a prescribed energy–momentum relationship by replacing the energy and momentum with appropriate operators and applying both sides of this relationship to a wave function. Thus, Schrödinger's formalism could be applied to determine the kind of de Broglie waves associated with a particle having any given energy–momentum relation. The de Broglie wave solutions then have the interpretation that their modulus square represents the intensity/probability density of particles in space at a given time. In special relativity, the energy–momentum relationship for a relativistic particle is a quadratic form in the energy and momentum unlike the Newtonian-nonrelativistic case where it is linear in the energy and quadratic in the momentum. Consequently the Schrödinger equation is a linear partial differential equation (PDE) that is of the first order in time and second order in spatial derivatives. The nonrelativistic Schrödinger equation as a consequence then leads to a unitary evolution of the wave function with time and with the additional pleasing property that evolution from time zero to time $t1 + t2$ can be expressed as the composition of evolution from time zero to time $t1$ followed by evolution from time $t1$ to time $t1 + t2$. In other words, Schrödinger evolution follows the semi-group property. This is in contrast to the case of special relativity wherein the wave equation is quadratic in time leading to the wave function at a given time being a superposition of a forward-propagating solution and a backward-propagating solution and the loss of unitarily of the evolution. Thus, the total probability will be conserved in the Schrödinger's nonrelativistic theory but not so in the relativistic theory. This difficulty was finally resolved by Paul Dirac who factorized the energy–momentum relationship of special relativity into linear factors in energy and momentum using 4×4 complex anti-commuting matrices and it led to Dirac's relativistic theory of the electron which along with Maxwell's equations is at the heart of quantum electrodynamics describing the interaction of light with matter. Dirac's equation being linear in the space, and time derivatives are a truly relativistic wave equation for quantum theory because first it respects Lorentz invariance in that space and time are treated as an equal footing and that we can find a spin or representation of the Lorentz group under which Dirac's equation remains invariant and second it is first order in time and therefore preserves the unitary semigroup property of the evolution, guaranteeing therefore the conservation of total probability. Light as we understand today is a second quantized electromagnetic field in which the vector potential components in the spatial frequency domain represent creation and annihilation operators of photons with different momentums and helicities. This is owing to the fact that the total energy of the electromagnetic field can in the Coulomb gauge be expressed as a quadratic functional of the spatial frequency components of the vector potential and each spatial frequency component of the vector potential, according to Maxwell's equations, evolves harmonically with time with a frequency being ω, where k is the wave vector. Such a picture is in perfect accord with the Hamiltonian theory of a collection of independent quantum harmonic oscillators. However, this model of light can actually be traced back to the work of Satyendranath Bose who proposed a statistical method for deriving Planck's law of black-body radiation wherein we are interested in distributing a total amount of energy E having N quanta at frequency ν so that $E = Nh\nu$, amongst p oscillators with the quanta being regarded as indistinguishable particles. When an oscillator has n such quanta, we say in the modern language of quantum mechanics that it has been excited to the nth energy level by an application of n creation operators to the vacuum. In this way, the entire field of black-body radiation is simply a collection of quantum harmonic oscillators and if we maximize the entropy, i.e., the total number of ways of distributing this energy then we end up with the famous Bose–Einstein statistics that gives us the relative number of photons at each frequency. This in turn enables us to determine the intensity of black-body radiation as a function of frequency. Summarizing the modern point of view, the entire photon field described as an operator electromagnetic four potential field is just a superposition of plane waves of different wave vectors whose coefficients are creation and annihilation operators at the different wave vectors and a given state of the photon field is actually a linear combination of number states wherein the number states is specified by the number of photons that occupy each state of definite wave-vector/momentum and definitive helicity/spin.

A particular kind of photon state called a coherent state is a state constructed by an appropriate linear combination of this kind that turns out to be an Eigen state for all the photon annihilation operators. This model for the quantum electromagnetic field implicitly contains both the particle nature and the wave nature of light. The particle nature is contained in the presence of the creation and annihilation operators while the wave nature is contained in the plane waves that act as carrier signal fields for creation and annihilation operators.

Second quantization means a quantization of a classical field theory. It is called second quantization for the following reasons: the first quantization is simply a wave equation like the three-dimensional wave equation, the Schrödinger wave equation for a single or an infinite number of quantum particles, the Klein Gordon equation or the Dirac equation. If such a classical wave equation is quantized, then it describes an infinite number of quantum particles. This can be seen clearly from the following example: take the Klein–Gordon equation that is the wave equation described earlier corresponding to Einstein's energy momentum relation with the energy and momenta replaced by appropriate operators. We expand the solution wave field as a three-dimensional Fourier series within a cube of side length L. The coefficient of this Fourier series then becomes quantum operators, each of which and its adjoint, i.e. Hermitian conjugate describe a single quantum particle. Thus, a second quantized field can equivalently be described by a countable infinite number of quantum particles. For Bosonic quantum fields like the Maxwell electromagnetic photon field, the state of the field can be described, e.g. by specifying how many particles are occupying each state of definite momentum and helicity or equivalently by specifying the momentum and helicity of each of the particles. In this case, there can be zero, one or more than one particle having a specified momentum and helicity. In the case of Fermionic fields like the second quantized Dirac field, we cannot have more than one particle having a definitive value of momentum and spin [20–22]. The output power level of cascade quantum laser can be 10 μW when the operating frequency approaches 1 THz. CATR techniques aiming at providing a feasible approach to THz antenna measurement. The terahertz DRAs have applications in sensing, spectroscopy, light emission enhancing the efficiency light emission, photo detection, on-chip communication, optical components, etc. Developing hardware initially may be a bit costly. This cost can be reduced one production in mass will be taken up. Fabrication steps of TDRA are dimensions for terahertz wavelength—nanofabrication process will take place as silver deposition-, electron-beam lithography for nano-patterning and Si deposition, annealing, spin-coating and curing of SiO_2 with subsequent peel from carrier substrate, plasma plume, dry etching Si resonators embedded in an SiO_2 scanning electron microscope. Here, we start with a silver layer deposited by electron beam evaporation on a standard, cleaned silicon wafer of 3″, coating and baking at 180° temperature, emersion to wash, lift off to remove extra and then anneal at 600° temperature.

2.3 Design of a terahertz dielectric resonator antenna

Figure 2.1 shows a frequency spectrum and Figure 2.2 shows a terahertz rectangular antenna, which is designed based on Table 2.1, which has all dimensions of feed, substrate and silicon-radiating elements. SPP phenomenon takes place at light matter interaction, and Gaussian beam is input given to feed. Silver nano-wave guide has been used as feed. It is well known that when a photon is incident on a semiconductor substrate, an electron may transit from the valence band to the conduction band if the band gap E_g is smaller compared to the energy of the photon. Light matter interaction, electron–hole pair are formed, with an electron in the conduction band and a hole in the valence band. This will introduce currents and antenna radiations (THz). Dipole formation and dipole moments are called as a wave vector, a second-order equation or time derivative or called retarded potentials at far fields.

2.4 Fabrication and testing

Terahertz time-domain spectroscopy (TDS) is used for testing. The terahertz TDS system operates both in transmission and reflection modes. SU-8 of 500–600 nm is used as glue Si DRA. It can be tested in reflection or transmission mode. Terahertz TDS in reflection mode is used for testing. To develop DRA of Si, process on a 3″ wafer can be used. The complex Si-binding process makes it hard. For terahertz, we use high resistivity Si (intrinsic, float zone Si). The materials like TiO_2 are used for optical frequencies applications. The VNA is used to directly probe and test at 5 THz; else free-space testing can also be an option to test THz antennas. To combine single-crystal Si with SiO_2, it is very complicated. To work at terahertz, the Si has to be intrinsic Si (no impurities), and so, cannot bond to SiO_2 by typical techniques (like anodic bonding) and needs a very thin cross-linking polymer (imagine it is like glue). This is the role of the thin SU-8.

2.5 Terahertz antenna far-field radiations: flowchart

Light in Gaussian pulse for phase matching.

⇓

The electrons **and positrons or the quantum matter field in a Fermionic coherent state.**

⇓

Define the **Fermionic annihilation operator.**

⇓

Describe the **Fermionic Dirac field.**

⇓

Compute the Dirac current.

⇓

Determine Eigen solution to the Dirac wave equation using 4×4 Dirac α-matrices satisfying the anti—commutation relation.

⇓

Express the **Dirac current density.**

⇓

Find **far-field magnetic vector potential.**

⇓

Determine **far-field radiation pattern.**

⇓

Berezin weight function.

⇓

Minimize the **quantum fluctuation correlation** in the pattern for possible matching the **average antenna pattern to the desired pattern.**

⇓

Electromagnetic radiated fields (Super directive)

Electrons dominate at THz generation. A laser-A 20-mW tungsten halogen lamp is selected as the white light source, incident power, biconvex lens, focal length plano-convex lens are used to collimate the broadly diverged light into a numerical aperture objective, and sheet polarizer focuses the light beam into a pinhole, sufficient signal power excites the DRA.

The single dielectric resonator operating in its fundamental electric dipolar mode. *e* Dielectric resonator operates on the displacement current, hence, displacement charge density, displacement current density, displacement charges, displacement current coupling of near fields with electric dipoles. Silver feed is inserted into an SiO_2 substrate. DRA proximity feed is used in a THz antenna. SPP is used to introduce radiations in silicon DRA. Gaussian beam is used for phase matching due to nonlinear phenomenon.

The fabrication part of terahertz DRAs can be taken up using silicon wafers. The materials of microwave DRAs built were Eccostock, Rogers TMM, RT Duroid, alumina (Al_2O_3), TiO_2, graphene, silicon and MOS_2.

2.6 Mathematical analysis of terahertz RDRA

Consider a cavity resonator of one angstrom size, i.e. a cube with each side of length $a = 10^{-10}$ m. The Maxwell equations in such a cube have solutions of the form

$$A_r(t,x,y,z) = \sum_{mnp} c(mnp,t)u_r, mnp(x,y,z), r = 1,2,3\ldots \tag{2.1}$$

where u_r, mnp are spatial functions obtained by integrating the electric field w.r.t.

$$\left\{\cos\left(\frac{m\pi x}{a}\right), \sin\left(\frac{m\pi x}{a}\right)\right\}$$

$$\otimes\left\{\cos\left(\frac{n\pi y}{a}\right), \sin\left(\frac{n\pi y}{a}\right) \otimes \left\{\cos\left(\frac{p\pi y}{a}\right), \sin\left(\frac{p\pi z}{a}\right)\right\}$$

Multiplied by some constants depending on the indices (m, n, p), we may, without any loss of generality, assume that these are normalized so that

$$\int_C u_r, mnp(r)\bar{u}_s, m'n'p'(r)d^3r = \delta_{rs}\delta_{mm'}\delta_{nn'}\delta_{pp'}$$

The dependence of $c(mnp, t)$ on t is $\exp(i\omega(mnp)t)$ where $\omega(mnp)$ are the characteristic frequencies of oscillation:

$$\omega(mnp) = \left(\frac{\pi c}{a}\right)\sqrt{m^2 + n^2 + p^2}, \; m, n, p = 1, 2, \ldots \tag{2.2}$$

which are of the order of magnitude:

$$\omega = \frac{\pi c}{a}$$

The electric field is

$$E_r = \partial_t A_r = \sum_{mnp} c(mnp, t)i\omega(mnp)u_r, mnp(r)$$

The magnetic field is

$$B = curlA$$

which is of the order of magnitude $|c(mnp, t)|/a$ where *by* $c(mnp, t)$ we actually mean its average in a coherent state. The total electric field energy within the cavity C is

$$U_E = \left(\frac{\epsilon_0}{2}\right)\int_C |E|^2 d^3r$$

which has components of the order of magnitude

$$\epsilon_0/\omega(mnp)c(mnp, t)|^2 a^3 = \epsilon_0\omega(mnp)^2 a^3 |c(mnp, t)|^2$$

The total magnetic field energy within the cavity is $U_B = (2\mu_0)^{-1}\int_c |B|^2 d^3r$

which has components of the order of magnitude:

$$\left|\frac{c(mnp, t)}{a}\right|^2 \frac{a^3}{\mu_0} = |c(mnp, t)|^2 \frac{a}{\mu_0}$$

The ration of the orders of magnitude of the electric field energy and the magnetic field energy within the cavity therefore has the order of magnitude:

$$\frac{U_E}{U_B} \approx \mu_0\epsilon_0\omega(mnp)^2 a^2 \approx \frac{\omega^2 a^2}{c^2} \approx 1$$

As expected, the canonical commutation relations are

$$[A_r(t,r), \partial_t A_s(t,r')] = \left(\frac{ih}{2\pi}\right)\delta^3(r-r') \tag{2.3}$$

These fields are

$$[c_r(mnp,t), \omega(mnp)c_s(m'n'p',t)^*] = \left(\frac{h}{2\pi}\right)\delta_{rs}\delta_{mm'}\delta_{nn'}\delta_{pp'}$$

So that the eigenvalues of $c_r(mnp,t)^*c_r(mnp,t)$ are positive integer multiples of $h/2\pi\omega(mnp)$. This means that the field energy within the cavity when a finite number of modes are excited. We assume eigenvalues that are of the same order of magnitude as positive integer multiples of $h\omega/2\pi$ as expected by Planck's quantum theory of radiation. This fact also fields the result that $|c(mnp,t)|$ is of the order of magnitude of $\sqrt{h/(2\pi\omega)}$.

Now we come to the question of computing the order of magnitude of the pointing vector power flux at a given radial distance R from the quantum cavity antenna caused by the surface current density induced by the magnetic field on the antenna surface. The magnetic field on the surface and hence the corresponding induced surface current density both have the order of magnitudes of $|c(mnp,t)|/a$, which is of the order $a^{-1}\sqrt{h/\omega}$. Therefore, the far-field magnetic vector potential at a distance R from the cavity is of the order of magnitude (use the retarded potential formula) $(a/R)\sqrt{h/\omega}$ and hence the corresponding far-field radiated magnetic field is of the order of magnitude $(w/c)(a/R)\ \sqrt{h/\omega}$ while the near-field magnetic field of the order of magnitude $(a/R^2)\sqrt{h/\omega}$. Actually, these expressions for the magnetic field must be multiplied by $\sqrt{N}$ where N is a positive integer corresponding to the largest modal eigenvalue of the operators $(2\pi\omega(mnp)/h)$ $c(mnp,t)^*c(mnp,t)$.

The far-field Poynting vector has the order of magnitude of $B^2c/2\mu_0$ that is of the order

$$\left(\frac{c}{2\mu_0}\right)\left(\sqrt{N}\cdot\left(\frac{\omega}{c}\right)\left(\frac{a}{R}\right)\cdot\sqrt{\frac{h}{\omega}}\right)^2 = \left(\frac{h}{2\mu_0}\right)N\cdot\left(\frac{\omega}{c}\right)\left(\frac{a^2}{R^2}\right)$$

And the total power radiated outward by this quantum antenna in the far-field zone is thus of the order of magnitude:

$$P = N = \left(\frac{h}{2\mu_0}\right)\left(\frac{a^2\omega}{c}\right)$$

Now we look at the order of magnitude of the power radiated in the far-field zone by the Dirac field of electrons and positrons within the cavity. The Dirac equation is

$$\left[i\gamma^\mu\partial_\mu - m\right]\psi(x) = 0$$

Or more precisely in arbitrary units,

$$\left[\left(\frac{ih}{2\pi}\right)\partial_t - c\left(\alpha, \left(-\frac{ih}{2\pi}\right)\nabla\right) - \beta mc^2\right]\psi(x) = 0$$

where the appearances of the constants h, m, c are explicitly shown. Now $|\psi(x)|^2 c$ is the probability density of the electron that must integrate to unity over the cavity volume. Thus, $\psi(x)$ is of the order of magnitude $a^{-3/2}$. The Dirac current density $J^\mu = e\psi^*\gamma^0\gamma^\mu\psi$ has the same order of magnitude as $e|\psi(x)|^2 c$ which is $ec/a^{3/2}$. Therefore, the far-field magnetic vector potential at a radial distance of R from the cavity is, in accordance with the retarded with the retarded potential theory of the order,

$$\left(\frac{ec}{a^{3/2}}\right)\cdot\left(\frac{a^3}{R}\right) = \frac{eca^{3/2}}{R}$$

The electric field in the far-field zone is then of the order

$$E \approx \omega \cdot \frac{eca^{3/2}}{R} \tag{2.4}$$

where ω is the characteristic oscillation frequency of the Dirac current. The magnetic field is of the order

$$B \approx a^{-1} \cdot \frac{eca^{3/2}}{R} = ec\sqrt{\frac{a}{R}}$$

If P is the characteristic momentum of the electrons and positrons in a given state, for example, P may be the average momentum of an electron in a given state, then according to de Broglie, P is of the order h/a since a is the order of the electron wavelength. Then the electron energy is of the order

$$E_e = c\sqrt{m^2c^2 + p^2} \approx c\sqrt{\frac{m^2c^2 + h^2}{a^2}} \tag{2.5}$$

And the characteristic frequency of oscillation of the Dirac wave field is then

$$\omega = \frac{E_e}{h}$$

The Poynting vector corresponding to the power radiated by the Dirac field in the far-field zone then has the order of magnitude:

$$S \approx c\left(\epsilon_0 E^2 + \frac{B^2}{\mu_0}\right) = c^3\epsilon_0\omega^2 ea^3 + \frac{e^2c^3a}{\mu_0} \tag{2.6}$$

Equations (2.1)–(2.6) have been derived for formulating rectangular quantum or optical antennas.

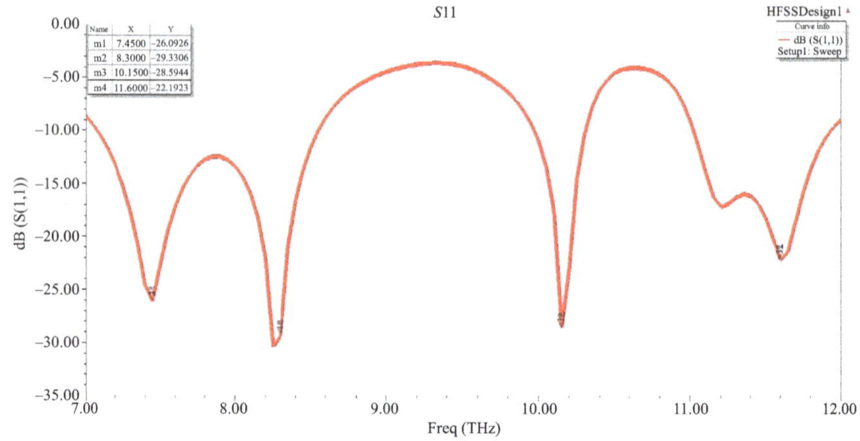

Figure 2.6 S11 of rectangular TDRA

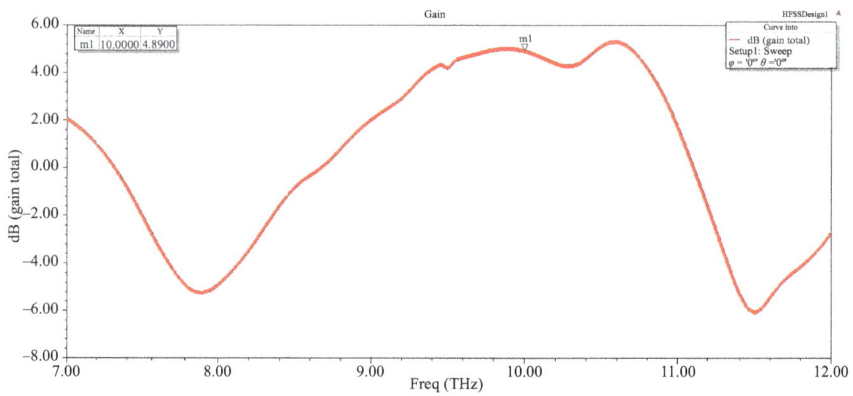

Figure 2.7 Gain of rectangular TDRA

2.7 Approximate analysis of a rectangular quantum antenna

The quantum antenna is assumed to be the cuboid region $[0, a] \times [0, b] \times [0, d]$. This rectangular cavity is assumed to comprise of photons, electrons and positrons. The exact equations governing the quantum fields corresponding to these particles are as follows:

1. The Maxwell equations for the four vector potential driven by the Dirac field current.
2. The Dirac field equations driven by an interaction between the Dirac field and the Maxwell field four vector potentials.

3. Define the Fourier components of the control classical laser generated electromagnetic field w.r.t. the cavity boundary conditions and the energy spectrum of the free Dirac field in the cavity.
4. The controllable part of the Dirac four current density is then given up to first-order perturbation terms by ($x = (t, r)$).
5. The far-field radiated electromagnetic potential generated by this controllable current field can be expressed in Hermitian quadratic form.
6. Let $A^c_\mu(x)$ denote the classical electromagnetic four potential from the laser and $J^c_\mu(x)$ the classical current density coming from the probe insertion.

$$A_\mu = -e\mu_0\psi * \alpha_\mu\psi + \mu_0 J^c_\mu$$

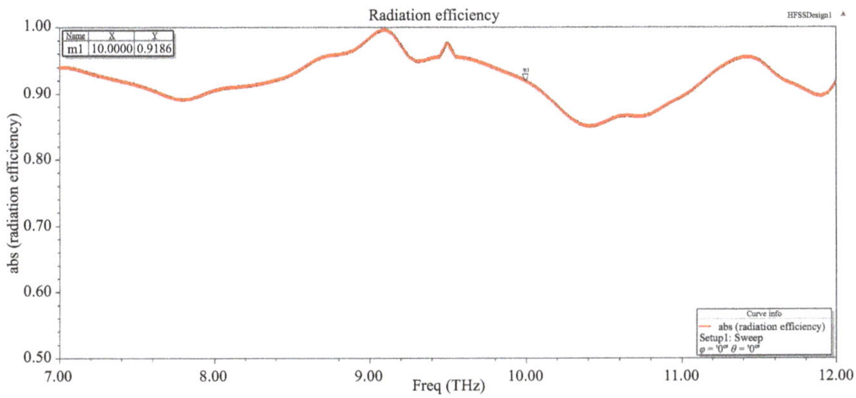

Figure 2.8 Efficiency of rectangular TDRA

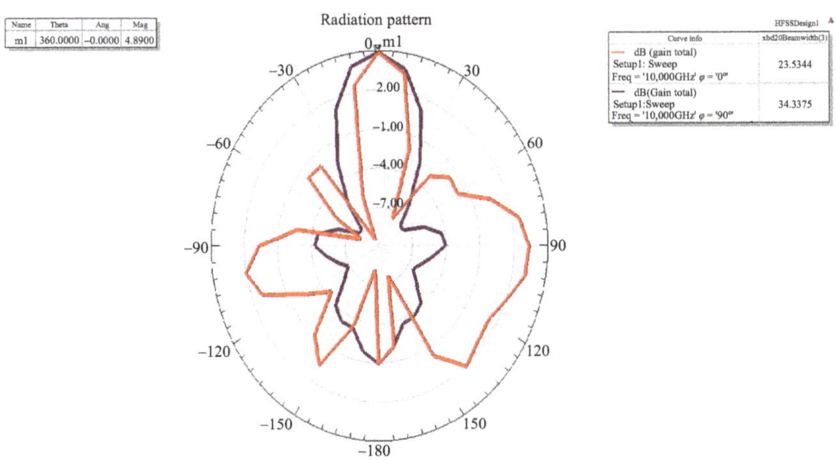

Figure 2.9 Radiation pattern of rectangular DRA

2.8 Terahertz DRA simulation results

A rectangular terahertz antenna has been simulated using high frequency structure simulator (HFSS) and results are presented in Figures 2.6–2.9.

2.9 Conclusion

Terahertz DRAs are futuristic and require high design precision fabrications technology. A laser machine can be used to fabricate computer simulation technology (CST) designs such .dfx or Gerber data files. These CST files can be directly exported to LPKF laser design and fabrication prototype machines. 3D printers can also be used to develop prototype polymer DRA models using specific inks as miniature 3D devices. DRAs can have multiple operating bands at terahertz. Rectangular DRAs at terahertz frequency have compact and simple designs. These are useful for high-speed communications. Terahertz DRAs are compact and efficient, and they can operate at an entire spectrum, i.e. microwave to optical regime. They have large design flexibility and customized aspect ratios. At optical frequency TDRAs can be used for retinal photo receptors and photonic energy harvesters. As compared to silver nano-wire waveguides, SP waveguides can give a much higher confinement for light field. Single-mode field patterns of the SP wave on a single interface between a metal and dielectric have been observed. The fabrication of terahertz device needs clean room for micromachining, dry and wet etching, angstrom engineering electron beam lithography, sputtering system, masking system, spin coater, bake plate, surface roughness tester, photo-resist chemical, microscope, spectrometer, signal generator, laser source, VNA and extension head (10 MHz to 500 THz). An SU 8 glue-type film or layer has 1-μm precision in fabrication. The GDS II design file is compatible with foundry. Hence, silicon cylindrical dielectric resonator antenna (CDRA) on a silver surface and an SiO_2 substrate at THz can be developed with the previous complex procedure and sophisticated instrumentations.

References

[1] R. D. Richtmyer, "Dielectric resonators," *J. Appl. Phys.*, vol. 10, pp. 391–398, 1939.

[2] S. A. Long, M. McAllister, and L. C. Shen, "The resonant cylindrical dielectric cavity antenna," *IEEE Trans. Antennas Propag.*, vol. AP-31, 1983.

[3] M. Gastine, L. Courtois, and J. J. Dormann, "Electromagnetic resonances of free dielectric spheres," *IEEE Trans. Microwave Theory Tech.*, vol. MTT15, no. 12, pp. 694–700, 1967.

[4] R. S. Yaduvanshi and H. Parthasarathy, *Rectangular Dielectric Resonator Antennas: Theory and Design*, Springer, New Delhi, 2016.

[5] J. Van Bladel, "On the resonances of a dielectric resonator of very high permittivity," *IEEE Trans. Microwave Theory Tech.*, vol. MTT-23, no. 2, pp. 199–208, 1975.

[6] R. S Yaduvanshi and V. Gaurav, *Nano Dielectric resonator for 5G applications*, CRC Press, Boca Raton, FL, 2020.
[7] K. M. Rajesh and B. Prakash, "Dielectric resonator antenna-a review and general design relation for resonant frequency and bandwidth," *Int. J. Microwave Millimetre Wave Comput. Aided Eng.*, vol. 4, no. 3, pp. 230–247, 1994.
[8] Y. M. Pan, K. W. Leung, and L. Guo, "Compact laterally radiating dielectric resonator antenna with small ground plane," *IEEE Trans. Antennas Propag.*, vol. 65, no. 8, pp. 4305–4310, 2017.
[9] Y. M. M. Antar, "Antennas for wireless communication: Recent advances using dielectric resonators," *IET Circuits Devices Syst.*, vol. 2, no. 1, pp. 133–138, 2008.
[10] A. Petosa and A. Ittipiboon, "Dielectric resonator antennas: A historical review and the current state of the art," *IEEE Antennas Propag Mag.*, vol. 52, no. 5, pp. 91–116, 2010.
[11] K. W. Leung, E. H. Lim, and X. S. Fang, "Dielectric resonator antennas," *Proceedings of the IEEE*, vol. 100, pp. 2181–2193, 2012.
[12] P. A. Martin and F. Rothen, *Many-Body Problems and Quantum Field Theory*, Springer, New York, NY, 2002.
[13] P. Bharadwaj and L. Novotny, "Spectral dependence of single molecule fluorescence enhancement," *Opt. Express*, vol. 15, pp. 14266–14274, 2007.
[14] L. Novotny, "Optical antennas: A new technology that can enhance light-matter interactions," *Front. Eng.*, vol. 39, no. 4, pp. 100–120, 2012.
[15] S. Lepeshov and A. Gorodetsky, "All-dielectric optically tunable metasurface for terahertz phase and amplitude modulation," *J. Phys. Conf. Ser.*, vol. 1461, p. 012203, 2020. IOP Publishing, doi:10.1088/1742-6596/1461/1/012203.
[16] M. Hayashi, *Quantum Information Theory*, Springer, Berlin Heidelberg, 2017.
[17] P. Bhardwaj, B. Deutsch, and L. Novotny, "Optical Antennas," *Adv. Opt. Photon.*, vol. 1, pp. 438–483, 2009.
[18] M. Humayun and L. Olmos de Koo, *Retinal Prosthesis—A Clinical Guide to Successful Implementation*, Springer, Cham, 2018.
[19] Y. Yang, Y. Yamagami, X. Yu, *et al.*, "Terahertz topological photonics for on-chip communication," *Nat. Photonics*, 2020.
[20] J. W. Choong, N. Nefedkin, and A. Krasnok, "Collectively driven optical nanoantennas," *Optics*, 2020.
[21] S. Weinberg, *The Quantum Theory of Fields*, vol. 1, Cambridge University Press, New York, NY, 1995.
[22] L. Zou *et al.*, "Dielectric resonator nanoantennas at visible Thank you for your attention! frequencies," *Opt. Express*, vol. 21, no. 1, pp. 1344–1352, 2013.

Chapter 3

Light–matter interaction in terahertz dielectric resonator antennas (DRA)

Abstract

Photons interact with atoms, molecules and other particles of metals and air particles in space or environment, out of which the phenomenon of scattering, absorption, polarization, coupling, propagation, creation and annihilation takes place. Transportation photons and particles takes place in temporal–spatial domains. Boltzmann coupled with Maxwell's equation can thus provide exact solution of wave propagation and quantum fields operators. Also, Schrödinger equation can provide solution to photon fields. Skin effect is dominant in photonic devices, and nonlinear phenomenon also gets introduced at optical spectrum. The solution of electromagnetic wave propagation and fields quantum is different from classical microwave regime. In this chapter, mathematical modeling of quantum fields generation, interaction and propagation has been developed and analyzed. The Dirac equation has been used to determine quantum wave fields. Human eyes have retinas with 5 million of cones of photonic wavelength in central part and 126 million rods in peripheral part to receive photons. They form vision by capturing any image from the real world. They are also known as photo receptors. In brief, light–matter interaction in retinal part takes place, conversion of phonic signal into electrical impulses takes place, now these electrical pulses get accumulated and travel through optical nerves and carry composite signal to brain, which further gets converted into pixels and thus image is formed. Light scattering and propagation solution is provided by Drude's model. Quantum entanglement is another important phenomenon to be used for quantum communication. This is also used for secure quantum communications.

3.1 Introduction

Light–matter interaction is nonlinear phenomenon, it can be expressed as behavior of optical spectrum, when, electromagnetic (e.m.) wave is propagating in space. Boltzmann Transport equation, Schrödinger equation and coupled Maxwell's equations can provide exact solution for this quantum phenomenon. These can be rays or waves or particles methods of solution. This chapter describes mathematical part of light–matter interaction and physics of optical antennas. The difference between microwave and quantum antennas has been developed. The solution of problems has been obtained based on classical and quantum approaches. Scattering and electrodynamics take place during light–matter interaction [1]. Orientation of propagation needs to be defined for both the cases separately. Maxwell's equation can provide solution to classical case and the Schrödinger equation is used for optical cases. Dipole moments and polarization are defined for both cases. Strong coupling for field enhancement is defined in optical case. Reflection and interference are two important parameters. Quality factor is defined for these cases. Three-dimensional control of light can be better achieved by spherical dielectric resonator antennas (DRAs). Due to skin effect, light is localized in spherical DRA. Absorption and field fluctuations are two important phenomena in light–matter interactions. The interaction of light with atoms, molecules and excitons takes place. Based on these, weak or strong coupling takes place. Thus, dispersion function is computed. Hamiltonian of photon and matter interaction is computed. Also, mean and covariance of e.m. fields are estimated.

3.2 Light–matter interaction theory in a quantum antenna

If we take an example of eye retina as quantum receive, only antenna, here, reflected light from an object falls on the surface of the eye cornea; it enters into eye lens, spherical vitreous humor and then to retina (rods and cones) in process to create an image of the object on the retina. Later the same signal of image is sensed by the brain. Plank's constant has a finite value of a quantum device, and absorption or emission of light particle takes place as photons. This process of image formation is usually described by modeling the surface of the eye lens to be spherical or more precisely of a parabolic shape so that using the standard Snell's laws of refraction, the path of a given light ray from the object can be broken into three parts: first, the part before it hits the lens; second, the part after it hits first surface of the lens and propagates within the lens till it hits the second surface; and third, the part after it emerges out of the second surface of the lens and till it hits the retinal screen. It can be shown that using Snell's laws of refraction that if u is the distance of the object from the lens and v is the distance of the image then $1/u + 1/v = 1/f$, where f is the lens focal length.

This is a classical ray description of image formation. The image of a point is formed on the retina if all the rays emanating from that point, after passing through the lens, meet at a single point on the retina. The condition for this to happen puts severe restrictions on the shape of the two surfaces of the lens. The wave description of light when it passes through the eye channel and till it hits the retina is however quite different. It involves modeling the eye channel as a cylindrical cavity resonator with the retinal disk being one surface and analyzing the wave fields within this cavity using the Maxwell equations. Specifically, using this formalism, we can calculate the e.m. wave field pattern on the retinal disk given its pattern at the other open disk end. This process just involves expressing the

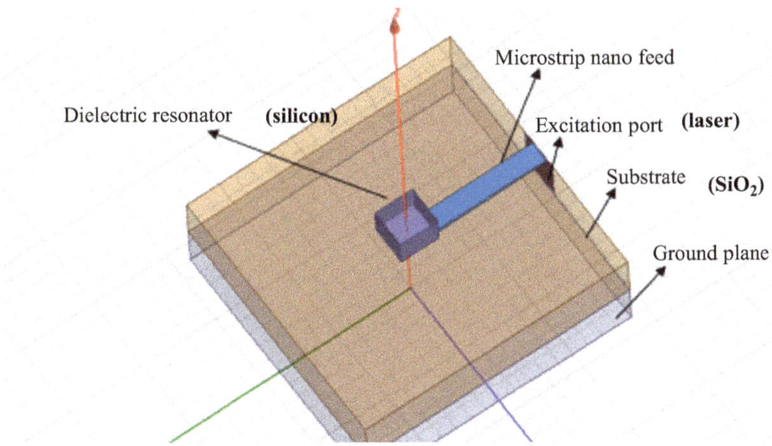

Figure 3.1 Light–matter interaction, SPP and radiated quantum states

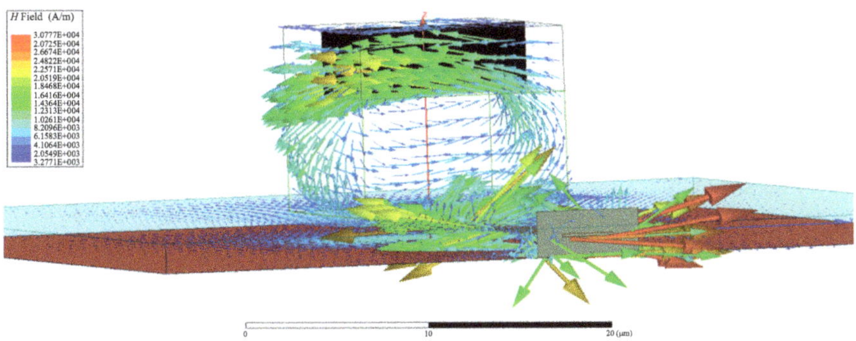

Figure 3.2 Effects of light–matter interactions at frequency 192.68 THz

transverse components of the *E*, *H* field in terms of the longitudinal components of the *E* and *H* fields and computing the coefficient of the longitudinal components using the specified fields pattern at the mouth of the cylindrical dielectric resonator antenna. Figure 3.1 shows rectangular DRA. Figure 3.2 shows *H*-fileds and Figure 3.3 shows radiated fileds at 192.68 THz frequency.

However, if we pass over to the quantum/nano-devices, even this wave field description is inaccurate. Indeed, today we know that the e.m. field is neither a ray nor a wave field. It is actually a quantum wave operator field. Quantum wave operator field analysis can provide exact solution to quantum fields. More precisely, it is an ensemble of an infinite number of quantum harmonic oscillators or in fact, a superposition of plane or constrained plane waves with coefficient being creation and annihilation operators in a boson Fock space. It is therefore not quite precise to talk about the wave field pattern on the retina but rather specify the

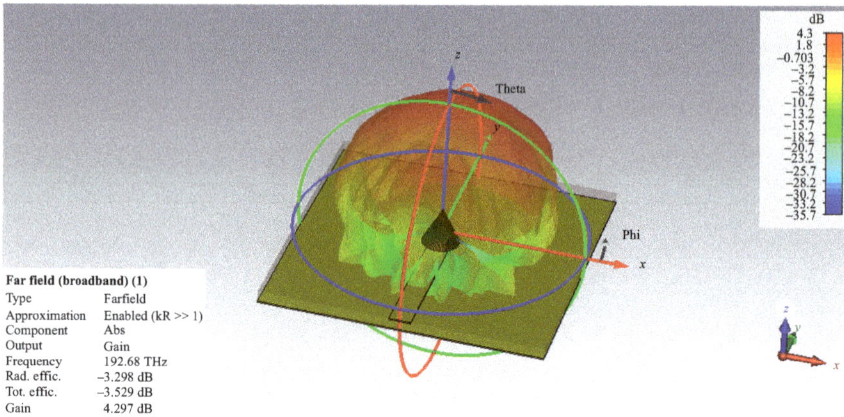

Figure 3.3 Light–matter interaction and radiation pattern of THz DRA at frequency 192.68 THz

quantum state of the wave field like a coherent state, or a number states, and then compute the statistics of the wave operator field.

On the retinal surface, the action of creation and annihilation operators on coherent states or on photon number states is used. After we do such an analysis, we can ask a more interesting question: how does light within the cylindrical cavity resonator interact with the matter field? Specifically, when light propagates within the cylindrical eye cavity, it interacts with electrons, atoms and molecules of the fluid medium of which the eye cavity is composed and its propagation therefore gets affected by this medium. In classical wave field theory, we would model this fluid medium as a plasma composed having a definitive permittivity, permeability and conductivity and would analyze the propagation of the e.m. field in this plasma using the Vlasov equations that are in fact the coupled Boltzmann equation for the particle distribution function of the plasma and the Maxwell e.m. field. The result of this analysis using linearized perturbation theory would be **dispersion relation between the oscillation frequencies of the plasma and the e.m. field and the wave vector.** However, at the quantum level the description of the interaction between light and matter is more subtle. A simplified analysis would be to model the matter as just the second quantized electron–positron field using Dirac's relativistic wave equation with the wave function being an operator field and then to include the photon interaction term in the usual way. In this way, the total Hamiltonian of the photon and matter field splits into three terms: one, the Hamiltonian of the e.m. field described as a quadratic form in the photon creation and annihilation operators within the cavity; two, the Hamiltonian of the Dirac field with cylindrical cavity boundary conditions on the wave function described as a quadratic form in the electron–positron creation and annihilation operators; and three, the Hamiltonian of the interaction between the photon and electron–positron field described as a quadratic form in the electron–positron creation–annihilation operators multiplied with a linear form in the photon

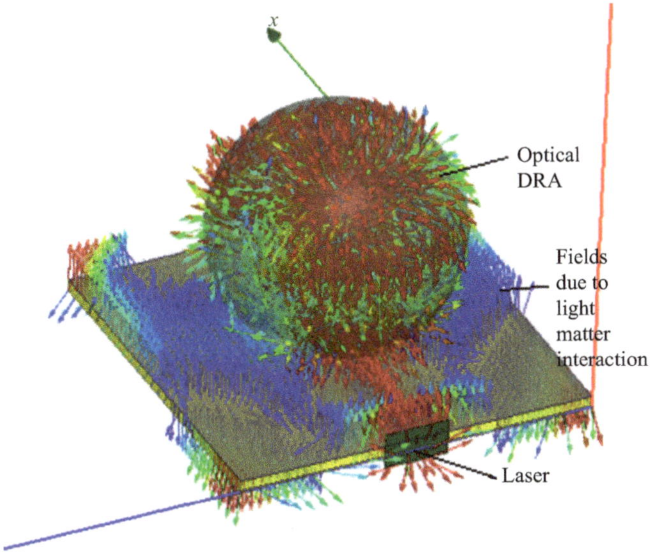

Figure 3.4 Light–matter interaction in optical DRA resulting into SPP and e.m. field radiations into space

creation–annihilation operators. This interaction Hamiltonian can be used to compute the amplitudes for scattering but since we are primarily interested in the statistics of the e.m. wave field pattern on the retinal surface, we shall describe this interaction using the Dirac current density expressed as a quadratic form in the electron–positron creation–annihilation operator fields that drives the photon field using the wave equation for the e.m. field in the presence of a current density. Thus, we shall by perturbation theory be able to calculate the change in the e.m. fields pattern on the retinal surface caused by the Dirac current in terms of operators and then by assuming a definitive state of the photon–electron–positron field, we shall be able to compute the mean and covariance of the e.m. field on the retinal screen surface in this state. Figure 3.4 is an example of light–matter interaction. Quantum field radiations in optical antennas or terahertz antennas occur due to light–matter interactions.

3.3 Theory of quantum entanglement

Entanglement is the process of generation of two coherent states of photons. These states are found to be synchronized. They are useful in secure quantum communication.

Radiated far fields are defined as single state, by making use of operator fields. The theory of entanglement has been discussed as follows:

$$\psi^{(0)}(x) = \sum_k [b(k)\chi_k(x) + c(k)^* \eta_k(x)] \tag{3.1}$$

The free cavity constrained e.m. four potential is defined as

$$A_{\mu}(x) = \sum_{k} [a(k)\theta_k(x) + a(k)^*\theta_k(x)^*] \tag{3.2}$$

where $a(k)a(k)^*$ are the photon annihilation and creation operators, while $b(k)$ $b(k)^*$ are the electron annihilation and creation operators and $c(k)c(k)^*$are the positron annihilation and creation operators. We get, for an approximate value of the Dirac current operator,

$$J^{\mu} = \left(\psi^{(0)} + \psi^{(1)}\right)^* \alpha^{\mu}\left(\psi^{(0)} + \psi^{(1)}\right) = J^{\mu(0)}(x) + \delta J^{\mu} \tag{3.3}$$

where $J^{\mu(0)} = \psi^{(0)*}\alpha^{\mu}\psi^{(0)}$,

$$\delta J^{\mu} = \psi^{(0)*}\alpha^{\mu}\psi^{(1)} + \psi^{(1)*}\alpha^{\mu}\psi^{(0)}$$

Now,

$$\begin{aligned} J^{\mu(0)} &= \sum_{k,m} (b(k)\chi_m(x) + c(k)^*\eta_k(x))^*\alpha^{\mu}(b(m)\chi_m(x) + c(m)^*\eta_m(x)) \\ &= \sum_{k,m} [b(k)^*b(k)\chi_k(x)^*\alpha^{\mu}\chi_m(x) + c(m)^*c(k)\alpha^{\mu}\eta_m(x)\eta_k(x)^* \\ &+ b(k)^*c(m)^*\alpha^{\mu}\eta_m(x)\chi_k(x)^* + c(m)^*c(k)\alpha^{\mu}\eta_m(x)] \end{aligned} \tag{3.4}$$

Which is the free Dirac current, i.e., in the absence of the interactions with the photon field? We have already indicated how to compute the far-field radiation pattern produced by this field and how to evaluate the moments of this field. Specifically, if $G(x-y)$ denotes the causal Green's function for the wave operator, then the electromagnetic four potentials produced by the Dirac current are given by

$$\begin{aligned} A_{\mu}(x) &= \int G(x-y)J^{\mu}(y)d^4y \\ &= \int G(x-y)J^{\mu(0)}(y)d^4y + \int G(x-y)\delta J^{\mu}(y)d^4y \\ &= A_{\mu}^{(0)}(x) + \delta A_{\mu}(x) \end{aligned} \tag{3.5}$$

Remark: We can consider a fermionic coherent state rather than a fermionic number state. Such a state is parameterized by a Grassmannian vector variable $\gamma = (\gamma_b(k), \gamma_c(k))$ and is denoted by $|\varnothing(\gamma)>$, and the action of electron and positron annihilation operators on the state is

$$b(k)|\varnothing(\gamma)> = \gamma_b(k)|\varnothing(\gamma)>,$$
$$c(k)|\varnothing(\gamma)> = \gamma_c(k)|\varnothing(\gamma)>$$

In order that the canonical anticommutation relations (CAR)

$$[b(k), b(m)]_+ = [c[k], c(m)]_+ = [b(k), c(m)]_+ = 0$$

holds good, we require that the Grassmannian parameters satisfy the following anticommutation rules:

$$\gamma_b(k)\gamma_b(m) + \gamma_b(m)\gamma_b(k) = 0,$$
$$\gamma_c(k)\gamma_c(m) + \gamma_c(m)\gamma_c(k) = 0,$$
$$\gamma_b(k)\gamma_c(m) + \gamma_c(m)\gamma_b(k) = 0,$$

Further, by our analogy with bosonic coherent states, we impose the requirement that

$$\left[\frac{\partial}{\partial\gamma_r(k)}, \gamma_s(m)\right]_+ = \delta(r,s)\delta(k,m), r, s = b, c$$

And that

$$b(k)^*|\varnothing(\gamma)> = \left(\frac{\partial}{\partial\gamma_b(k)}\right)|\varnothing(\gamma)>,$$
$$c(k)^*|\varnothing(\gamma)> = \left(\frac{\partial}{\partial\gamma_c(k)}\right)|\varnothing(\gamma)>$$

We then get

$$= \varnothing(\gamma)<b(k)^*b(m)|\gamma_b(m)|\varnothing(\gamma)> = <\varnothing(\gamma)|b(k)^*\gamma_b(m)|\varnothing(\gamma)>$$
$$= <b(k)\varnothing(\gamma)|\gamma_b(m)|\varnothing(\gamma)> = \gamma_b(k)^*\gamma_b(m)$$

On the one hand, while on the other

$$\langle\varnothing(\gamma)|b(m)b(k)^*|\varnothing(\gamma)\rangle$$

$$= \langle\varnothing(\gamma)\left|\left(\frac{\partial}{\partial\gamma_b(k)}\right)b(m)\right|\varnothing(\gamma)\rangle$$

$$= \langle\varnothing(\gamma)\left|\left(\frac{\partial}{\partial\gamma_b(k)}\right)b(m)\right|\varnothing(\gamma)\rangle$$

$$= \langle\varnothing(\gamma)\left|\left(\frac{\partial}{\partial\gamma_b(k)}\right)\gamma_b(m)\right|\varnothing(\gamma)\rangle$$

$$= \delta(k,m) - \langle\varnothing(\gamma)\left|\gamma_b(m)\left(\frac{\partial}{\partial\gamma_b(k)}\right)\right|\varnothing(\gamma)\rangle$$

$$= \delta(k,m) - \gamma_b(m)\langle\varnothing(\gamma)|b(k)^*|\varnothing(\gamma)\rangle$$

$$= \delta(k,m) - \gamma_b(m)\langle b(\gamma)\varnothing(\gamma)|\varnothing(\gamma) \geq \delta(k,m) - \gamma_b(m)\gamma_b(k)^* \tag{3.6}$$

This is in agreement with the CAR

$$[b(m), b(k)^*]_+ = \delta(m, k)$$

Provided that we assume that CAR

$$[\gamma_b(m), \gamma_b(k)^*]_+ = \delta(m, k)$$

and likewise,

$$[\gamma_c(m), \gamma_c(k)^*]_+ = \delta(m, k)$$

By imposing such restrictions, we can calculate easily the momentum of the current density field and hence of the radiated field is a state that is jointly coherent for the bosons (i.e. photons) and for the Fermions. We observe that the perturbation of the current density of the Dirac field caused by the interactions between the electron–positron field and the photon field is given up to first order in the photon field and second order in the Fermion field by an expression of the form

$$\begin{aligned}\delta J^\mu = {} & \psi^{(0)*}\alpha^\mu\psi^{(1)} + \psi^{(1)*}\alpha^\mu\psi^{(0)} \\ = {} & -e\psi^{(0)*}(x)\alpha^\mu \int S_e(x-y)\gamma^\mu A_\mu(y)\psi^{(0)}(y)d^4y + h\cdot c.\end{aligned} \tag{3.7}$$

This expression is manifestly trilinear in the operators. Specifically, it is quadratic in the electron–positron field and linear in the photon field, totally yielding a trilinear term. It can be expressed as

$$\delta J^\mu(x) = \sum_{k,m,q} \big(F_1^\mu(x)|k, m, q)b(k)^*b(m) + F_2^\mu(x|k\cdot m\cdot q)b(k)^*c(m)^*$$

$$+F_3^\mu(x|k\cdot m\cdot q)c(k)b(m) + F_4^\mu(x|k\cdot m\cdot q)c(k)c(m)^*a(q) + h\cdot c\big) \tag{3.8}$$

It should be noted that the photon operators $a(q), a(q)^*$ commute with all the electron–positron operators $b(q), b(q)^*, c(q), c(q)^*$. From this expression, it is clear that the photon operators tend to couple the other modes of the electron–positron field and hence produce additional terms in the far-field radiation pattern. If we have a state $|\psi>$ of the electron–positron field in which there are $n_e(k) = 0, 1$ electrons with momentum-spin index k, $n_p(k) = 0, 1$ positron with momentum-spin index k and $n_{ph}(k) = 0, 1, 2, \ldots$ photon with momentum–helicity index k for $k = 1, 2, \ldots$, then we can calculate easily the moments of the currents fluctuation field $\delta J^\mu(x)$ in this state by simply applying the following rules:

$$\begin{aligned}&b(k)|n_e(k) = 0> = 0, b(k)|n_e(k) = 1> = |n_e(k) = 0>,\\ &c(k)|n_p(k) = 0> = 0, c(k)|n_p(k) = 1> = |n_p(k) = 0>,\\ &b(k)^*|n_e(k) = 0> = |n_e(k) = 1>, b(k)^*|n_e(k) = 1> = 0,\\ &c(k)^*|n_p(k) = 0> = |n_p(k) = 1>, c(k)^*|n_p(k) = 1> = 0,\end{aligned}$$

in view of the Pauli-exclusion principle and likewise, for the photon numbers

$$a(k)\left|n_{ph}(k)\right> = \sqrt{\left|n_{ph}(k)\right>}\left|n_{ph}(k)-1\right>,$$
$$a(k)^{*}\left|n_{ph}(k)\right> = \sqrt{\left|n_{ph}(k)+1\right>}\left|n_{ph}(k)+1\right>$$

more precisely, we can evaluate the moments

$$< \psi|\delta J^{\mu_1}(x_1) \otimes \cdots \otimes \delta J^{\mu_m}(x_m)|\psi >$$

By noting that the quantity

$$\delta J^{\mu_1}(x_1) \otimes \cdots \otimes \delta J^{\mu_m}(x_m) \tag{3.9}$$

Is a homogeneous polynomial of degree 3 in the electron–positron–photon operators with the photon operator appearing with a total degree of m and the electron–positron operators appearing with the total degree of 2. We can also evaluate the previous moment in joint coherent state $\left|\varnothing_{ep}(\gamma)\otimes\varnothing_{ph}(u)\right>$.

3.4 Conclusion

Quantum computing and quantum communications are two different fields using quantum entanglement. Entanglement can have wide applications in secure communication. Coherent states are secured and utmost important in quantum communications. Quantum mechanics can provide analytical solution for quantum communications. Light–matter interaction can increase confinement. This confinement is very useful at terahertz frequencies for healthcare and security purpose. Terahertz scanners and spectroscopy are useful as biosensors to the mankind. Protein, skin and fat testing sensors can be developed based on light confinements technologies at terahertz frequencies.

Reference

[1] S. Lepeshov and A. Gorodetsky, "All-dielectric optically tunable metasurface for terahertz phase and amplitude modulation," *J. Phys.: Conf. Ser.* vol. 1461, p. 012203, 2020 IOP Publishing. doi:10.1088/1742-6596/1461/1/012203.

Chapter 4
Terahertz dielectric resonator antennas design and modeling

Abstract

Theory of operation of a terahertz (THz) dielectric resonator antenna (DRA) is explained and mathematical expressions for rectangular DRA, spherical DRA, cone DRA and cylindrical DRA are broadly given in this chapter. THz emission is produced solely by the photo-induced current in the silicon (Si) DRAs, also telecom wavelength of 1,550-nm DRA is presented. Our finding is that THz waveforms are lower than those for pumping at 1,100 nm, due to the nonuniform absorption of different parts of the excitation spectrum that is also called plasmon frequency. DRAs exhibit an infinite number of resonant modes. In this chapter, the development of high-efficiency DRAs at optical frequencies and THz frequencies has comprehensively explained using mathematically equations for optical currents densities and quantum fields.

4.1 Introduction to terahertz DRA

Terahertz (THz) cavity that can be modeled as given in Figures 4.1–4.3 is spherical, cylindrical and conical shape, respectively. Since light is in fact a quantum electromagnetic (e.m.) field, when it falls on cavity, it starts immediately interacting with the matter within the cavity. A coarse approximation of this matter is to regard as a **sea of electrons and positrons** and then to apply the formalism of **quantum electrodynamics** to describe its interaction with the **incident photons**, both before the light hits the cavity matter directed based on boundary conditions. The field in the spherical, cylindrical and conical cavity is actually a quantum e.m. field in certain state and we wish to calculate the probability that such field will be formed. The penetration of light wave into silver nano waveguide takes place to generate surface plasmon polytrons (SPP). **The e.m. response is an ensemble of collective electron oscillations (plasmons) coupled plasma or gaseous form of metal**s. Optical antennas are also not typically driven with galvanic transmission lines—localized oscillators are instead brought close to the feed point of the antennas, and electronic oscillations are driven capacitive coupling, their properties may be strongly shaped and materials dependent owing to surface plasmon resonances

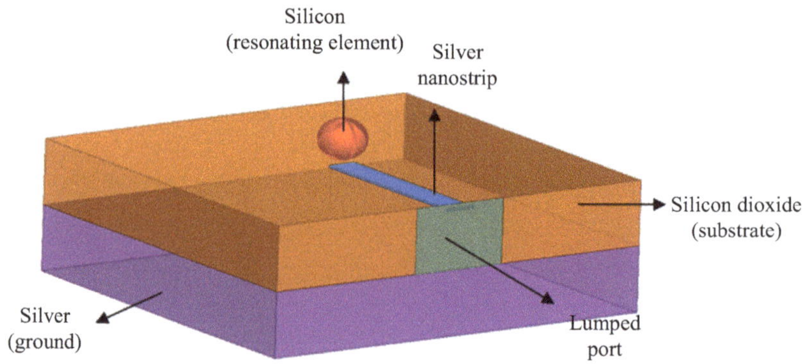

Figure 4.1 Spherical terahertz DRA

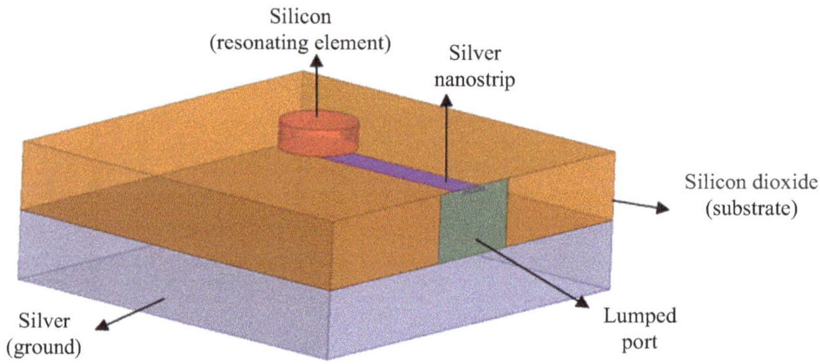

Figure 4.2 Cylindrical terahertz DRA

[1–6]. The optical dielectric resonator antennas (DRAs) are fed by localized light emitters, not by real currents, i.e., local density of e.m. states, which can be expressed in terms of Green's function tensor *G*.

The cellular communication in the 5G and beyond can be in the sub-THz band, i.e., 100–350 GHz, because the data rate of 20–100 Gbps is needed. Here, atmospheric path loss in THz frequency can be more than 100 dB/km. There is a need to operate in frequency windows below 1 THz, where the atmospheric attenuation is relatively low, and the high-gain antennas can be good solution for this. Hence, solid-state communication sources with a power of over 10 mW are required [1]. The gain required for such antennas is minimum of 15.27–20.71 dB for the frequency of 100–350 GHz for a data rate of 20 Gbps. There is need to maintain effective cellular wireless communication at a 0.1–10 THz band to develop powerful sources, and efficient detectors for cellular wireless communications systems, for expanding the transmission range more than 100 m. The hybrid antenna array for beam forming massive multiple-input and multiple-output can significantly improve the spectrum efficiency [2]. Here, the study of optical antennas is still in

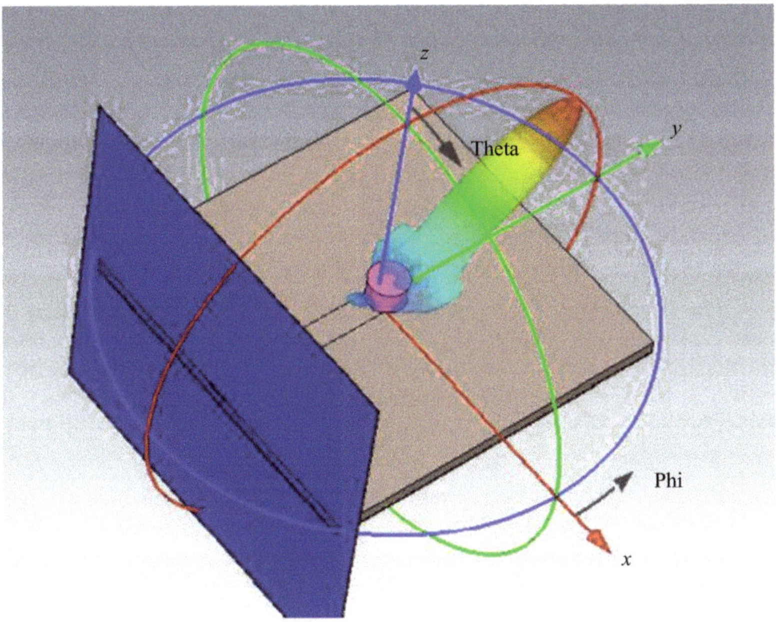

Figure 4.3 Super directive far field radiations in terahertz CDRA

its infancy and some properties directly derive from classical and microwave antenna theory, the direct downscaling of antenna designs into the optical regime is not possible because radiation penetrates into metals and gives rise to plasma oscillations. An optical antenna is designed to increase the interaction area of a local absorber or emitter with free radiation, thereby making the light–matter interaction more efficient. In the excitation of certain resonant modes, the DRAs may be used as resonant cavities or efficient radiators. With knowledge of its resonant modes, one can qualitatively predict the antenna behavior and can estimate the produced far field during design stage itself [3]. The generation of short vertical electric dipole and horizontal magnetic dipole is the origin of radiation. Their design resonance frequencies depend on the dimensions, geometry and permittivity of ceramics used. At optical, the nanoscale version of the microstrip line is known as nanostrip waveguide. We have numerically investigated the basic elements of photonics (DRAs) and plasmonic (nanostrip waveguide). By observing the fundamental antenna parameters, the performance of DRAs at optical frequencies have been explained. The Tx mode of DRA by means of the propagation vector orientation is also explained. The magnetic field vectors of the fundamental mode of the nanostrip waveguide show their compatibility for modal coupling to DRA. Nano waveguide dispersive properties have been described by the Drude model. The coupling of the magnetic field coming from the nanostrip's fundamental mode to DRA's takes place. **The optical communication bands are 185–205 THz**: L-band ($1.625-1.565\ \mu m = 0.06\ \mu m$), C-band ($1.565-1.530\ \mu m = 0.35\ \mu m$) and

S-band (1.530−1.460 μm = 0.70 μm). The cylindrical dielectric resonator antenna (CDRA) is made from silicon materials. The structure is multilayer, i.e., ground plane is made from silver, then the substrate is SiO_2 materials and DRA is made from silicon materials. The whole DRA structure is excited via a nano waveguide made from a noble silver metal (Ag) whose conductive properties are calculated using the Drude model. DRA dielectric is made of Si, with $\varepsilon_r = 11.9$ and estimated loss tangent, $\tan\delta = 0.0025$. It is excited by microlasers such as microdisks and photonic crystal lasers. The length of the transmission lines is characterized to the wavelengths (λ) of incoming and outgoing radiations [4]. However, while working at the optical frequencies, most of the incident light is transparent through the metals. This gives rise to plasmonic free electrons; thus the feed line is analyzed considering shorter effective wavelength (λ_{eff}), which depends on the material properties. The properties of the noble metals are explained as per Drude's model [5]. The radiating power flows perpendicular to the feed waveguide's plane. However, depending on the excited radiation resonating mode, nano dielectric resonator antennas (NDRAs) can radiate either end-fire or broadside.

To perform **cavity quantum e.m. field computation, we must solve the joint Maxwell–Dirac equations within the cavity, and with the appropriate boundary conditions**, these are corresponding to what shape of the cavity we are using. The cavity shape can be cone, cylindrical or spherical. This means that we expand the Maxwell field in terms of basis functions, which vanish on the boundary**. Then the coefficients in this expansion will be the photon creation and annihilation operator fields. Likewise, we must expand the Dirac field within the cavity using the same orthogonal basis functions, whose coefficients will now be the electron–positron creation and annihilation operator fields** [6].

The resulting Hamiltonian of these fields will then comprise three components: one, the Hamiltonian of the free photon field expressed as a quadratic form in the photon creation and annihilation operators; two, the Hamiltonian of the free electron–positron field expressed as a quadratic form in the electron–positron creation and annihilation operators; and finally three, the interaction Hamiltonian between the electron–positron field and the photon field.

It is this last interaction term that is of importance in scattering theory of light and the formation of fields. Using the interaction picture of Dyson, we can expand the solution to the Schrödinger evolution operator for this quantum system as a Dyson series in powers of the interaction Hamiltonian. Then, by using the technology of Feynman diagrams, we can calculate various terms of the scattering matrix to determine the probability amplitudes. The basic physics behind this computation does not depend on what shape we choose for the cavity. No matter what shape we choose, we can in principle determine a complete orthonormal basis of test functions that satisfy the appropriate boundary conditions for the given cavity and then expand the Maxwell and the Dirac field in terms of these test functions.

Before the interaction between light and matter, the system was in an initial state and after this interaction, including the interaction of light with the matter, the state of the whole system changes with time and this change can be determined by the action of the unitary evolution operator on the initial state with the unitary evolution operator constructed from the total Hamiltonian of the interacting electrons, positrons and photons using **the Dyson series**. The final state of this system is then equivalently determined once we know the matrix elements of the evolution operator between any given initial state and a final state of the particles. From this final state, the probability of getting a given field is computed from the basic rules for quantum probability.

Thus, we shall by perturbation theory be able to calculate the change in the e.m. fields pattern caused by the Dirac current in terms of operators and then by assuming a definitive state of the photon–electron–positron field, we shall be able to compute the mean and covariance of the e.m. field in this state. Their operating wavelengths are very small, mostly in the nanometer size. Optical antennas that work on Drude's theory have the nature of nonlinearity. Optical antennas have far field with finite values of Planck's constant.

The physics in optical antennas involves the dominance behavior of skin effects, SPP and light–matter interaction (creation and annihilation). It is well known that light can penetrate through metallic walls at optical frequencies. Light–matter interaction at optical domain is studied to design quantum devices. In optical DRA, SPP phenomenon occurs due to strong light–matter interactions. A simplified analysis would be to model the matter as just the second quantized electron–positron field using Dirac's relativistic wave equation. In this process, wave function is an operator field, and then the process includes the photon interaction term in the usual way. In this way, the total Hamiltonian of the photon and matter field splits into three terms:

1. One, the Hamiltonian of the e.m. field is described as a quadratic form in the photon creation and annihilation operators within the cavity.
2. Two, the Hamiltonian of the Dirac field with cylindrical cavity (in cylindrical DRA) boundary conditions on the wave function is described as a quadratic form in the electron–positron creation and annihilation operators.
3. And three, the Hamiltonian of the interaction between the photon and electron–positron field described as a quadratic form in the electron–positron creation–annihilation operators multiplied with a linear form in the photon creation–annihilation operators.

This interaction Hamiltonian can be used to compute the amplitudes for scattering, but since we are primarily interested in the statistics of the e.m. wave field pattern, we shall describe this interaction using the Dirac current density expressed as a quadratic form in the electron–positron creation–annihilation operator fields that drive the photon field using the wave equation for the e.m. field in the presence of a current density.

The previous quantum theory can be applied to understand the concept of quantum physics as stated in the following.

Antennas have characteristic dimensions of the order of a wavelength of light, demanding fabrication accuracies better than 10 nm. The advent of nano-science and nanotechnology provides access to this length scale with the use of novel top-down nanofabrication tools (e.g. focused ion beam milling and electron–beam lithography) and bottom-up self-assembly schemes. The fabrication of optical antenna structures is an emerging opportunity for novel optoelectronic devices. In light-emitting devices, an electron and a hole pair combines to emit a photon. The antenna serves to make both the excitation and the emission more efficient.

It can be expressed in terms of the system's dyadic Green's function *G*. This reflects the fact that the decay from the excited state happens in response to the emitter's own field. Thus, the excited state lifetime of the quantum emitter is determined by the Green function *G* of the system in which the emitter is embedded.

Rigorously treating the metal as a strongly coupled plasma is required, which leads to a reduced effective wavelength. This effective wavelength is related to the external (incident) wavelength. The plasma wavelengths of the metal $n1$ and $n2$ are constants that depend on the geometry and dielectric parameters of the antenna. The effective wavelength is roughly a factor of 2–6 shorter than the free-space wavelength.

Laser Gaussian input pulse for nonlinear phase matching (coherent states) of metals at optical frequency. The absorption by infrared-active optical phonons takes place beyond 7 THz. The mode-locked fiber lasers operate at wavelengths of 1.1 and 1.55 μm with pulse repetition rates of 10 and 20 MHz at 10-mW power. The compact THz photonic devices operate up to multi-THz frequencies that are compatible with Si CMOS technology. The spectrum of potential applications depends on the available spectral bandwidth, signal-to-noise ratio and data acquisition speed. In general, the techniques for THz generation and detection exploit either photoconductivity or optical nonlinearity. Photoconductive techniques for THz emission and detection are widely used due to their simplicity, compactness and possibility of direct coupling to fiber optics. THz emission from photoconductivity was first demonstrated using Si. These DRAs are investigated at optical frequencies (C-band). Here, another potential application of optical nanoantennas in wireless broadcasting and wireless optical links is used between two points at the nano- and microscale. Here, we develop the idea of wireless links using optical nanoantennas, inspired by the concept of microwave links at radio frequencies.

4.2 Mathematical formulations used to describe working of quantum DRA

In order to describe the propagation of e.m. waves a spherical cavity, we must express the Maxwell equation in the frequency domain using spherical polar coordinates. An elegant formulation of this is based on the theory of multipole moments, where we account one component of the *E* field that is

$$\underline{E}_{lm}(\mu, \underline{r}) = f_l(r)\underline{L}\gamma_{lm}(\hat{r})$$

where $\underline{L} = -\underline{i}\vec{r} \times \vec{\nabla}$*is* the vector angular momentum operator when used in quantum mechanics, and $\gamma_{lm}(\hat{r})$ are the standard spherical harmonics.

Likewise, we assume that one component of the magnetic field is

$$\underline{H}_{lm}(u,r) = gl(r)\underline{L}\gamma_{lm}(\hat{r})$$

$\hat{r}.\underline{E}_{lm} = 0, \hat{r} \cdot \underline{H}_{lm} = 0$; these are longitudinal components.

We select f_l at g_l to that $\underline{E}_{lm}$ and $\underline{H}_{lm}$ satisfy the Helmholtz equation (i.e. the ware equation in the frequency domain):

$$(\nabla^2 + k^2)\begin{bmatrix}\underline{E}_{lm}\\ \underline{H}_{lm}\end{bmatrix} = 0$$

This result in f_l and g_l satisfies modified D Bessel differential equations that have far fields behavior in the form of spherical waves propagating radially outward. These components in addition satisfy Maxwell equations:

$$\begin{aligned}&\text{div}\underline{E}_{lm} = 0,\\ &\text{div}\underline{H}_{lm} = 0\end{aligned}$$

Hence, we can build up all lollops for spherical wave propagation using $\underline{E}_{lm}$ and $\underline{H}_{lm}$. The coefficients that we choose for expanding the e.m. field are that of $\underline{E}_{lm}\underline{\nabla} \times \underline{H}_{lm}, \underline{H}_{lm}, \nabla \times \underline{E}_{lm}$that determine the Hamiltonian (field energy of the e.m. field) within the cavity as a quadrature form in these coefficients and hence these coefficients upon second quantization become the creation and annihilation operators of the e.m. field.

The surface current density in the spherical surface generated by the boundary conditions of these field operators generates a quantum e.m. field in space, where mean-square fluctuations may be determined in the coherent state of the field defined by the joint Eigen states of the annihilation operators. Finally in order to discuss the interaction of the matter within the spherical DRA cavity with the quantum e.m. field, we solve the following stationary Dirac equation:

$$((\alpha - i\nabla) + \beta m)\psi = E\psi$$

For the four-component wave function ψ by expanding, ψ as a linear combination of the basis function obtained earlier and denote the solved

By $\pm E_n, '\psi_n, \chi_n,$ we care to express the general time dependent on wave function as

$$\psi(t,\underline{r}) = \sum_n \left[a(n)e^{-iEnt}\psi_n(\underline{r}) + b(n)^+ e^{iEnt}\chi_n(\underline{r})\right)$$

where $\{a(n)\}$ are the electron annihilation operators and $b(n)^+$ are the positron creation operators. The interaction energy operator between electron–positron and the photon field is then

$$Hlm^{(t)} = \int \psi^+\left(t,\underline{\vec{r}}\right)\left(\vec{\alpha}A,\ e\vec{A}\left(t,\vec{r}\right)\right)\psi\left(t,\vec{r}\right)d^3\vec{r}$$

Therefore, in a polynomial of third-order electron–position–photon operators and their interdependence on perturbation theory or the Dyson series, we can use $H_{lm}(t)$, the effect of the e.m. field on matter and the effect of matter on fields formation.

For e.m. fields and changed matter within a **cylindrical cavity**, a detailed study is given in the following.

In a spherical cavity, the vector solutions to Helmholtz equation with boundary conditions were

$$\begin{aligned} E_{lm}(\underline{r}) &= f_l(r)\underline{L}Y_{lm}(\hat{r}) \\ H_{lm}(\underline{r}) &= g_l(r)\underline{L}Y_{lm}(\hat{r}) \end{aligned}$$

With boundary conditions,

$$\hat{r} \times \left[E_{lm}(\underline{r}) + \frac{1}{j\omega\varepsilon} \text{curl } H_{lm}(\underline{r}) \right] \Big|_{r=R} = 0$$

and

$$\hat{r} \cdot [\text{curl } E_{lm}(\underline{r})]|_{r=R} = 0$$

correspond to the vanishing of the tangential components of the electrified and the normal component of the magnetic field on the surface.

The boundary conditions determine the characteristic frequencies ω of oscillation.

4.3 Cylindrical terahertz DRA

In the core of a cylinder cavity, we solve Helmholtz equation in cylindrical coordinates to obtain the basis functions.

For expanding the e.m. fields and the Dirac fields specifically, the basic functions satisfy

$$\left(\nabla^2 + k^2\right) \begin{bmatrix} E_t \\ H_t \end{bmatrix} \leq 0$$

With boundary conditions determined by the few that $E_\perp$ vanishes to $t = 0$, and $t = \text{d}$, $H_\perp$ vanishes to $t = 0$ and $t = \text{d}$, and E_t vanishes to $\rho = R(\partial H_t/\partial \rho)$ and to $\rho = R$.

Doing the results in an expansion of the e.m. field as

$$\begin{aligned} \vec{E}(t, \vec{r}) \quad = \quad & R_e \sum c_E(n, m) \underline{\chi}_{Enm}(\underline{r}) e^{j\omega_E(n,m)t} \\ & + R_e \sum d_E(n, m) \underline{\gamma}_{Enm}(\underline{r}) e^{j\omega_H(n,m)t} \end{aligned}$$

And likewise, for $\vec{H}(t, \vec{r})$.

The boundary conditions completely determine the characteristic frequencies $\omega_E(n, m), \omega_H(n, m)$ of oscillations as well as the basis function χ_E, γ_E, etc.

(For that matter it should be noted that giving a boundary of any shape whether spherical or cylindrical or any other, in order get the correct basis function for expanding with e.m. fields or the Dirac field, we express the Laplacian in the Helmholtz equation in a coordinate say q_1 is such that q_1 = constant that is defining the boundary of the surface.)

To solve the Dirac equation by the same method, we choose any of their basis functions, say $\{u_n(\underline{r})\}$, that vanish on the cylindrical boundary and expand the Dirac wave function as

$$\underline{\Psi}(\underline{r}) = \sum_{n=1}^{\infty} \underline{c}_n u_n(\underline{r})$$

where $\underline{c}_n \epsilon C^4$ and resultant into the time-independent Dirac equation:

$$\left(\left(\vec{\alpha} - i\vec{\nabla}\right) + \beta m\right)\underline{\Psi}(\underline{r}) = E\Psi(\underline{r})$$

We then multiply this equation by $u_m(\underline{r})$ and integrate over the cylindrical volume to obtain Eigen-matrix equation for $\{\underline{C}_n\}$ and E; thus our term-dependent Dirac wave function has the form

$$\underline{\Psi}(t, \underline{r}) = \sum_{n,k} \exp(-iE_k t) c_n^{(k)} u_n(\underline{r})$$

And in the second quantization process, $c_n^{(k)}$ become the certain and annihilation operators of electron and positron. Then the interaction energy

$$H_I(t) = e \int \psi(t, r)^T (\alpha^{-1}) \vec{A}\left(t, \vec{r}\right) \psi(t, r) d^3\vec{r}$$

defines a cylindrical form in the electron–positron photon creation–annihilation operators to which the theory of the Dirac series, Feynman diagram, etc. can all be applied to calculate mathematical properties or the quantum fields. The quantization process for the Dirac operator $c_n^{(k)}$ involves introduction to the canonical anticommutation relation (CAR):

$$\{\psi(\underline{r}),\ \psi^+(\underline{r}')\} = \delta^3(\underline{r} - \underline{r}')$$

Since $\sum u_n(\underline{r}) u_n(\underline{r}') = \delta(\underline{r} - \underline{r}')$, it follows that there (AR') are $\left\{c_n^{(k)}, c_m^{(l)^+}\right\} \overset{n}{=} \delta_{kl}\delta_{n,m}$ in show, where an expression for the interaction Hamiltonian between the e.m. field of the photons and the Dirac fields of electron and positron comparing of bosonic and fermionic operators are the canonical commutator rules of the bosonic operators and the CAR(anticommutator) of the fermionic operators can be used to calculate all the matrix elements of this interaction Hamiltonian and hence all the quantum statistical properties of the image field formed. This interaction Hamiltonian may also be used to calculate radiative correlation to the Dirac wave function and hence to the Dirac current field and correlation to the far-field radiation e.m. field pattern, and hence correlation to the mean fluctuations of the radiation quantum field in any state.

Remarks: The Eigen functions of ∇^2 for a cylindrical boundary are

$$J_m\left(\frac{\alpha_m[n]\rho}{R}\right)\cdot(\cos(m\varphi),\sin(m\varphi))\cdot\sin\left(\frac{p\pi t}{d}\right)$$

or

$$J_m\left(\frac{\alpha_m[n]\rho}{R}\right)\cdot(\cos(m\varphi),\sin(m\varphi))\cdot\cos\left(\frac{p\pi t}{d}\right)$$

where *m, n, p* is curve integers $\alpha_m[n]$ are either the zeros of the $J_m(n)$ or $J_m^{'}(n)$, with $J_m(n)$ being the Bessel function. The corresponding Eigen value determines the oscillation frequency:

$$\omega(mnp)=\left(\left(\frac{p\pi}{d}\right)^2+\frac{\alpha_m[n]^2}{R^2}\right)^{1/2}$$

The frequencies of the Dirac equation obtained using their basis function will however be different.

4.4 Conical terahertz DRA

In fact, it is the conical cavity that most accurately represents the shape of the eye retina photo receptors (Figure 4.4). The conical cavity is regarded as a section of the sphere. Let us to denote the cone angle. Then the appropriate basis functions for expansion are obtained by solving the Helmholtz equation:

$$(\nabla^2+k^2)\psi=0$$

in spherical polar coordinates in the region $0\le r\le R, |\theta|\le\alpha, 0\le\varphi\le 2\pi$ with vanishing boundary conditions on the surfaces $\{r=R, |\theta|\le\alpha, 0\le\varphi\le 2\pi\}$.

And $\{0\le r\le R, \theta=\alpha, 0\le\varphi\le 2\pi\}$.

These Eigen functions can be obtained by forming linear combinations of those for the spherical cavity case and the choosing the linear combination. Coefficients to that the previous boundary conditions, the previous cases yield the possible Eigen-frequencies oscillations.

4.5 Conclusion

In this chapter, THz DRAs of various shapes and geometries have been explained, i.e., rectangular, spherical, cylindrical and conical DRAs at optical frequencies. Their coherent states have been given more importance. The current density

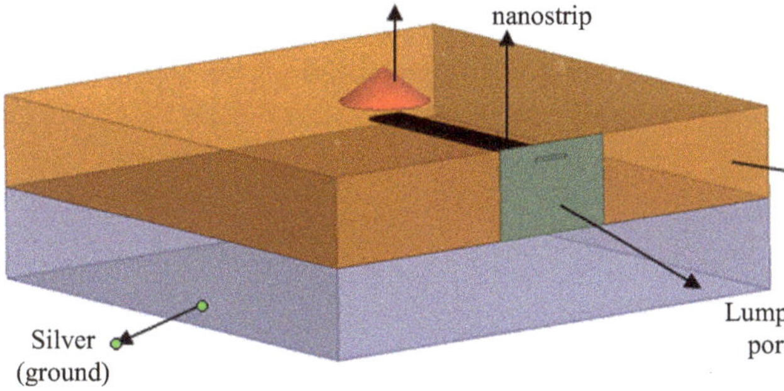

Figure 4.4 Conical terahertz DRA

fluctuations have controlled to its optimized threshold value for getting desired far fields. These quantum antennas can be named THz nanoantennas or optical antennas. These are similar to radio waves and microwave antennas, and their purpose is to convert the energy of free propagating radiation to localized energy and vice versa. Optical antennas exploit the unique properties of metal nanostructures, which behave as strongly coupled plasmas at optical frequencies. Applications are optical sensors, lasers, solid-state lighting, photovolatics, scanner and imaging and in telecommunication, short (S), conventional (C) and long (L) bands of the optical communication spectrum. The inter-chip wireless communication, including infrared and THz technologies, can be used. Here, properties of the dielectric materials were assumed as constant and the Drude model was considered for the metal used (Ag), transmit/couple optical energy from/into plasmonic circuits. The time-dependent current across the full depth of the Si wafer is calculated by numerically solving the pulse-propagation equation. The coherent polarizations occur due to the simultaneous generation of heavy-hole–light-hole wave packets. Nanoantennas enable optical wireless links, which would present much less absorption losses, largely outperforming conventional plasmonic waveguides. Replacing waveguide networks by nano-antenna chip-to-chip and intra-chip links also provide more on-chip space that can be used to house other circuitry, hence enabling further miniaturization.

References

[1] P. Bharadwaj, B. Deutsch, and L. Novotny, "Optical antennas," *Adv. Opt. Photon.*, vol. 1, no. 3, pp. 438–483, 2009.

[2] G. N. Malheiros-Silveira, G. S. Wiederhecker, and H. E. Hernández-Figueroa, "Dielectric resonator antenna for applications in nanophotonics," *Opt. Express*, vol. 21, no. 1, pp. 1234–1239, 2013.

[3] W. T. Sethi, H. Vettikalladi, H. Fathallah, and M. Himdi, "Dielectric resonator nantennas for optical communication," *Optical Communication Technology*, Pedro Pinho, IntechOpen, doi: 10.5772/intechopen.69064.
[4] G. N. Malheiros-Silveira and H. E. Hernández-Figueroa, "Wireless optical coupling evaluation in a dielectric resonator nanoantenna," *OSA Continuum*, vol. 1, no. 3, pp. 805–811, 2018, https://doi.org/10.1364/OSAC.1.000805.
[5] A. Alù and N. Engheta, "Wireless at the nanoscale: Optical interconnects using matched nano antennas," *Phys. Rev. Lett.*, vol. 104, no. 21, p. 213902, 2010.
[6] D. M. Solís, J. M. Taboada, F. Obelleiro, and L. Landesa, "Optimization of an optical wireless nanolink using directive nanoantennas," *Opt. Express*, vol. 21, no. 2, pp. 2369–2377, 2013.

Chapter 5
Surface plasmon polytrons (SPP) into terahertz DRA

Abstract

Surface plasmonic resonance occurs at interface of metal and substrate at terahertz frequencies. The terahertz's wavelength is located between the microwave and the infrared region of the electromagnetic spectrum. Optical wavelength falls after infrared region. Terahertz DRA (TDRA) makes use of surface plasmon polytron (SPP) phenomenon for the excitation of dielectric resonator antenna (DRA). The Drude model helps to characterize TDRA, when switching from microwave to terahertz frequencies. Drude's scattering is function of frequency, resistance and conductance at photonic wavelength. Here, photon's generation is classified by bosonic fields or fermionic fields. The nonlinearity of photons is described by Dirac second quantized field equations. Feynman path integral gives know-how about particle's moment fluctuations. Correlation coefficient manipulation provides proper beam formation. Photon's spin is the main cause of nonlinearity in terahertz antennas. Optical spectrum has frequencies from 10^{13} to 10^{15} Hz. Optical DRA can couple optical energy into plasmonic resonance and vice versa. Plasmon resonance thus becomes the main cause of excitation to TDRA. Terahertz antennas that involve SPP phenomenon for radiations can provide highly directive radiation pattern and high gain.

5.1 Introduction

In 1939, Richtmyer found that dielectric objects can resonate and radiate into free space based on boundary conditions air–dielectric interface [1–5]. In 1985, SA Long experimentally proved. Hence, dielectric resonators (DRs) are very efficient radiators. DR materials are commercially available with vast range of permittivity ranging from 10 to 1,600 F/m. An antenna designer can exercise choice for exciting various higher order modes depending on material's permittivity and dimensions of DR at optical frequency [6–11].

It consists of electrons and positrons that are solved by Dirac second quantized field equations based on quantum electrodynamics to describe photon spin. Here, electromagnetic (e.m.) field is produced by linear superposition of creation and annihilation operators of photon fields [12–16]. The electron positron creation and

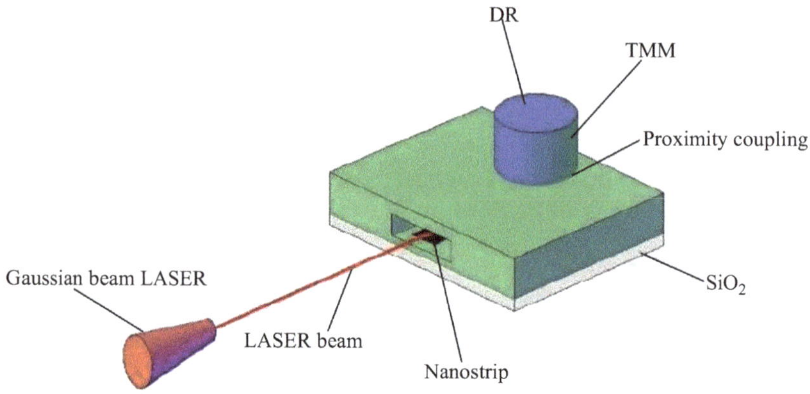

Figure 5.1 Terahertz DRA design with CDRA, laser and SPP

annihilation operator fields plus photon's creation and annihilation operators' fields are total fields produced in terahertz DRA (TDRA) as shown in Figure 5.1. The current densities of fields are obtained by quadratic functions of the Dirac field operators. This way, the current density in turn produces quantum e.m. fields. These fields can be described by retardation potentials or Dirac wave function [21,22]. Hence computations of mean and mean-square fluctuations of quantum e. m. fields produced by TDRA plus free photon e.m. fields in any state shall provide the complete solution (Figures 5.2 and 5.3).

The terahertz RDRA (TRDRA) is excited by plasmonic resonance (Figure 5.4). Terahertz input is given into an SiO_2 substrate thorough silver nano waveguide. Energy band gap of semiconductor materials is targeted to accelerate by input excitation. Transient photocurrents are thus produced [23–25]. Accelerated charge carriers are thus generated. Thus, oscillations in TRDRA also known as resonance are created at THz frequency. This results into surface plasmon polytron (SPP) and radiations into space, and it takes place on the basis of boundary conditions of TRDRA. Thus, these radiations are governed by the rule of plasmonic. Using higher order modes concept, dimension scaling can become possible. It is a device that can convert terahertz radiations into localized energy and vice versa. Proximity feed is used to excite TDRA. The Poynting theorem describes power dissipated by a time harmonic system. Nano waveguide (silver) is inserted into a substrate to create SPP and SPR (surface plasmon resonance). Then through capacitive coupling, near field is coupled to TDRA for generating far-field radiations.

5.2 Working principle of TDRA

1. Light in, light out.
2. Input excitation by semiconductor LASER.
3. SPP and SPR plasmonic resonances take place into interface of substrate and ground.

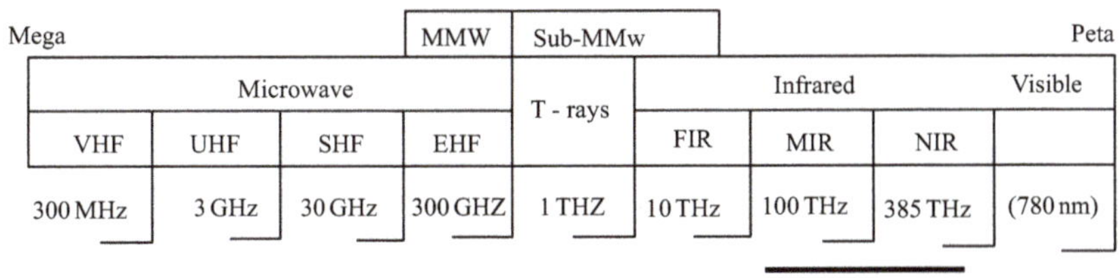

Figure 5.2 Frequency spectrum used in terahertz DRA

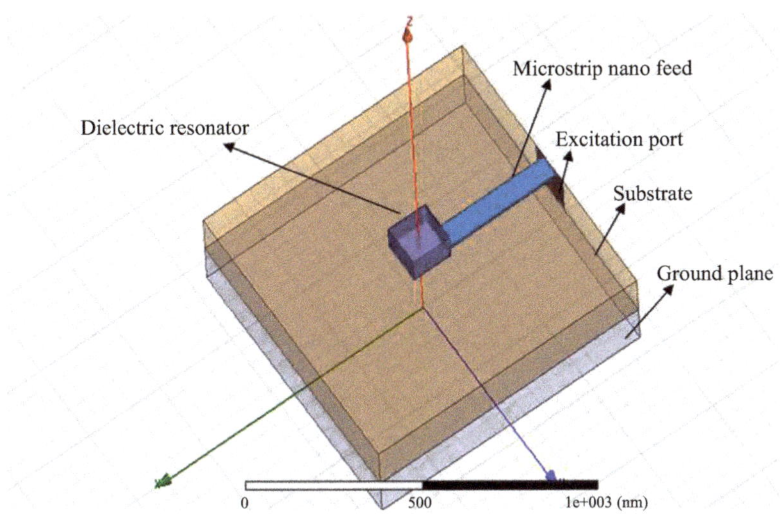

Figure 5.3 Terahertz rectangular DRA with Gaussian input

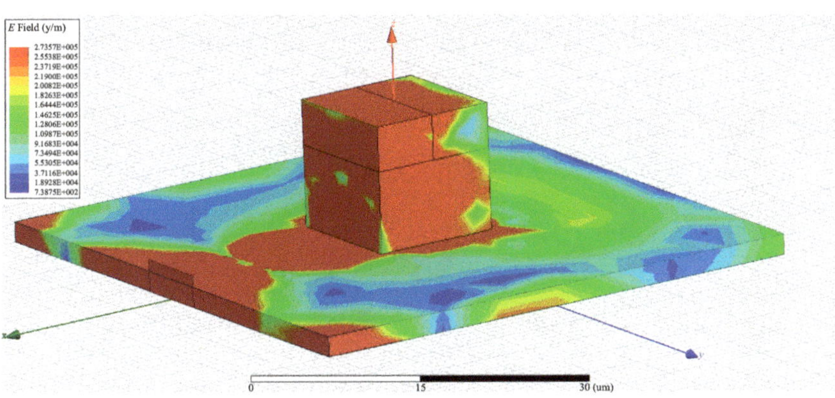

Figure 5.4 Terahertz RDRA fields

4. Electromagnetic coupling takes place to TRDRA.
5. Capacitive coupling takes place due to proximity feed. It forms photon clouds and fluctuations due to spin effect of photons.
6. Screening of electronic–hole takes place near band gap due to accelerated charge carriers.
7. Initially very small band-gap energy was observed.
8. Laser provided higher photon energy than initial band gap energy.
9. Laser accelerated photons due to excitation known as plasmonic resonance.
10. Laser created clouds of charge carrier (screening of fields takes place).
11. Quick change in field takes place thus generated transient currents.

5.3 Terahertz CDRA design and simulations

Design of TCDRA is given in Figure 5.5.

Resonant frequency (f) formulations for TDRA (cylindrical) is given in the following:

$$f = \frac{6.324c}{2\pi d\sqrt{\varepsilon_r + 2}}\left[0.27 + 0.36\left(\frac{d}{2h}\right) + 0.002\left(\frac{d}{2h}\right)^2\right]$$

Skin effect plays an important role at terahertz frequencies. The penetration of e.m. fields into metal ground plane takes place. Election oscillations cause skin effect. Drude's scattering is a function of frequency, resistance and conductance. Dielectric resonator antenna (DRA) is an efficient radiator. It can excite an infinite number of resonant modes. Knowledge of resonant modes can predict antenna behavior. Dissipation factor or dielectric loss can be predicted based on Drude's model. Photons generation can be called as fermionic or bosonic fields. Beam formation can be worked on the basis of correlation coefficients. Radiation or emissions are based on Dirac second quantized field equation. Feynman path integral is used to define photon density and moments fluctuations. Photon spin is the main cause of nonlinearity in quantum antennas. Beaming of these nonlinear photons fields requires the treatment of quantum mechanics for getting a solution of transmitting

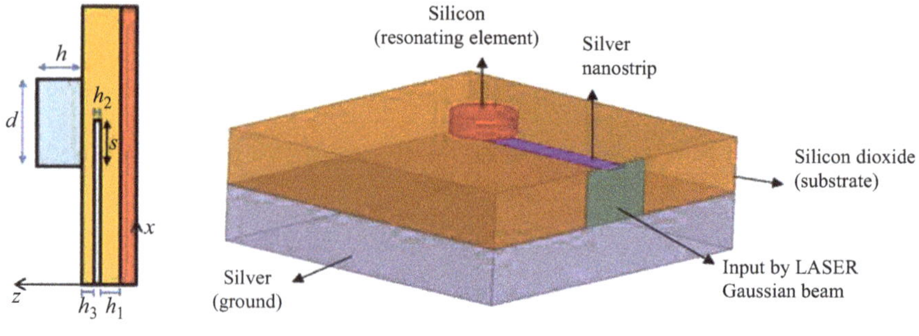

Figure 5.5 TCDRA with Gaussian input to silver nano waveguide

Table 5.1 Dimensions terahertz CDRA

S. no.	Parameter	Value
1	Radius r	0.250 μm
2	Height h	0.325 μm
3	Permittivity	11.9 F/m
4	Input frequency	473 THz
5	Radiated frequency	192 THz
6	Gain	10.5 dB
7	Substrate	Arlon
8	Dimension of the substrate	4.6×4.5 μm

quantum states from one location to another. The Dirac equation can provide a generation of photons and its fields to constitute currents. This current density becomes the main factor to the introduction of quantum e.m. field. This field can be estimated on the basis of the solution of retarded potential. Solution of resonant modes can be worked out making the use of linear super position of photon fields by creation and annihilation. Here, we apply standard commutation rule for photons and anticommutations rule for electron, positron/hole. Figures 5.1–5.10 shows terahertz DRA radiation parameters. Table 5.1 shows design dimension of cylindrical DRA.

5.4 Terahertz DRA main features

1. Resonant and light scattering.
2. Antenna structures and resonant frequency.
3. Higher order resonances and fundamental resonance.
4. Near field and far fields.
5. Focused electron beam.
6. Feed gap between waveguide and DRA (proximity feed).
7. The feed gap and impedance match (source and DRA).
8. Proper impedance-match, high-energy transfer between a localized source and the TDRA.
9. The collective excitation corresponds dipole moment (super directive and super radiance).
10. Physical dimensions are reduced by the square root of the dielectric constant, i.e., an ε_r of DR increased to 100, which can reduce dimensions by a factor of 10.
11. These DRAs operate via displacement current in a low-loss high-permittivity dielectric, resulting in reduced energy dissipation in the resonators.
12. At optical regime, radiation penetrates into metals and gives rise to plasma oscillations.
13. Best efficiency the internal energy dissipation of any antenna must be minimized.
14. Radio wave and microwave technologies predominantly make use of antennas to manipulate e.m. fields.

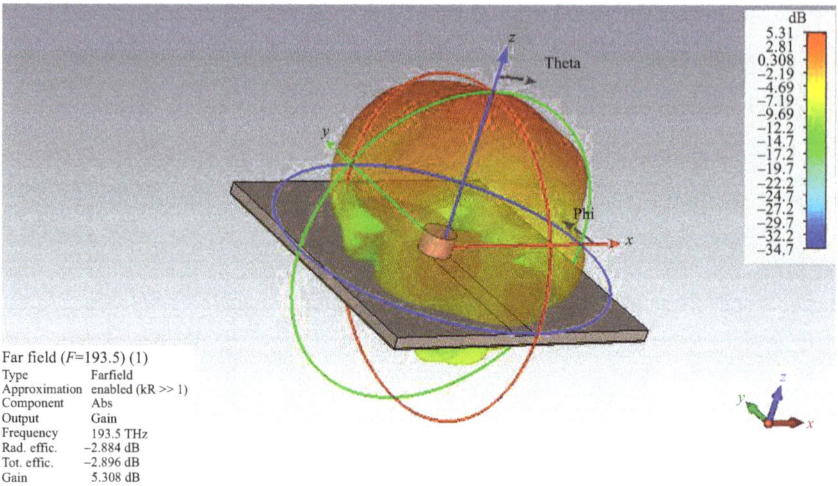

Figure 5.6 Radiation in TCDRA

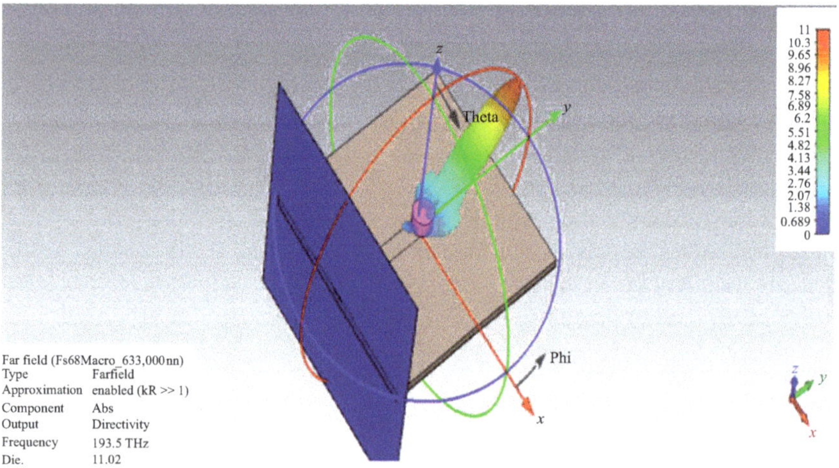

Figure 5.7 Radiation pattern of TCDRA (super directive)

15. THz interfacing efficiently between propagating radiation and localized fields.
16. The fabricated device comprising an arrangement of TiO_2 near resonance introduces a progressive phase delay to the wave front, resulting in a clearly observable beam deflection.
17. The use of low-loss DRs at THz to concentrate radiation from far-field modes into a volume opens a new paradigm alternative to purely plasmonic

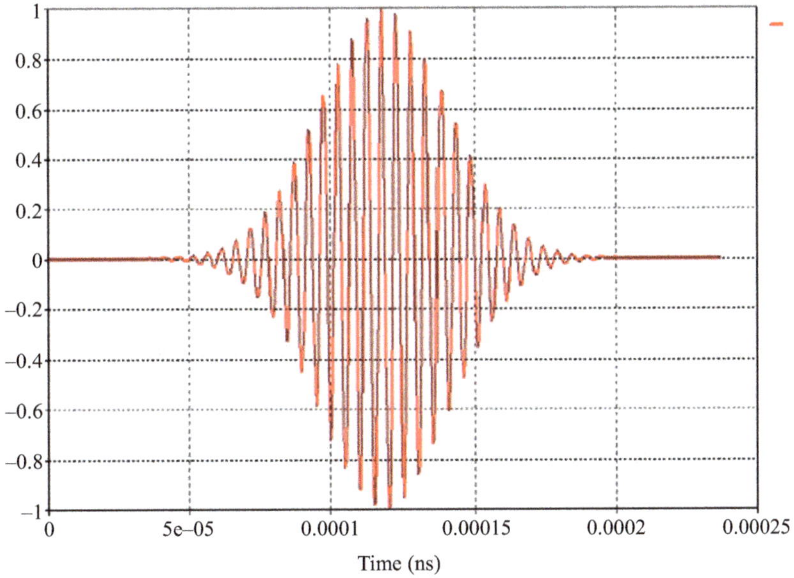

Figure 5.8 Gaussian beam input to TCDRA

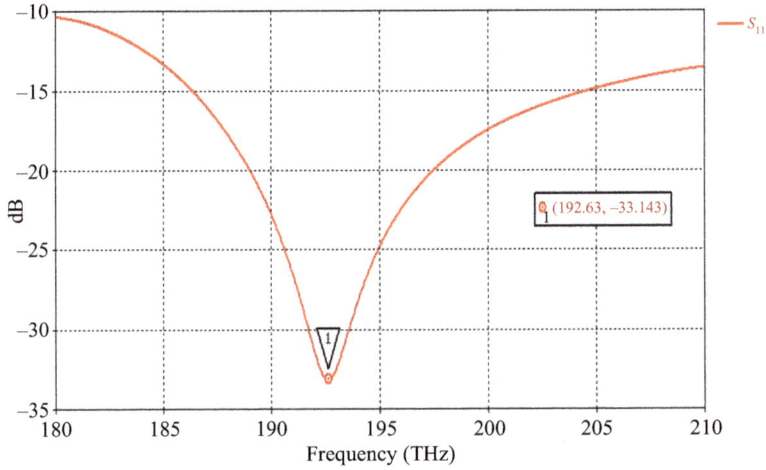

Figure 5.9 Reflection coefficient of nano-CNDRA at 193 GHz

(metallic) antennas. This is the precursor to purely DRAs with minimal energy dissipation.

18. DRA dimensions are in order of a wavelength of light, demanding fabrication accuracies better than 10 nm.

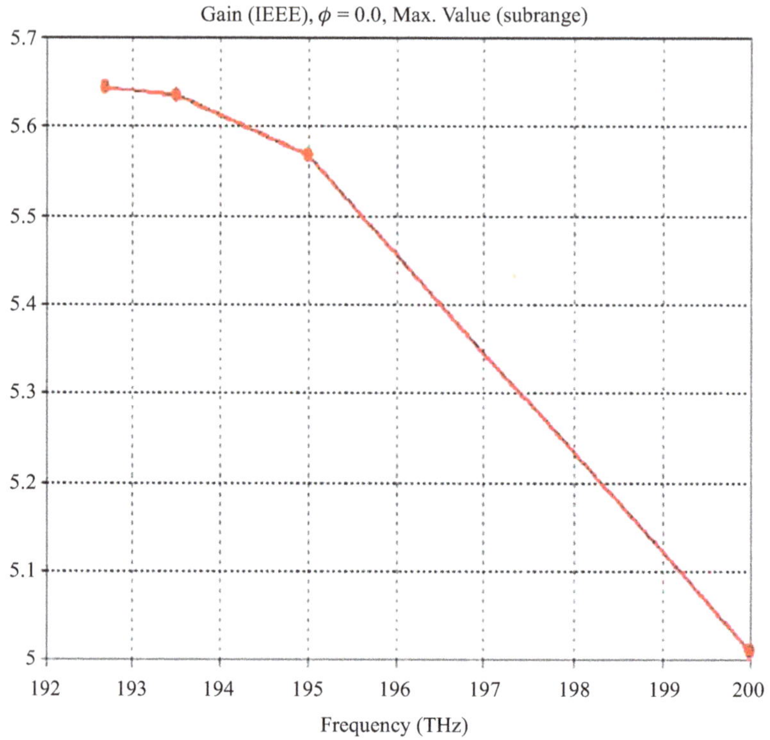

Figure 5.10 Gain 5.6 dBi of nano-CDRA at 193 GHz

19. THZ DRA has applications short-range optical communications, ultrafast computation and molecular sensing.
20. Here CDRA with a diameter of 180 nm, a height of 50 nm is used.
21. Optical DRA 65% of the power is absorbed by the array, as plasmonic and dielectric loss.
22. Antenna radiation is due to the resonance of surface waves that propagate on the antenna.
23. SPR is the resonant oscillation of conduction electrons at the interface between negative and positive particles, ω_p is the frequency of bulk longitudinal electron excitations, the plasma frequency.
24. The fundamental limit is given by the tunneling time (SPP at interface).
25. A resonant plasmonic antenna contributes to the local density of states in the junction.
26. Here, h is Planck's constant yielding a time limit on the order of 4 fs, the limit is given by the circuit time constant RC.
27. The emitted power as a function of the current fluctuations.
28. The emitted power is on the order of 10 pW.
29. The coupling between the antenna mode and the propagating plasmon mode has to be optimized.

30. The figure of merit is quality measure of TDRA.
31. Surface plasmon polaritons are mixed electronic and e.m. waves.
32. The magnetic field of an SPP is parallel to the surface and perpendicular to its direction of propagation.
33. The surface plasmon polaritons are solutions of Maxwell's equations in which the effects of retardation, non-propagating, collective vibrations of the electron plasma near the metal surface.
34. Scanning near-field optical microscopy did it become possible to measure the surface polariton field.
35. If the contribution of the antenna mode is larger than the contribution of the non-radiative modes, it becomes possible to avoid quenching (absorption) and therefore to increase the radiative efficiency.
36. Antenna radiation is due to the resonance of surface waves that propagate on the antenna.
37. Sensor development using SPP and studies about kinetics of interaction between biomolecules are the most investigated applications in the field.
38. The contributions from SPPs and from other surface waves to the antenna radiation. It is found that for antennas with long arms that support higher order resonances, SPPs provide a dominant contribution to the antenna radiation.
39. For sensor, measure the surface polariton field at interface.
40. Generation of surface plasmon polaritons with a focused laser beam.
41. Get electric-field intensity distribution on a silver surface.
42. Polytrons at a spot size of the focused beam.
43. In antenna, contribution to the local density of states.
44. Spatially averaged enhancement factor in the junction barrier. The antenna produces a narrow emission spectrum and provides an enhancement of the efficiency.
45. The efficiency defined as the number of emitted photons due to plasmon leakage per electron.
46. The fields emitted by a fluctuating current density in the tunneling gap. The power spectral density of the current fluctuations is given as $\langle I^2 \rangle$.
47. Optical antennas deal with light emission, based on surface plasmons.
48. Photon emission in the visible by electrically driven optical antennas has been reported demonstrating a spectral control and also a directional control.
49. SPR-based sensor.
50. The TDRA is modeled as a resonant plasmonic cavity.
51. $\in(\omega) = 1 - \frac{\omega_p^2}{\omega^2}$,

    k_{spp} is the wave number

$$k_{sp} = \frac{\omega}{c}\left[\frac{\in_1 \in(\omega)}{\in_1 + \in(\omega)}\right]^{1/2}$$

$$\omega_{sp} = \frac{\omega_p}{\sqrt{\in_1 + 1}}$$

5.5 Mathematical formulations used in TDRA

Following sequence of activities takes place in quantum antennas:

1. Light in (LASER).
2. Input excitations is Gaussian beam from LASER (nonlinear).
3. Nano waveguide feed is used at optical frequency (light–matter interaction).
4. SPP takes place at an SiO_2 substrate and ground metal interface.
5. Nano-DRA is excited by SPP through proximity feed due to capacitive coupling.
6. Radiated frequency is lower to input laser frequency ($f_p <$ LASER f_r for propagation).
7. Electrons, positrons and photons fields become a radiated beam of quantum antenna at terahertz frequencies with clouds of photons spins field that becomes as input excitation (SPP) to nano-DRA.

This is a very important and exciting phenomenon in quantum antenna. The mathematical aspect of its working principle is input excitation to radiated power of quantum. With the illumination of femtosecond laser pulses, electron-hole pairs are created in the photoconductive material when the laser pulses have higher photon energy than the band-gap energy of the photoconductive material, and the quick change in the electric field at the gap results in a transient current and finally an e. m. pulse in the THz frequency range is radiated. Mean of photons along with spin field included during radiation needs to develop quantum mechanics control for desired radiation orientation characteristics. Silver nanostrip or nano waveguide is used in the model. For coherently coupled photon sources whose emission direction can be dynamically steered by phase control. Feed gap is used for impedance matching. Optical antenna operates on "light in light out" mode:

$$\epsilon(\omega) = \epsilon'(\omega) + i\epsilon''(\omega)$$
$$\epsilon(\omega) = 1 - \frac{\omega_p}{\omega^2 + i\omega'}$$
$$\omega_p = \frac{\sqrt{n \cdot e^2}}{m \cdot \epsilon_o}$$

$$\nabla \times \nabla \times E(r,\, \omega) - \frac{\omega^2}{c^2}\epsilon(r,\, \omega)E(r,\, \omega) = 0$$

$$\rho = \epsilon o E_z e^i\left(k_x^x - \omega t\right)$$

$$\text{Mean/fluctuations} = \frac{\langle\psi|A_u(x)A_v(x')|\psi\rangle}{\langle\psi|A_{u_1}(x_1)A_{u_2}(x_2)\cdots A_{u_n}(x_n)|\psi\rangle}\text{(controlled to threshold)}$$

where $A_u(x) = \frac{\mu o}{4\pi}\int\frac{\hat{J}_u(p)e^{ipx}}{p^2 - io}dp^4$; $h \cdot c/ = E$, where is the photon energy; ω_P is the plasma volume frequency; Γ is the damping constant; ω is the center frequency where $\omega > \omega_P$.

Drude's model is used to characterize metal transport properties when switching from microwave to optical frequency. It is dispersive in nature. Loss tangent is a dielectric dissipation factor:

$$\tan\delta = \frac{\epsilon''}{\epsilon'}$$

At THz frequency, the penetration of e.m. fields into metal ground plane takes place.

This causes electron oscillations or plasmonic resonance or surface wave plasmonic resonance.

Skin depth is a function of frequency, resistance and conductance.

$$\delta = \sqrt{\frac{2\rho}{\omega\mu}}\sqrt{\sqrt{1 + (\rho\omega\epsilon)^2} + \rho\omega\epsilon}$$

$$Z_o = \sqrt{\frac{j\omega_o\mu_o\mu_{vz}}{\sigma 0}};\ \text{surface impedance}$$

$$Z = \frac{P}{|I|^2};$$

Silver nano waveguide dimensions are given as $L(\lambda_o) = \frac{m\lambda_{eff}}{2}$, m is the resonant modes order.

$$\lambda_{eff} = \frac{2L(\lambda_o)}{m} = \frac{\lambda_o}{\eta_{eff}}$$

The length of nano waveguide will be shorter at optical frequency as compared to microwave. Transcendental equation gives solution to determine resonant frequency.

$k_z \tan(k_z z/2) = \sqrt{(\epsilon_r - 1)k_0^2 - k_z^2}$; when propagating in the z direction.

Resonant frequency is given by the following transcendental equation:

$$k_o^2 = \sqrt{k_x^2 + k_y^2 + k_z^2};$$

where $k_o^2 = \sqrt[\omega]{\mu_o\epsilon_o}$ and $k_x = \frac{m\pi}{a}$; $k_y = \frac{n\pi}{b}$; $k_0 = 2\pi/\lambda$;

$$\sigma = \epsilon_o\epsilon_z e^i\left(k_x^x - \omega t\right).$$

Surface charge density is given as follows:

$$\rho(t,r) = -e|\psi(t,r)|^2$$

Current density is given as follows:

$$\vec{J}(t,r) = -\frac{ieh}{2m}\left[\psi(t,r)\underline{\nabla}\vec{\psi}(t,r) - \vec{\psi}(t,r)\underline{\nabla}\psi(t,r)\right];$$

$\psi(t,r) \rightarrow$ wave function

$$i\hbar\frac{\partial\psi(t,r)}{dt} = \frac{\hbar^2}{2m}\nabla^2\psi(t,r) - Z\frac{e^2}{r}\psi(t,r)$$

Electric scalar vector potential is given as follows:

$$\varnothing\,(t,r) = -\frac{e}{4\,\pi\varepsilon 0}\int\frac{|\psi|\left(t-\left|\frac{\underline{\underline{r-r'}}}{C}\right|,\,r'\right)}{|\underline{r-r'}|}d^3r'$$

Magnetic vector potential:

$$A(t,r) = -\frac{ieh\mu_o}{8\,\pi m}\int|\psi|\left(\pm\frac{\underline{\underline{r-r'}}}{C},\,\underline{r}\right)$$

$$\times\,\nabla\bar{\psi}\left(t-\frac{\underline{\underline{r-r'}}}{C}r'\right)\nabla'_{\psi}\left(t\frac{\underline{\underline{r-r'}}}{C}\right)d^3_{r'}$$

where $\psi(t,r)$ satisfies Schrödinger's equation that gives a solution of wave function for single electrons:

$$\frac{\partial\psi(t,r)}{dt} = \frac{h^2}{2m}\nabla^2_{\psi}(t,r) - \frac{Ze^2}{r}\psi(t,r)$$

This provides discrete dynamic solution. If there are N electrons in joint wave function then wave function can be described as

$$\frac{\partial\psi}{\partial t} = -\frac{h^2}{im}\sum_{1=1}^{N}\nabla^2\psi - \sum_{1=1}^{N}\frac{Ze^2}{|\overline{n}|}\psi$$

Hence, current density can be expressed as

$$J(x) = e\psi(x)^*\lambda^{\mu}\psi(s)$$

where $\lambda^{\mu} = \Upsilon^{o}\Upsilon^{\mu}$

This satisfies Dirac second quantized field equations. This is obtained in a quadratic function of the Dirac field operators. This current density produces quantum e.m. field.

Quantum-retarded potential:

$$A(t,r) = \int\frac{J^{\mu}(t-|r-r'|,\,r'}{4\pi|r-r'|}d^3r'$$

where $A(t,r)$ is the bosonic field. It is expressed as quadratic function field operators.

Quantum antenna radiated fields:

$$\langle\psi|A_u(x)A_v(x')|\psi\rangle \text{ fluctuation fields}$$

$$A_u(x) = \frac{\mu_o}{4\pi}\int\frac{\hat{J}_{\mu}(p)e^{ipx}}{p^2-10}d^4_P$$

5.6 Terahertz DRA applications

There are many advanced applications at THz frequencies as follows:

1. Imaging for pattern recognition with a higher resolution of images, biosensors spectroscopy.
2. Large band width for 5G and beyond, 6G, IOT, LIDAR and sensors.
3. Compact size for wearable, secured communication due to encryption.
4. Higher order modes for dimension scaling, i.e. miniaturization.
5. Defense and space application, i.e. detection of size of aircraft.
6. On-chip communication (chip-to-chip high speed communication).
7. It can detect any metallic object carried by any person at airport without manual search.
8. At airport advanced landing system during a bad weather.
9. Rectenna for energy harvesting, i.e. renewable energy source. Nanoantenna can receive direct optical energy.
10. Biological sensor detection and biomedical applications in imaging.
11. Comprehensive integrated border management system.
12. The advantages of nanoantenna rectennas are as follows: it can work whole day (day and night), and not sensitive to weather conditions, drones, e-cars and sensors.

5.7 Conclusion

TDRA has very sharp beam width, super directive and high gain. They are having compact size and very high operating frequency, i.e. THz frequency. They are good candidates for high-resolution imaging, scanning, LiDAR and sensors. Necessity of fabrication and testing facilities at such a high frequency in THz range is needed. Correlation coefficient manipulation is used in beam formation in quantum antenna. It is observed that the spin of photons becomes a main function of non-linearity at optical frequency. There is strong need to establish relationship between radiated frequencies, SPPs resonance frequencies and input excitation frequency. SPP wavelength is ten times smaller to free-space wavelength. Hence, radiated frequency shall be smaller to input excitation frequency in plasmonic antennas. Terahertz medical sensing and ultrafast data transfer and ultrawide-band antenna. Plasmon frequency must be smaller to input excitation frequency for propagation. Used for high-speed chip-to-chip communications and biomedical sensing.

References

[1] R. S. Yaduvanshi and H. Parthasarathy, *Rectangular Dielectric Resonator Antennas: Theory and Design*, Springer, New Delhi, 2016.

[2] M. Hayasi, *Quantum Information Theory: Mathematical Foundation*, Springer, Berlin Heidelberg, 2017.

[3] C. A. Balanis, *Antenna Theory: Analysis and Design*, Wiley, New York, NY, 2005.

[4] A. Öchsner and H. Altenbach (Editor), *Properties and Characterization of Modern Materials (Advanced Structured Materials)*, Springer, Singapore, 2017.

[5] S. A. Long, *Dielectric Resonator Antenna Handbook*, Artech House, Norwood, MA, 2007.

[6] P. Bhardwaj, B. Deutsch, and L. Novotny, "Optical antennas," *Adv. Opt. Photon.*, vol. 1, pp. 438–483, 2009.

[7] D. H. Youn, M. Seol, J. Y. Kim, *et al.* "TiN nanoparticles on CNT–graphene hybrid support as noble-metal-free counter electrode for quantum-dot-sensitized solar cells," *ChemSusChem*, vol. 6, no. 2, pp. 261–267, 2013.

[8] R. P. Feynman and A. R. Hibbs, *Quantum Mechanics and Path Integration*, McGraw-Hill, New York, NY, 1965.

[9] L. Novotny and N. Van Hulst, "Antennas for light," *Nat. Photonics*, vol. 5, no. 2, p. 83, 2011.

[10] W. T. Sethi, H. Vettikalladi, and H. Fathallah, "Dielectric resonator Nano antenna at optical frequencies," In Information and Communication Technology Research (ICTRC), 2015 International Conference on, IEEE, pp. 132–135, 2015.

[11] F. J. Rodríguez-Fortuño, A. Espinosa-Soria, and A. Martínez, "Exploiting metamaterials, plasmonics and nanoantennas concepts in silicon photonics," *J. Opt.*, vol. 18, no. 12, p. 123001, 2016.

[12] L. Zou, W. Withayachumnankul, C. M. Shah, *et al.* "Dielectric resonator nanoantennas at visible frequencies," *Opt. Express*, vol. 21, no. 1, pp. 1344–1352, 2013.

[13] X. Wu, F. Tian, W. Wang, J. Chen, M. Wu, and J. X. Zhao, "Fabrication of highly fluorescent graphene quantum dots using L-glutamic acid for in vitro/in vivo imaging and sensing," *J. Mater. Chem. C*, vol. 1, no. 31, pp. 4676–4684, 2013.

[14] A. E. Krasnok, D. S. Filonov, C. R. Simovski, Y. S. Kivshar, and P. A. Belov, "Experimental demonstration of superdirective dielectric antenna," *Appl. Phys. Lett.*, vol. 104, no. 13, p. 133502, 2014.

[15] N. Shinohara and H. Matsumoto, "Experimental study of large rectenna array for microwave energy transmission," *IEEE Trans. Microwave Theory Tech.*, vol. 46, no. 3, pp. 261–268, 1998.

[16] G. Varshney, A. Verma, V. S. Pandey, R. S. Yaduvanshi, and R. Bala, "A proximity coupled wideband graphene antenna with the generation of higher order TM modes for THz applications," *Opt. Mater.*, vol. 85, pp. 456–463, 2018.

[17] W. Chen, R. L. Nelson, D. C. Abeysinghe, and Q. Zhan, "Optimal plasmon focusing with spatial polarization engineering," *OPN Opt. Photonics News*, pp. 36–41, 2009.

[18] P. Biagioni, J.-S. Huang, and B. Hecht, "Adult teratoma of the testicle metastasizing as adult teratoma," *Rep. Prog. Phys.*, vol. 75, pp. 1–40, 2012.

[19] Y. Zhao and A. Alu, "Optical nanoantennas and their applications," *IEEE Radio Wirel. Symp. RWS*, pp. 58–60, 2013.

[20] S. Abadal, E. Alarcón, A. Cabellos-Aparicio, M. Lemme, and M. Nemirovsky, "Graphene-enabled wireless communication for massive multicore architectures," *IEEE Commun. Mag.*, vol. 51, no. 11, pp. 137–143, 2013.

[21] M. Schnell, P. Alonso-González, L. Arzubiaga, *et al.*, "Nanofocusing of mid-infrared energy with tapered transmission lines," *Nat. Photonics*, vol. 5, no. 5, pp. 283–287, 2011.

[22] J. Wen, S. Romanov, and U. Peschel, "Excitation of plasmonic gap waveguides by nanoantennas," *Opt. Express*, vol. 17, no. 8, p. 5925, 2009.

[23] J.-S. Huang, T. Feichtner, P. Biagioni, and B. Hecht, "Impedance matching and emission properties of optical antennas in a nanophotonic circuit," *Nano Lett.*, vol. 9, no. 5, pp. 1897–1902, 2008.

[24] L. Zou, W. Withayachumnankul, C. M. Shah, *et al.*, "Efficiency and scalability of dielectric resonator antennas at optical frequencies," *IEEE Photonics J.*, vol. 6, no. 4, pp. 1–10, 2014.

[25] G. N. Malheiros-Silveira, G. S. Wiederhecker, and H. E. Hernández-Figueroa, "Dielectric resonator antenna for applications in nanophotonics," *Opt. Express*, vol. 21, no. 1, pp. 1234–1239, 2013.

Chapter 6

Terahertz conical dielectric resonator antenna—design, simulation and implementations

Abstract

Terahertz conical dielectric resonator antennas (DRAs) have been designed, simulated and analyzed at a terahertz spectrum, i.e. 10 THz. A lower terahertz frequency band is useful for high data speed communication, whereas a terahertz frequency upper band is used for optical sensors or sensing applications. The free-space wireless communications also use terahertz DRAs (TDRAs). A unique and novel geometry for terahertz dual-band operations has been investigated with mathematical formulations. The comprehensive analyses in a terahertz regime have been worked out with a reflection coefficient (S_{11}), a voltage standing wave ratio, a radiation pattern and efficiency using a conical TDRA. The conical TDRA is designed at 10 THz frequency with 4.9 dBi, gain and bore sight radiation pattern. The equivalent circuit of conical TDRA has been developed at given frequencies.

6.1 Introduction

Optical antennas such as dipoles, graphene patch, silver and gold patch antennas and many other metallic nano antennas have been theoretically demonstrated by many research groups such as Bhardwaj, A Alu and Novotny [1–3]. Dielectric materials for use of antennas have been available since 1939 developed by Ritchmeyer. Dielectric resonator antenna (DRA) has high efficiency, temperature stability, ease of fabrication and low losses at microwave and optical frequencies. They have been used so far to develop classical antennas at microwave frequencies for applications in mobile personal communication, radar, satellite and sensors. Many researchers such as SA Long, MM Antaar, A Kishk and R Mongia, Guha and Ravi have contributed in this fields since 1980 [4–6]. These dielectric materials are ceramics having permittivity, ε_r, ranging from 9 to 100 (dielectric material permittivity). Reduced dimensions and efficient radiations are two very important properties of DRAs. DRA's main characteristics are a wide range of operations from microwave to optical, simple and efficient, built with high dielectric constant ceramics for miniaturization, ease of fabrication and highly stable at high operating temperature. Optical DRAs have been trending in current era research and still they are infant in photonics science [2,3]. Nanophotonics is a branch of science that deals with transmission and reception of signals at terahertz and optical frequencies and falls in nanometer dimensions [4,5]. A nanoantenna has two important features, miniaturization and super directivity [6–8]. So far researchers have worked on metallic antennas such as Yagi–Uda, dipole antenna and Bowtie antenna, but they suffer from losses at terahertz and optical frequencies. Nanotechnology is a multidisciplinary area of science that merges physics, chemistry, materials, biology, medicine and engineering branches for innovations on current trends of optical DRAs in a nano regime [7,9,10]. The physics in optical antennas involves a dominance behavior of skin effects, surface plasmon politrons (SPP) and light–matter interaction (creation and annihilation of photons) [11–13]. It is well known that light can penetrate through opaque objects such as clothes, plastic suitcases at optical frequencies; hence detection and scanning becomes possible in concealed items [14]. Light–matter interaction at an optical domain is an important parameter in optical devices, and their mathematical parts are described in the next section.

In a terahertz DRA (TDRA), an SPP phenomenon occurs due to strong light–matter interactions [15–17]. Total electric dipole moment in scattering can be computed using tensor analysis. This way radiations of desired wave vector with polarization, scattering, suppression, forward and backward scattering can thus be computed mathematically. Optical electromagnetic fields are realized by quantum wave operator fields. It can be precisely defined as an ensemble of an infinite number of quantum harmonic oscillators. It can be analyzed as a superposition of constant plane waves with coefficients being creation and annihilation operators in a boson Fock space. At an optical spectrum, it is better to talk about quantum state of wave field like a coherent state of photons. The light interacts with electrons, atoms and molecules of a material medium it travels. At terahertz and optical frequencies, the material medium is treated as a plasma. Having a definite permittivity, permeability and conductivity of the medium diffraction, reflection and refraction is solved based on linear and nonlinear properties [1,18,19]. Nonlinear dispersion relation can be developed between an oscillation frequency of plasma, electromagnetic fields and wave vector. The frequency of radiation by DRA and frequency of plasma oscillations are different. The plasmon frequency is lesser than input excitation frequency due to the absorption of photonic phenomenon. In optical antennas, radiated resonant frequency is always lesser than input excitation frequency. The light–matter interaction at optical frequency is more subtle. The simple way to model shall be a treating matter as a second quantized electron positron field using Dirac's relativistic wave equation with wave function being an operator field and then to include photon interaction with it [20–22]. An SPP phenomenon can be better described by a Drude model at a metal dielectric interface as metal behaves as a plasma or gas at optical frequency. The total Hamiltonian of the photons and matter fields can thus be defined by creation and annihilation operators, Dirac's fields and electron–positron–photons creation and annihilation. Thus interaction Hamiltonian is used to compute the amplitudes of scattering and electromagnetic wave fields patterns based on surface current density. Mean and covariance of electromagnetic fields shall be based on state of photon–electron positron fields due to Dirac currents in terms of operators. The resonant frequency of cone optical DRA will depend upon the volume and shape of DRA. Resonant modes excitation shall be mainly governed by aspect ratio, cavity dimensions, boundary conditions and feed position of DRA [23–25].

Since light is in fact a quantum electromagnetic field, when it falls on conical cavity, it starts immediately interacting with the matter within the conical cavity. A coarse approximation of this matter is to regard as a **sea of electrons and positrons** and then apply the formalism of **quantum electrodynamics** to describe its interaction with the **incident photons**, both before the light hits the conical cavity matter directed based on boundary conditions. The field in the cavity is actually a planar quantum electromagnetic field in a certain state and we wish to calculate the probability that such a field will be formed. The penetration of radiation into silver metals gives rise to SPP. **The electromagnetic response due to collective electron oscillations (Plasmon's) takes place. Plasmons also become excitation to cone DRA.** SPPs are surface modes in silver metal nanostructures at interface. Optical antennas are compact in size in RF and microwave. There are many differences in classical and optical antenna concepts. Although metals are not perfect conductors at optical frequencies, free electron gas or plasma is. The optical antennas are fed by localized light emitters, not by real currents, also known as a local density of electromagnetic states (LDOS). These LDOS can be solved using a Green's function tensor. Developing these devices hardware are still in infant stage and involves huge cost with limited fabrication facilities worldwide.

At optical frequency DR design becomes compact and it can be useful for imaging, scanning and other biomedical healthcare sensor applications such as a UV sensor, environment sensor and oxygen saturation sensor. A conical optical DRA integrated with photodiodes can work as a retinal implant as it can provide all functionalities of cone-type photo receptors used in a retinal surface of human eye. The study has revealed that photons are received by retina using available arrays of cones and rods present in a retina at central part and periphery. These DRAs can very well operate at a vision spectrum of human eyes, which lies in between 430 and 750 THz. Similarity optical DRAs can set new trends of large bandwidth for high-speed terahertz communication and wireless optical communication. Being small in size, DRAs can fit into integrated circuits for chip-to-chip communications in system-on-chip and network-on-chip circuits and Internet-of-Things sensors for healthcare. Also 0.1–10 THz is lying as a vacant frequency band; hence it can be used for high-speed communication. Other application areas are light detection and ranging, three-dimensional holography, THz technology, sensing, photovoltaics, optical processing, security and biotechnology [23–29].

6.2 Design structure of conical THz DRAs

The conical DRA design structures are shown in Figure 6.1 at 10 THz using silicon ($\varepsilon_r = 11.9$) as radiating elements, i.e. DRAs. DRAs are placed at the top of the substrate with optimized design dimensions. The proposed conical DRAs are

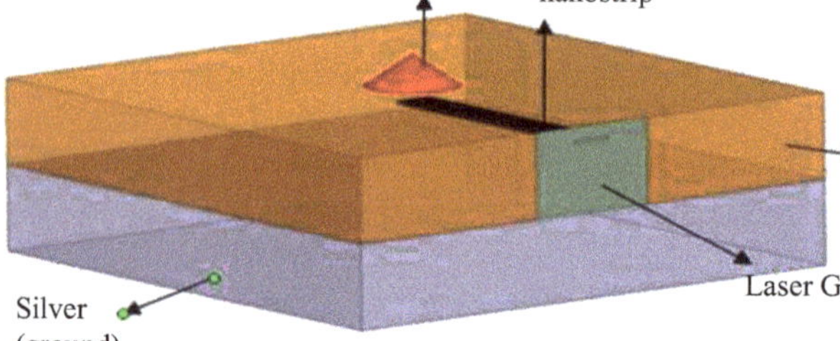

Figure 6.1 Conical terahertz DRA design architecture

excited using a rectangular nano silver feed line of width (w) with the stub of desired dimensions. The optimum value of the stub length has been chosen for good impedance match. The dispersive properties of silver used in the nano waveguide are selected using the Drude model as $\epsilon_{\text{Ag}} = \epsilon_o \left[\epsilon_\infty - \left(f_p^2 / \{ f(f + i\gamma) \} \right) \right]$. Here, f is the operating frequency, ϵ_{Ag} is the permittivity of silver, f_p is the plasma frequency, γ is the collision frequency and it is the inverse of relaxation time, ϵ_∞ is the offset of real part of the dielectric constant and $\epsilon_o = 8.85 \times 10^{-12}$ F/m. The cone DRA is designed with a ground plane and it will look like a biconical DRA due to image effect. The volume of a cone DRA is given as $= \pi r^2 (h/3)$. The finite value of Planck's constant in a TDRA becomes a source of radiation, in terms of quanta. In conical cavity, this matter is to regard as a sea of electrons and positrons and then applies the formalism of quantum electrodynamics to describe its interaction with incident photons, both before the light hits the cavity matter directed based on boundary conditions. In bosonic creation and annihilation operators, they satisfy the canonical communication relations and finally compute the mean and covariance of the electromagnetic field in a state. Hence, due to Planck's constant in quantum antennas, radiation energy comes in quanta. The mean-square fluctuations are at least of the order of magnitude of Planck's constant. The mathematical derivation of quantum antenna states has been done to differentiate it from classical antennas. These conical DRAs can also be used in biomedical applications, sensors, chip-to-chip communications and high-speed telecommunications.

Thus, discrete particle distributions of electrons, positrons and photons do matter and this discreetness is also responsible for producing quantum fluctuation and formation of particular state or statistical moments. The polynomial in the photon and electron, positron creation and annihilation operators hence determine their statistical moments, i.e. mean, variance, etc. in any given state of the electron positron photon field in the case of optical DRAs.

6.3 Model-1 multiband conical TDRA

Resonant frequency $f_r = 10$ THz, $\lambda = c/f$

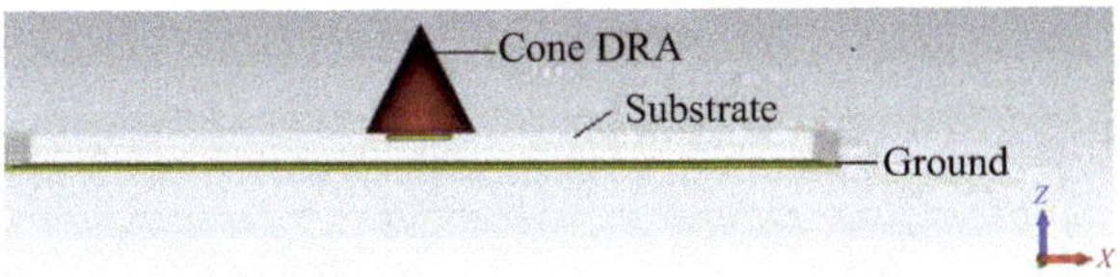

Figure 6.2 Conical 10-THz DRA with Gaussian beam laser input excitation (silicon radiating element at 10 THz)

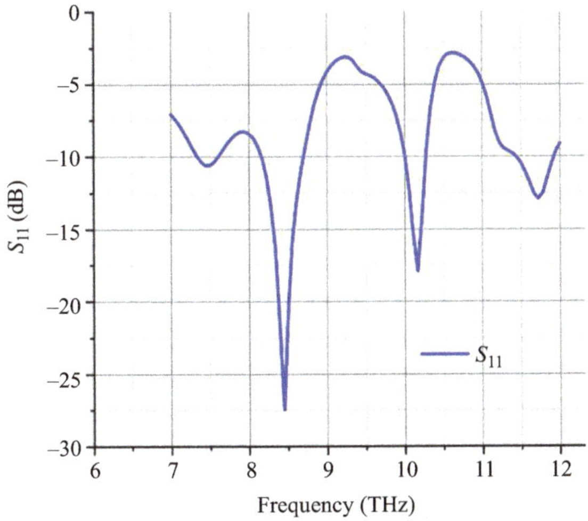

Figure 6.3 S_{11} simulated conical terahertz DRA

Substrate (SiO_2): $1.5*\lambda_o = 1.5*30 = 45\ \mu m$
and height of cone DRA $= \lambda_o/3 = 30/3 = 10\ \mu m$
Substrate dimensions: length*breadth*height $= 45*45*10\ \mu m^3$
Ground plane (silver): $1.5*\lambda_o = 1.5*30\ \mu m = 45\ \mu m$
and height of a DRA (silicon) $= \lambda_o/3 = 30/3 = 10\ \mu m$
Nanostrip line or waveguide (silver Ag): length*width*height $= 22.5*3*0.3\ \mu m^3$
Cone DRA (silicon): $h = \lambda_o/10 = 30/10 = 3\ \mu m$
(height of cone DRA),
Radius of base in cone DRA $(r) = 3\ \mu m$
TDRA aspect ratio $(h/r) = 1$

Figure 6.2 shows conical DRA with substrate and ground. The aspect ratio of cone DRA is used to excite resonant modes, i.e. fundamental or higher order modes; here it is chosen in between 0.5 and 2.5. Thus by varying aspect ratio, higher order modes can be excited. Here, this TDRA has dual band applications as shown in Figure 6.3. Figure 6.4 has a **radiation pattern of TDRA with gain that is 4.9 dBi**

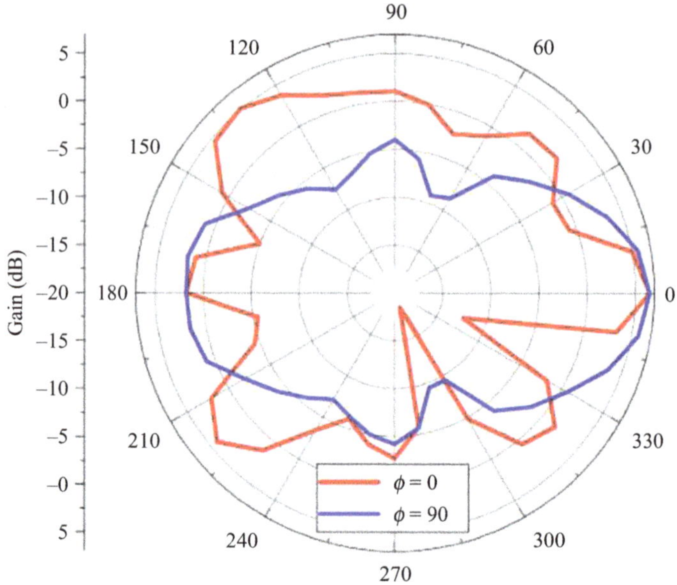

Figure 6.4 Radiation pattern of conical TDRA

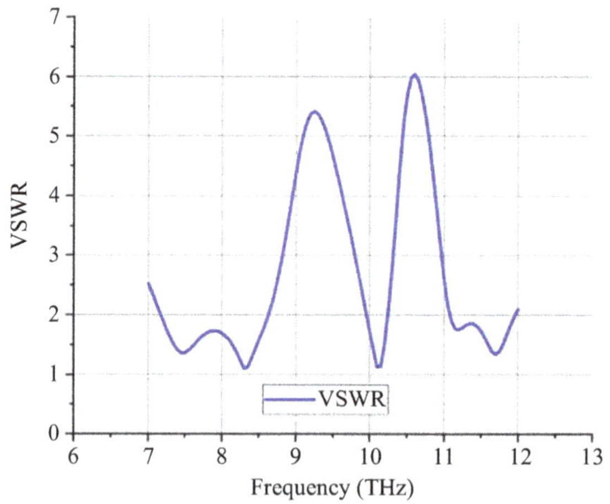

Figure 6.5 VSWR of conical TDRA

broadside. Figures 6.5 and 6.6 are voltage-standing wave ratio (VSWR) and input impedance at operating frequencies. Figure 6.7 shows a better efficiency of TDRA. Figure 6.8 shows current density *J*. Figure 6.9 shows *E*-fields. The ground plane material is assigned as silver (Ag). The substrate of the material, silicon dioxide was used and placed on silver ground plane (Ag). The loss tangent material for a

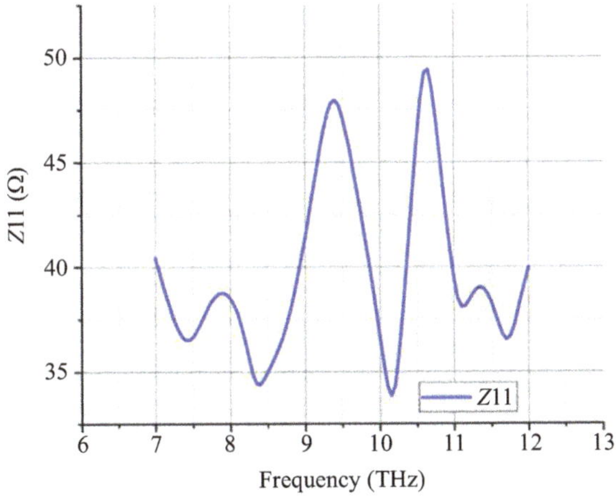

Figure 6.6 Z11 of conical TDRA with multibands

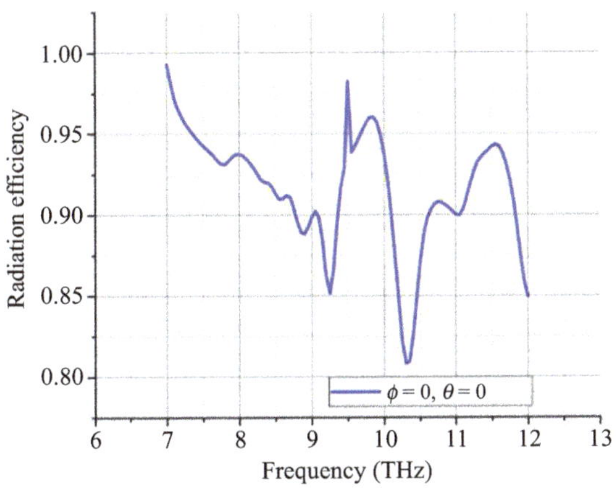

Figure 6.7 Radiation efficiency of TDRA

substrate was chosen to minimize the losses. Input excitation as Gaussian pulse from laser was assigned as input and coupled to silver nano waveguide for proximity feed. This silver nano waveguide is made up of silver nano wire. For perfect match, we designed nanostrip waveguide (Ag), whose lower part is connected to ground plane and top part is inserted into substrate (height h_1) near to DRA, as proximity feed. The width, length and height of the nano feed (h_2) have been chosen for matching of the design to 50-Ω impedance. A silicon cone DRA was placed on the top of a substrate as a radiating element. The silver nano waveguide is inserted into a substrate to create SPP and surface plasmon resonance (SPR). Light–matter interaction shall result into a generation of plasmonic frequency due to proximity feed. The plasmon frequency, f_p, is responsible for the excitation of resonant modes into DRA [30,31]. This will

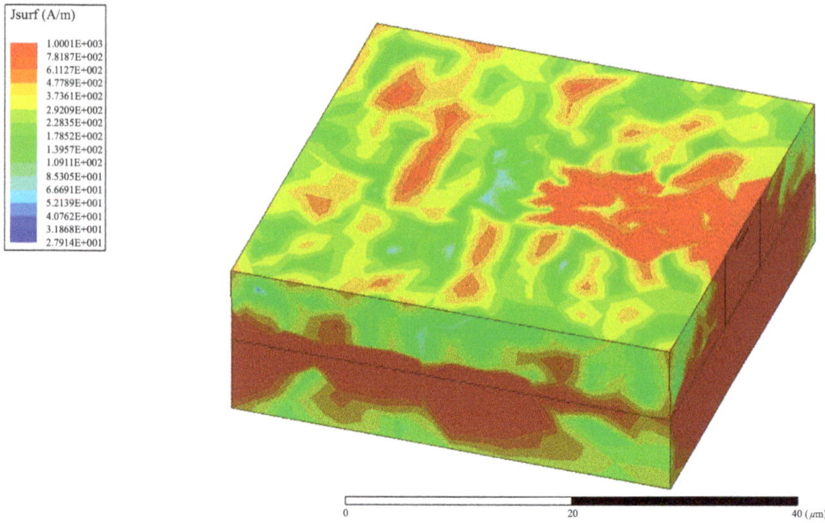

Figure 6.8 E-Fields J current density

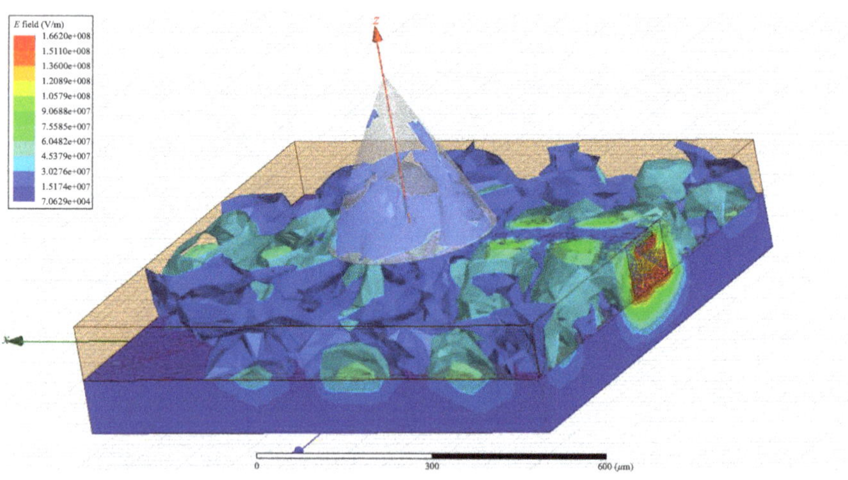

Figure 6.9 E-Fields of TDRA

cause DRA to radiate at resonant frequency. The quantum-mechanical model is required to show radiated states of cone DRA at terahertz *l* frequencies. Silver nano waveguide is used for the excitation of conical DRA with laser Gaussian beam. Figures 6.1–6.9 have shown simulated design and corresponding results obtained (Table 6.1).

Table 6.1 Comparison table

Reference	Mathematical analysis	Antenna geometry	Frequency	Feeding technique	Dimensions	ε_r	Multiband	Bandwidth
[3]	No	Conical	8.3 GHz	Microstrip *line*	$h = 5$ mm $r = 7$ mm	10	No	0.9 GHz
[2]	No	Conical annular ring	2.72–11.36	Microstrip line	$h = 10$ $r = 6$	20	Yes	8.64 GHz
Proposed work	**Yes**	**Conical TDRA**	**10 THz**	**Silver nano waveguide**	$\boldsymbol{r = 3\ \mu m}$ $\boldsymbol{h = 3\ \mu m}$	**11.9**	**Yes**	**High, dual band 300 GHz**

6.4 Mathematical modeling of terahertz conical DRA

The theory of terahertz conical DRA has been developed with (6.1)–(6.24), which describe the concept of terahertz radiations when excited with Gaussian beam input. In these equations, the case have been theoretically analyzed for statistical moments or terahertz radiated states are formed in conical TDRAs.

Let h be the height of the cone and r, its radius. We may thus assume the conical region to be specified by the following equation:

$$0 \leq z \leq d, x^2 + y^2 \leq a^2 z^2, \ a = \tan\left(\frac{\alpha}{2}\right)$$

where α is the cone angle, and this cone has its apex at origin and its surface is a revolution of a line passing through the origin making an angle $\alpha/2$ with the z-axis. Using spherical polar coordinate system to expand the field as

$$E(r, \theta, \varphi) = E_r(r, \theta, \varphi)\hat{r} + E_\theta(r, \theta, \varphi)\hat{\theta} + E_\varphi(r, \theta, \varphi)\hat{\varphi} \tag{6.1}$$

likewise, for the magnetic field the boundary conditions are

$$E_r = E_\varphi = 0, \ H_\theta = 0, \ \theta = \frac{\alpha}{2}$$

current into the PEC boundary conditions, the vanishing of the tangential component of the electric field and the normal component of magnetic field on the conical surface. Further if the top of the cone at $z = h$ is again PEC, then the boundary conditions corresponding to this is

$$E_\theta = E_\varphi = 0, \ H_z = 0, \ z = h$$

$$\text{Note that } \hat{z} = \cos(\theta)\hat{r} - \sin(\theta)\hat{\theta} \tag{6.2}$$

Hence, the boundary condition can also be expressed as

$$H_r \cos(\theta) - \sin(\theta) H_\theta = 0, r \cdot \cos(\theta) = h \tag{6.3}$$

We now write down the Maxwell curl equations in the spherical polar coordinate system.

Quantum image processing via the Belkin filter,

$$\begin{aligned} \text{curl } E &= \det\begin{pmatrix} \hat{r}/r^2 \sin(\theta) & \hat{\theta}/r \sin(\theta) & \hat{\varphi}/r \\ \dfrac{\partial}{\partial r} & \dfrac{\partial}{\partial \theta} & \dfrac{\partial}{\partial \varphi} \\ E_r & rE_\theta & r\sin(\theta)E_\varphi \end{pmatrix} \\ &= j\omega\mu\left(H_r\hat{r} + H_\theta\hat{\theta} + H_\varphi\hat{\varphi}\right) \end{aligned} \tag{6.4}$$

Similarly,

$$\begin{aligned}\text{curl}\,H &= \det\begin{pmatrix}\hat{r}/r^2\sin(\theta) & \hat{\theta}/r\sin(\theta) & \hat{\varphi}/r\\ \dfrac{\partial}{\partial r} & \dfrac{\partial}{\partial \theta} & \dfrac{\partial}{\partial \varphi}\\ H_r & rH_\theta & r\sin(\theta)H_\varphi\end{pmatrix}\\ &= j\omega\epsilon\left(E_r\hat{r}+E_\theta\hat{\theta}+E_\varphi\hat{\varphi}\right)\end{aligned}$$

Writing out the components explicitly gives

$$\begin{aligned}&(\sin(\theta)\,H\varphi)_{,\theta}-H_{\theta,\varphi}=-j\omega\,\mu r\sin(\theta)H_r\\ &E_{r,\varphi}-\sin(\theta)\left(r\cdot E_\varphi\right),r=-j\omega\,\mu r\sin(\theta)H_\theta\\ &(rE_\theta),r-E_{r,\theta}=-j\omega\mu\,rH_\varphi\end{aligned}\tag{6.5}$$

Hence, its dual with $E\rightarrow H$, $H\rightarrow -E$ and $\varepsilon\leftrightarrow\mu$. Also these Maxwell curl equations imply the Helmholtz equation:

$$\left(\nabla^2+k^2\right)(E,H)=0,\ k^2=\omega^2\mu\epsilon\tag{6.6}$$

Helmholtz equations in spherical polar coordinate for that, we note that

$$\begin{aligned}&\frac{\partial\hat{r}}{\partial r}=0,\ \frac{\partial\hat{r}}{\partial\theta}=\hat{\theta},\frac{\partial\hat{r}}{\partial\varphi}=\ \sin(\theta)\hat{\varphi},\\ &\frac{\partial\hat{\theta}}{\partial r}=0,\ \frac{\partial\hat{\theta}}{\partial\theta}=-\hat{r},\frac{\partial\hat{\theta}}{\partial\varphi}=\cos(\theta)\hat{\varphi},\\ &\frac{\partial\hat{\varphi}}{\partial r}=0,\ \frac{\partial\hat{\varphi}}{\partial\theta}=0,\frac{\partial\hat{\varphi}}{\partial\varphi}=-\hat{x}\cos(\varphi)-\hat{y}\sin(\varphi)=\ -\hat{\rho}\\ &=-\hat{r}\sin(\theta)-\hat{\theta}\cos(\theta)\\ &\text{Thus, }\nabla^2(E_r\hat{r})=(\nabla^2E_r)\,\hat{r}+2(\nabla E_r,\nabla)\hat{r}+E_r\nabla^2\hat{r}\\ &\text{With }(\nabla E_r,\nabla)\hat{r}=r^{-2}E_{r,\theta}\hat{\theta}+r^{-2}\cdot(\sin\theta)^{-1}E_{r,\theta}\hat{\varphi}\end{aligned}\tag{6.7}$$

And

$$\begin{aligned}&\frac{\partial^2\hat{r}}{\partial\theta^2}=\frac{\partial\hat{\theta}}{\partial\theta}=-\hat{r}\\ &\frac{\partial^2\hat{r}}{\partial\varphi^2}=\ \sin\theta\frac{\partial\hat{\varphi}}{\partial\varphi}=-\ \sin\theta\ \sin\theta\hat{r}+\cos(\theta)\hat{\theta}\end{aligned}$$

show that

$$\begin{aligned}\nabla^2r&=\frac{\cot\theta}{r^2}\frac{\partial^2\hat{r}}{\partial\theta^2}\ +\frac{1}{r^2}\frac{\partial^2\hat{r}}{\partial\theta^2}+\frac{1}{r^2\sin(\theta)}\frac{\partial^2\hat{r}}{\partial\varphi^2}\\ &=\frac{\partial^2\hat{r}}{\partial\varphi^2}\hat{\theta}-\frac{1}{r^2}\hat{r}+\frac{1}{r^2}(-\hat{r}-\cot\theta)\hat{\theta}\\ &=-\frac{2\,\hat{r}}{r^2}\end{aligned}\tag{6.8}$$

Thus,

$$\nabla^2(E_r\hat{r}) = \hat{r}\left((\nabla^2 E_r) - \frac{2E_r}{r^2}\right) + \frac{2}{r^2}E_{r,\theta}\hat{\theta} + \frac{2}{r^2\sin(\theta)}E_{r,\varphi}\hat{\varphi} \tag{6.9}$$

Next,

$$\nabla^2(E_\theta\hat{\theta}) = (\nabla^2 E_\theta)\hat{\theta} + 2(\nabla E_{\theta,}\nabla)\hat{\theta} + E_\theta\nabla^2\hat{\theta}$$

And

$$(\nabla E_{\theta,}\nabla)\hat{\theta} = -\frac{2E_{\theta,\theta}}{r^2}\hat{r} + \frac{E_{\theta,\varphi}}{r^2(\sin(\theta))^2}\cos(\theta)\hat{\varphi}$$

And further,

$$\begin{aligned}\nabla^2\hat{\theta} = &\ \frac{\cot\theta}{r^2}\frac{\partial\hat{\theta}}{\partial\theta} + \frac{1}{r^2}\frac{\partial^2\hat{\theta}}{\partial\theta^2} + \frac{1}{r^2(\sin(\theta))^2}\frac{\partial^2\hat{\theta}}{\partial\varphi^2}\\ = &\ \frac{\cot\theta}{r^2}\hat{r} - \frac{1}{r^2}\hat{\theta} + \frac{1}{r^2(\sin(\theta))^2}\left((-\sin(\theta)\cos(\theta))\hat{r} - (\cos(\theta))^2\hat{\theta}\right)\\ = &\ -\frac{2\cot\theta}{r^2}\hat{r} - \frac{1}{r^2(\sin(\theta))^2}\hat{\theta}\end{aligned} \tag{6.10}$$

Thus,

$$\begin{aligned}\nabla^2(E_\theta\hat{\theta}) = &\ (\nabla^2 E_\theta)\hat{\theta} - \frac{2}{r^2}E_{\theta,\theta}\hat{r} + \frac{2}{r^2(\sin(\theta))^2}E_{\theta,\varphi}\cos(\theta)\hat{\varphi} + E_\theta\frac{2\cot\theta}{r^2}\hat{r} - \frac{1}{r^2(\sin(\theta))^2}\hat{\theta}\\ = &\ 2\hat{r}\left(\frac{E_{\theta,\theta}}{r^2} + \frac{E_\theta\cot\theta}{r^2}\right) + \hat{\theta}\left(\nabla^2 E_\theta - \frac{E_\theta}{r^2(\sin(\theta))^2}\right) + \hat{\varphi}\left(\frac{2\cot\theta}{r^2\sin(\theta)}\right)E_{\theta,\varphi}\end{aligned}$$

Finally

$$\nabla^2(E_\varphi\hat{\varphi}) = (\nabla^2 E_\varphi)\hat{\varphi} + E_\varphi\nabla^2\hat{\varphi} + 2(\nabla E_\varphi\nabla)\hat{\varphi} \tag{6.11}$$

Now

$$(\nabla^2 E_\varphi, \nabla)\hat{\varphi} = -\frac{1}{r^2}E_{\varphi,\varphi}(\hat{r}\cdot\sin(\theta) + \hat{\theta}\cos(\theta))$$

And $\nabla^2\hat{\varphi} = \dfrac{\hat{\varphi}}{r^2(\sin(\theta))^2}$

Thus

$$\nabla^2(E_\varphi\hat{\varphi}) = \hat{r}\left(-\frac{1}{r^2}E_{\varphi,\varphi}\sin(\theta)\right) + \hat{\theta}\left(-\frac{1}{r^2}E_{\varphi,\varphi}\cos(\theta)\right) + \hat{\varphi}\left(\nabla^2 E_\varphi - \frac{E_\varphi}{r^2(\sin(\theta))^2}\right) \tag{6.12}$$

The Helmholtz equation for E gives, on equating the $\hat{r}$, $\hat{\theta}$ and $\hat{\varphi}$ components, the following three equations:

$$\nabla^2 E_r - \frac{2E_r}{r^2} - 2\left(\frac{E_{\theta,\theta}}{r^2} + \frac{E_\theta \cot\theta}{r^2}\right) - \frac{E_\varphi}{r^2} + k^2 E_r = 0,$$
$$\nabla^2 E_\theta + \frac{2E_{r,\theta}}{r^2} - \frac{E_\theta}{r^2(\sin(\theta))^2} - \frac{E_{\varphi,\varphi}}{r^2} + k^2 E_\varphi = 0,$$
$$\nabla^2 E_\varphi - \frac{E_\varphi}{r^2(\sin(\theta))^2} + \frac{2}{r^2\sin(\theta)} E_{r,\varphi} + \frac{2\cot\theta}{r^2\sin(\theta)} E_{\theta,\varphi} = 0$$

Likewise, with E replaced by H, it is hard to solve this Helmholtz equation taking into account the earlier Maxwell curl equations. Before delving into the special case of these equations, we simplify matters assuming that the entire electromagnetic field is replaced by a single scalar field $\psi(r,\theta,\varphi)$ within the cone.

The Helmholtz equation is

$$\frac{2}{r}\psi_{,r} + \psi_{,rr} + \frac{1}{r^2}\left(\cos(\theta)\psi_{,\theta} + \psi_{,\theta\theta}\right) + \frac{1}{r^2}\psi_{,\varphi\varphi} + k^2\psi = 0 \tag{6.13}$$

With the boundary conditions that ψ vanishes on all the surface of the cone, we can equivalently express this equation as

$$\frac{2}{r}\psi_{,r} + \psi_{,rr} - L^2\frac{\psi}{r^2} + k^2\psi = 0$$

where

$$L^2 = -\left(-\frac{1}{\sin(\theta)}\frac{\partial}{\partial\theta}\sin(\theta)\frac{\partial}{\partial\theta} + \frac{1}{(\sin(\theta))^2}\frac{\partial^2}{\partial\varphi^2}\right) \tag{6.14}$$

is the angular part of the Laplacian or what is known in quantum mechanics as the square of the angular momentum operator separation of variable gives us

$$\Psi(r,\theta,\varphi) = R(r)Y_{lm}(\theta,\varphi)$$

where Y_{lm} are the spherical harmonics that satisfy the eigen relations:

$$L^2 Y_{lm} = l(l+1)Y_{lm},\ L_z Y_{lm} = mY_{lm}, L_z = -i\frac{\partial}{\partial\varphi} \tag{6.15}$$

Thus, the radial component of the wave function $R(r)$ satisfies

$$2rR'(r) + r^2R''(r) + \left(k^2r^2 - l(l+1)\right) + R(r) = 0$$

We substitute it into this equation:

$$R(r) = r^aF(r)\text{to get}$$
$$2r\left(ar^{a-1}F(r) + r^aF'(r)\right) + r^2\left(a(a-1)r^{-2}F(r) + 2ar^{a-1}F'(r) + r^aF''(r)\right) +$$
$$(k^2r^2 - l(l+1)r^aF(r)) = 0$$

Or equivalently,

$$r^2F''(r) + F'(r)(2r + 2ar) + F(r)\big(a(a+1) + k^2r^2 - l(l+1)\big) = 0$$

Choosing $a = -1/2$ gives us

$$r^2F''(r) + r\frac{F'(r)}{r} + F(r)\left(k^2r^2 - \left(l + \frac{1}{2}\right)^2\right) = 0 \tag{6.16}$$

which is the Bessel equation of the order $l + 1/2$. Its solution is

$$F(r) = j_{l+1/2}(kr)$$

And hence, $R(r) = r^{-1/2}j_{l+1/2}(kr)$

The general solution to the three-dimensional Helmholtz equation then can be expressed as a superposition:

$$\Psi(r,\theta,\varphi) = \sum_{i\geq 0,|m|\leq l} c(l,m)r^{-1/2}j_{l+1/2}(kr)\,Y_{lm}(\theta,\varphi)$$

The first boundary condition is

$$0 = \left(r, \frac{\alpha}{2}, \varphi\right) = \sum_{lm} c(l,m)j_{l+1/2}(kr)Y_{lm}\left(\frac{\alpha}{2}, \varphi\right) = 0$$

Or equivalently writing the spherical harmonics in terms of modified Legendre polynomials:

$$Y_{lm}(\theta,\varphi) = P_{lm}(\cos\theta)\exp(im\varphi) \tag{6.17}$$

This Boundary condition gives us

$$\sum_{l} c(l,m)P_{lm}\left(\cos\frac{\alpha}{2}\right)j_{l+1/2}(kr) = 0,\ \forall\, m \in z, 0 \leq z \leq L$$

where $L = \sqrt{h^2 + a^2}$ is the length of the side of cone. The other boundary condition is Ψ that vanishes on the circular lid at z=h, $0 \leq \theta \leq \frac{\alpha}{2}$:

$$0 = (h \cdot \tan\theta, \theta, \varphi)$$

So that

$$\sum_{l} c(l,m)P_{lm}(\cos\theta)j_{l+1/2}(kh\tan\theta) = 0,\ \forall\, m \in z, 0 \leq \theta \leq \frac{\alpha}{2} \tag{6.18}$$

By multiplying these equations by linearly independent function of r and θ, respectively, and integrating over the appropriate range, we get a sequence of linear homogeneous equations for $c(l,m)$, $l \geq 0$foreach m that therefore have a nontrivial solution if the corresponding infinite determinant vanishes this, which gives as a nonlinear equation for k from which one can in principle obtain the characteristics frequencies of oscillation.

We now come back to the electromagnetic field case and analyze it in a different way using the multiple-expansion method based on vector-valued functions.

Let electric field be given by $E_{lm} = f_l(r)LY_{lm}(\theta, \varphi)$
where

$$L = -ir \times \nabla$$

is the usual angular momentum vector in quantum mechanics then

$$\hat{r} \cdot E_{lm} = 0. \tag{6.19}$$

Also, the Helmholtz equation for E_{lm} and the fact that ∇^2 commutes with L implies that

$$\frac{2}{r}f_l^{'}(r) + f_l^{''}(r) - \frac{l(l+1)}{r^2}f_l(r) + k^2 f_l(r) = 0$$

Or equivalently

$$r^2 f_l^{''}(r) + 2rf_l^{'}(r) + \left(k^2r^2 - l(l+1)\right)f_l(r) = 0$$

which has solutions

$$f_l(r) = h_l(kr)$$

where h_l are the Hankel functions. E_{lm} also satisfies the Gauss equation:

$$\text{div}E_{lm} = 0 \tag{6.20}$$

Since

$$\hat{r}.L = 0$$

And

$$\nabla \cdot L\psi = -i\nabla \cdot (r \times \nabla\psi) = i\nabla \cdot (\nabla \times (r\psi)) = 0$$

Since $\nabla \times r = 0$, we can repeat the same analysis for magnetic field by defining

$$h_{lm} = h_l(kr)LY_{lm}(\theta, \varphi)$$

Thus, the general solution to the Helmholtz equation in spherical polar coordinate can be expressed as

$$\begin{aligned} E(\omega, r) &= \sum_{lm} [c(l,m)h_l(kr)LY_{lm}(\hat{r}) + d(l,m)\text{curl}(h_l(kr)LY_{lm}(\hat{r}))/j\omega\epsilon] \\ H(\omega, r) &= \sum_{lm} [d(l,m)h_l(kr)LY_{lm}(\hat{r}) - c(l,m)\text{curl}(h_l(kr)LY_{lm}(\hat{r}))/j\omega\mu] \end{aligned} \tag{6.21}$$

It is a simple matter to verify that these fields satisfy Maxwell's curl equations:

$$\mathrm{curl}E = -j\omega\mu H, \mathrm{curl}H = j\omega\epsilon E$$

By making use of the relations,

$$\mathrm{div}(h_l(kr)LY_{lm}(\hat{r})) = 0, \ \left(\nabla^2 + k^2\right)(h_l(kr)LY_{lm}(\hat{r})) = 0$$

For our cone problem, the coefficient $c(l,m)$, $d(l,m)$ and characteristics oscillation frequency $\omega = k/\sqrt{\mu\varepsilon}$ are obtained from the following boundary conditions:

$$\begin{aligned} &E_r\left(r, \frac{\alpha}{2}, \varphi\right) = 0, \ H_\theta\left(r, \frac{\alpha}{2}, \varphi\right) = 0, \ 0 \le r \le L, 0 \le \varphi \le 2\pi \\ &E_z(h \cdot \tan\theta, \theta, \varphi) = 0, 0 \le \theta \le \frac{\alpha}{2}, 0 \le \varphi \le 2\pi \end{aligned} \tag{6.22}$$

Denote the characteristics frequencies by $\omega[n]$, $n = 1, 2, 3\ldots$ and for each n, we have eigen vector $(c(l,m,n))_h, (d(l,m,n))_n$. Then the general solution in a time domain can be expressed as

$$E(t,r,\theta,\varphi) = R_e\left[\sum_{lm,n} [c(l,m,n)h_l(k[n]r)LY_{lm}(\hat{r})\exp(-j\omega[n]t)\right.$$

$$\left. + d(l,m,n)curl(h_l(k[n]r)LY_{lm}(\hat{r})\exp(-j\omega[n]t))/j\omega[n]\epsilon]\right]$$

$$\psi^{n+1}(x) = -\sum_{k=0}^{n} \int S(x-y)\gamma^\mu \psi^k(y) A_\mu^{n-k}(y) d^4y$$

$$A^{\mu(n+1)}(x) = \sum_{k=0}^{n} \int G(x-y)\psi^k(y)^* \alpha^\mu \psi^{n-k}(y) d^4y \tag{6.23}$$

For $n = 0, 1, 2\ldots$, $\psi^{(0)}(x)$, $A^{\mu(0)}(x)$ describe the free fields, i.e. in the absence of electron–positron and photon interaction and for $n = 1, 2, 3\ldots$, $\psi^{(n)}(x)$, $A^{\mu(n)}(x)$ describe the higher order interaction terms. In this way, continuing up to any order N of accuracy, photon radiation field can be expressed as

$$A^\mu(x) \cong \sum_{n=0}^{N} e^n A^{\mu(n)}(x) \tag{6.24}$$

Equations (6.1)–(6.24) have been developed to impart a complete view of quantum radiations in conical DRAs at higher frequencies (terahertz and optical frequencies). Table 6.1 shows a comparative analysis of the previous work. The proposed model has good bandwidth and mathematical analysis as novelty.

6.5 Equivalent electrical circuit of conical terahertz DRA

This frequency-dependent resistance is also called a dynamic resistance of the circuit. Figure 6.10 is built as an equivalent electrical circuit of conical TDRA at

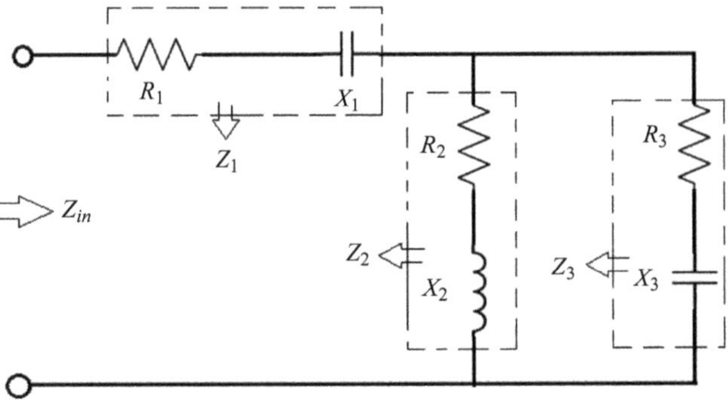

Figure 6.10 Equivalent circuit of conical terahertz DRA

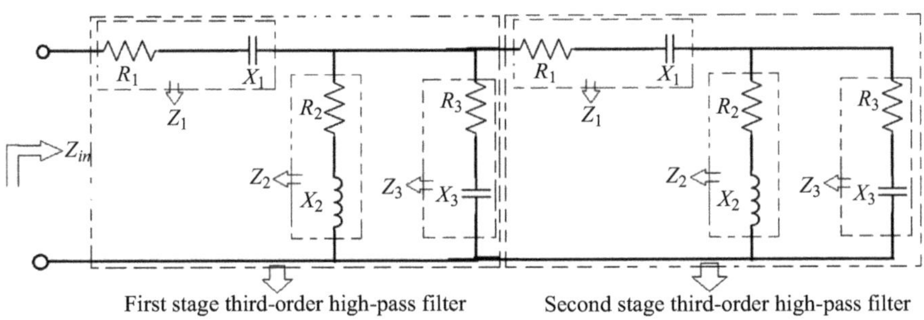

Figure 6.11 Expanded form of equivalent electrical circuit of conical terahertz DRA

fundamental mode and Figure 6.11 is developed for showing higher order resonant modes.

The conical TDRA input impedance of the circuit:

$$Z_{in} = Z_{inreal} + Z_{inimaginary}(\omega)$$
$$Z_{inreal} \neq f(\omega)$$

where Z_{inreal} means only real part of input impedance that is called only resistive part of input impedance

Conical TDRA-resistive part of input impedance is a function of frequency dependent:

$$Z_{inreal} = \text{resistance} = f(\omega)$$

Resonant frequency of this circuit depends on R, L and C but input impedance of this circuit is a function of R, L, C and frequency:

$$Z_{in} = f(R, L, C, \omega)$$

The impedance depends only on the real part of Z_{in} and real part of Z_{in} is dependent only on the frequency of the signal. The quality factor has been estimated from the equivalent circuit; hence bandwidth, resonant frequency, components values can thus be obtained at terahertz frequency.

$$\text{Quality factor } (Q) = 2\pi \times \frac{\text{maximum energy stored per cycle}}{\text{power dissipatetd per cycle}}$$

$$Q = \frac{\omega_r}{(\omega_2 - \omega_1)}$$

$$= \frac{\sqrt{\dfrac{-\left(\frac{L_2}{C_3^2} - \frac{R_3^2}{C_1} - R_2^2\left(\frac{1}{C_1} + \frac{1}{C_3}\right) - \frac{2}{C_1}\left(R_2R_3 - \frac{L_2}{C_3}\right)\right) \pm \sqrt{\left(\frac{L_2}{C_3^2} - \frac{R_3^2}{C_1} - R_2^2\left(\frac{1}{C_1} + \frac{1}{C_3}\right) - \frac{2}{C_1}\left(R_2R_3 - \frac{L_2}{C_3}\right)\right)^2 + 4\left\{L_2R_3^2 - L_2^2\left(\frac{1}{C_1} + \frac{1}{C_3}\right)\right\}\frac{1}{C_1C_3^2}}}{2\left\{L_2R_3^2 - L_2^2\left(\frac{1}{C_1} + \frac{1}{C_3}\right)\right\}}}}{\sqrt{\dfrac{1}{L_2C_3 + R_2R_3C_1C_3 + \{L_2C_3 + R_1C_1C_3(R_2 + R_3)\}}} - \sqrt{\dfrac{R_1C_1 - R_2C_1}{R_1C_1L_2C_3 - R_3C_1L_2C_3}}}$$

6.6 Conclusion

The proposed terahertz conical DRA has highest bandwidth as shown in comparative Table 6.1 and excellent gain. Hence, it is best suitable for high-speed communication requirements. Gaussian beam input excitation was used to compensate phase nonlinearity with specifications as beam power: $P1=1.0$ W, wavelength: 632.47 nm, propagation vector: (0/0/1), focus point distance from source center: 1.27E−06 m and linear polarization. Conical TDRA can be used to generate higher order modes depending upon a proper selection of aspect ratio; hence higher gain can be obtained at higher order modes as compared to fundamental mode. The design, dimensions, radiation pattern of THz conical DRAs, along with S_{11} and other antenna parameters have been obtained in Figures 6.1–6.9, and Figures 6.10 and 6.11 are drawn as equivalent R, L, C circuit based on which dynamic impedance is found to be frequency dependent. They can provide wide bandwidth and high speed data rate. TDRA can be used for optical communication. New geometry-based terahertz antenna has opened up new space for fast speed communication antennas. The equivalent circuit of TDRA is a novel field of analysis for terahertz or optical antennas.

References

[1] A. A. Kisk, Y. Yin and A. W. Glisson "Conical DRA for wireless applications," *IEEE J.*, 2002.

[2] V. Gaurav, R. S. Yaduvanshi, and V. S. Pandey, "Conical shape DRA for ultra wide band applications," IEEE Conference, 2015.

[3] S. Dash, T. Khan, and B. Kanaujia, "Conical dielectric resonator antenna with improved gain and bandwidth for X-band applications," *International Journal of Microwave and Wireless Technologies*, vol. 9, no. 8, pp. 1749–1756, 2017.

[4] L. Novotny and N. van Hulst, "Antennas for light," *Nat. Photonics*, vol. 5, no. 2, pp. 83–90, 2011.

[5] P. Biagioni, J.-S. Huang, and B. Hecht, "Nanoantennas for visible and infrared radiation," *Rep. Prog. Phys.*, vol. 75, p. 024402, 2012.
[6] L. Zou, W. Withayachumnankul, C. Shah, *et al.*, "Dielectric resonator nano antennas at visible frequencies," *Opt. Express*, vol. 21, no. 1, pp. 1344–1352, 2013.
[7] G. N. Malheiros-Silveira, G. S. Wiederhecker, and H. E. Hernández-Figueroa, "Dielectric resonator antenna for applications in nano photonics," *Opt. Express*, vol. 21, no. 1, pp. 1234–1239, 2013.
[8] K. M. Luk and K. W. Leung, *Dielectric Resonator Antennas*, Research Studies Press LTD, England, 2003.
[9] Y. Zhao, N. Engheta, and A. Alù, "Effects of shape and loading of optical nano antennas on their sensitivity and radiation properties," *J. Opt. Soc. Am.*, vol. B28, no. 5, pp. 1266–1274, 2011.
[10] E. Ozbay, "Plasmonic: merging photonics and electronics at nanoscale dimensions," *Science*, vol. 311, pp. 189, 2006.
[11] C. Balanis, *Antenna Theory: Analysis and Design*, 3rd ed., Wiley Interscience, New York, NY, 2005.
[12] A. A. Kishk and Y. M. M. Antar, "Dielectric resonator antennas," *Antenna Engineering Handbook*, 4th ed., John L. Volakis, New York, NY.
[13] A. Petosa, *Dielectric Resonator Antennas Handbook*, Artech House, Norwood, MA, 2007.
[14] A. Bonakdar and H. Mohseni, "Impact of optical antennas on active optoelectronic devices," *Nanoscale*, vol. 6, pp. 10961–10974, 2014.
[15] R. K. Mongia and P. Bhartia, "Dielectric resonator antennas—are view and general design relations for resonant frequency and bandwidth," *Int. J. Microwave Millimeter Wave Comput. Aided Eng.*, vol. 4, no. 3, pp. 230–247, 1994.
[16] S. A. Maier, *Plasmonic—Fundamentals and Applications*, Springer, 2007.
[17] A. Mehmood, O. H. Karabey, and R. Jakoby, "Dielectric resonator antenna with tilted beam," *IEEE Antennas Wirel. Propag. Lett.*, 2016.
[18] S. Fakhte, H. Oraizi and L. Matekovits, "High gain rectangular dielectric resonator antenna using uniaxial material at fundamental mode," *IEEE Trans. Antennas Propag.*, vol. 65, no. 1, pp. 342347, 2017.
[19] R. S. Yaduvanshi and H. Parthasarathy, "Coupled solution of Boltzmann transport equation, Maxwell's and Navier–Stokes equations," *IJACSA*, 2010.
[20] R. Cicchetti, A. Faraone, E. Miozzi, R. Ravanelli, and O. Testa, "A high gain mushroom-shaped dielectric resonator antenna for wideband wireless applications," *IEEE Trans. Antennas Propag.*, vol. 64, no. 7, pp. 2848–2861, 2016.
[21] L. Zou, W. Withayachumnankul, C. Shah, *et al.*, "Efficiency and scalability of dielectric resonator antennas at optical frequencies," *IEEE Photonics J.*, vol. 6, pp. 1–10, 2014.
[22] J. D. Jackson, *Classical Electrodynamics*, Wiley, New York, NY, 1962.
[23] R. S. Yaduvanshi and H. Parthasarathy, *Rectangular Dielectric Resonator Antennas: Theory and Design*, Springer, New Delhi, 2016.

[24] R. S. Yaduvanshi and V. Gaurav, *Nano Dielectric Resonator for 5G Applications*, CRC Press, Boca Raton, FL, 2020.
[25] S. Fakhte, "High gain rectangular dielectric resonator antenna using uniaxial material at fundamental mode," *IEEE Trans. Antennas Propag.*, 2017.
[26] M. Humayun and O. de Koo, *Retinal Prosthesis—A Clinical Guide to Successful Implementation*, Springer, Cham, 2017.
[27] P. Mühlschlegel, H.-J. Eisler, O. J. F. Martin, B. Hecht, and D. W. Pohl, "Resonant optical antennas," *Science*, vol. 308, no. 5728, pp. 1607–1609, 2005.
[28] E. Cubukcu, E. A. Kort, K. B. Crozier, and F. Capasso, "Plasmonic laser antenna," *Appl. Phys. Lett.*, vol. 89, no. 9, p. 093120, 2006.
[29] R. K. Mongia and P. Bhartia, "Dielectric resonator antenna—A review and general design relations for resonant frequency and bandwidth," *Int. J. Microwave Millimeter Wave Comput. Aided Eng.*, vol. 4, pp. 230–247, 1994.
[30] A. Petosa and A. Ittipiboon, "Dielectric resonator antennas: A historical review and the current state of the art," *IEEE Antennas Propag. Mag.*, vol. 52, pp. 91–116, 2010.
[31] S. K. K. Dash, T. Khan, and A. De, "Dielectric resonator antennas: An application oriented survey," *Int. J. RF Microwave Comput. Aided Eng.*, vol. 27, no. 3, 2017.

Chapter 7
Cylindrical terahertz and optical DRA—design and analysis

Abstract

Terahertz cylindrical dielectric resonator antennas (DRAs) have been studied and implemented using computer simulation technology. The theoretically analyses have also been performed at terahertz frequency. Mathematical formulation for the Poynting vector and far fields radiated pattern for terahertz antennas have been developed. Their simulation results along with theoretical concepts have been presented. The work carried out in this chapter is much useful for high-speed communication applications and retinal artificial photoreceptors applications. This optical DRAs work can also be used for sensing, scanning, imaging and chip-to-chip communications. The cylindrical optical DRAs have been built at 521 THz with a gain of 4.04 dBi. Another optical DRA has also been developed at 10 THz with 6-dBi gain. Simulated results on both optical DRAs have been included in this chapter. Antenna parameters such as reflection coefficient (S_{11}), radiation pattern, voltage standing wave ratio, impedance ($Z11$) plots along with other antenna results are included in this chapter. These are compact antennas and suitable for 5G and beyond networks for providing better connectivity with lower RF exposer levels.

7.1 Introduction

The radiation pattern is based on input to feed, **current density fluctuations** and retardation potentials. **The fluctuations in quantum antennas must be controlled to a minimum threshold value. The radiated field is a state of jointly coherent for the bosons (i.e. photons) and for the fermions (positrons–electrons).** These are basic concepts of radiations into dielectric resonator antennas (DRAs) at terahertz and optical frequencies. Hence, interaction between these **bosonic and fermionic fields causes the electromagnetic (e.m.) field to change** and induce surface currents that radiate out in space. **This change in e.m. fields must be optimized to a threshold value, i.e. to** obtain a desired **variance of quantum radiation pattern fluctuations,** controlling **current density fluctuations**. These DRAs are useful for high-speed communications and optical sensors for biomedical applications.

Terahertz and optical cylindrical dielectric resonator antennas (CDRAs) are simple, efficient, fascinating, futuristic, promising and an interdisciplinary subject, whose applications are in mobile and wireless communications that have capability to deliver ultrahigh-speed data rates [1]. It can be used for energy harvesting, sensing, imaging, microscopy and scanning in biomedical applications. They can be very useful to harvest energy from natural sources such as sunlight. Nanoantennas, light antennas, photonic antennas, optical antennas and quantum antennas are all synonyms used for nanometer-size antennas that operate at optical wavelength and their physics is different from classical microwave antennas [2]. Light–matter interaction is a well-known phenomenon in photonic antennas. Scattering light particle takes place, and mathematical solution for scattering e.m. waves can be provided with coupled Boltzmann–Maxwell's equation or using Drude's model theory [3]. This phenomenon of light–matter interaction is used in photonic or optical antennas. Photonic antennas are nanocircuits and thus can be used for chip-to-chip communication in electronic printed circuit boards and for in-house communications. Photonic antennas can operate in coherent states. Mario Agio and Andrea Alù have accomplished tremendous work in the field of optical antennas [4–7]. They can be well described based on Drude's model, where metal behavior is different at optical frequencies. e.m. Signal at optical frequency can partially penetrate through metals. SPP (surface plasmon polytrons) are emissions from light–matter interactions and dominated by florescence as coherent absorptions–emissions as dipole moment scattering, which is described by Green's function tensor.

7.2 Model 2 TCDRA at 10-THz resonant frequency

7.2.1 Design computations

1. Here, all the measurements are given in terms of free-space wavelength (λ_o)
   Resonant frequency = 10 THz
   Impendence = 50 Ω
   We know,

$$\lambda = \frac{c}{f}$$

   where c is the speed of light in free space = 3×10^8 m/s

$$\lambda_o = \frac{3 \times 10^8}{10 \times 10^{12}} = 0.3 \times 10^{-4}\ \text{m}$$
$$= 30\ \mu\text{m}$$

2. Substrate (SiO_2) → $1.5 \times \lambda_o$
   $= 1.5 \times 30\ \mu\text{m} = 45\ \mu\text{m}$

i.e. length and breadth of substrates (by the help of basic model given we know the dimensions of substrate, ground, nanostrip waveguide and nano-DRA).

The height of substrate is given by $= \lambda_o/3 = 30/3\ \mu$m

$= 10$ μm (use double height of substrate for high gain)

So, length × breadth × height

$= 45 \times 45 \times 10\ \mu m^3$

3. Ground (silver) → $1.5 \times \lambda_o$

$= 1.5 \times 30$ μm $= 45\ \mu$m

i.e. length and breadth of silver ground.

We use an equal height of ground and substrate (given by basis model) = 10 μm

So, length × breadth × height

$= 45 \times 45 \times 10\ \mu m^3$

4. Nanostrip waveguide (silver): →

Length of nanostrip = 1/2 × length of substrate or ground

$= 1/2 \times 45 = 22.5\ \mu$m

Width and height of nanostrip waveguide is according to matching, but the width of nanostrip should be less than or equal to width of nano-DRA used.

5. Nano-DRA (silver) →

Here, the fixed height of dielectric resonator in terms of wavelength (λ_o)

$= \lambda_o/10 = 30/10\ \mu$m

$= 3\ \mu$m (height of DRA)

Now, by the variation of the diameter (D) of the dielectric resonator we can achieve resonance in this nanoantenna for resonance, we have to put the nano-DRA having an aspect ratio between 0.5 and 2.5 (height of DRA/radius of DRA).

We know the relation between resonance frequency and dimension of DRA.

$$F_{\text{mnp}} = \frac{C}{2\sqrt{\in r\mu r}}\sqrt{\left(\frac{4}{3a}\right)^2 + \left(\frac{p}{h'}\right)^2}$$

$p = 1$ (for fundamental)

$a =$ side of DRA and $h' = 2 \times$ height of DRA

According to basic model, height is fixed at $\lambda_o/10 = 3\ \mu$m

As an aspect ratio between 0.5 and 2.5, we select the radius of cylinder = 4 μm

And the diameter of cylinder $= a = 8\ \mu$m

The height of cylinder $= 3\ \mu$m; so, $h' = 2 \times$ height of DRA $= 6\ \mu$m

$\geq$ aspect ratio of DRA $= 3/4 = 0.75$

We use silicon DRA with $\varepsilon_r = 11.9$

$$F_{\text{mnp}} = \frac{3 \times 108}{2\sqrt{11.9}}\sqrt{\left(\frac{4}{3 \times 8 \times 10^{-6}}\right)^2 + \left(\frac{1}{6 \times 10^{-6}}\right)^2}$$

$\simeq$10 THz (resonant frequency)

7.3 Terahertz antennas detailed description

Both **microwave and photonic antennas** have common features such as resonant frequency, impedance, current, rower, gain, reflection coefficient (S_{11}), scattering parameters, voltage-standing wave ratio (VSWR), bandwidth, beam width, directivity, polarization, permittivity, permeability, dielectric constant, efficiency, quality factor, resonant modes, axial ratio, aspect ratio, frequency ratio, low profile, directive gain (DG), magnetic dipole, electric dipole, dipole moment, higher order modes, Poynting vector, magnetic vector potential, wave number, boundary conditions, image effect, Bessel's function, Green's function, frequency spectrum, conductivity, wavelength ($\lambda = C/f$); group delay, loss tangent, tan (∂_o), free–spaceimpedance$\left(\eta = \sqrt{\mu_o/\epsilon_o} = 377\ \Omega\right)$, end-fire radiation pattern, broadside radiation pattern, power gain, multiband, correlation, circular polarization, RHCP and LHCP (right-hand circular polarization and left-hand circular polarization), MIMO (multi-input–multi-output), envelopment correlation coefficient, DG, mean effective gain, channel capacity loss, total active reflection coefficient, surface current, volume current and bore sight radiation pattern.

This process in microwave antennas just involves expressing the transverse components of the *E–H* fields in terms of the longitudinal components of the *E* and *H* fields. Computing the coefficients of the longitudinal components using the specified field pattern of the cylindrical and cone-shaped microwave antennas is essential to determine operating resonant modes.

Indeed photonic case, today we know that the e.m. field is neither a ray nor a wave field. It is actually a quantum wave operator field. The mechanism of the photonic antenna is described by scattering. The distortion of the surface current is due to the source of the light scattering. The surface current on the surface is proportional to the total tangential magnetic field and the scattering amplitude is also proportional to the total tangential magnetic field on the *f* surface antennas, which are solved based on volume currents instead of surface currents. Displacement currents are introduced due to dynamic inductances and resistance becomes a function of frequency in photonic antennas. Drude's model solution is obtained by taking all scattering parameters of SPPs. In currents induced by e.m. fields, it is easy enough to expand the whole scattering process by calculating the reemitted light with the well-known formula for the vector potential. The second-order resonance shows the familiar quadruple emission pattern.

More precisely, it is an ensemble of an infinite number of quantum harmonic oscillators or in fact, a superposition of plane or constrained plane waves with coefficients being creation and annihilation operators in a boson Fock space. It is therefore not quite precise to talk about the wave-field pattern on the photonic antennas but rather specify the quantum state of the wave field like a coherent state, or a number state and then compute the statistics of the wave operator field. On the surface, the action of creation and annihilation operators on coherent states or on photon number states is used. After we do such an analysis, we can ask the more interesting question: how does light within the cylindrical cavity resonator interact with the matter field? Specially, when light propagates within the cylindrical eye cavity, it interacts with electrons, atoms and molecules of the fluid medium of which the cavity is composed of fluid and plasma medium.

In classical wave field theory, we would model this fluid medium as a plasma composed of a definite permittivity, permeability and conductivity and would analyze the propagation of the e.m. field in this plasma using the Vlasov equations that are in fact the coupled Boltzmann equation for the particle distribution function of the plasma and the Maxwell equation for e.m. field. The result of this analysis using linearized perturbation theory would be dispersion relation between the oscillation frequencies of the plasma and the e.m. field and the wave vector. However, at the quantum level, the description of the interaction between light and matter is more subtle. A simplified analysis would be to model the matter as just the second quantized electron–positron field using Dirac's relativistic wave equation with the wave function being an operator field and then to include the photon interaction term in the usual way.

In this way, the total Hamiltonian of the photon and matter field splits into three terms: one, the Hamiltonian of the e.m. field described as a quadratic form in the photon creation and annihilation operators within the cavity; two, the Hamiltonian of the Dirac field with cylindrical cavity boundary conditions on the wave function described as a quadratic form in the electron–positron creation and annihilation operators; and three, the Hamiltonian of the interaction between the photon and electron–positron field described as a quadratic form in the electron–positron creation–annihilation operators multiplied with a linear form in the photon creation–annihilation operators. This interaction Hamiltonian can be used to compute the amplitudes for scattering but since we are primarily interested in the statistics of the e.m. wave field pattern on the retinal surface, we shall describe this interaction using the Dirac current density expressed as a quadratic form in the electron–positron creation–annihilation operator fields that drives the photon field using the wave equation for the e.m. field in the presence of a current density.

Thus, we shall by perturbation theory be able to calculate the change in the e. m. field pattern on the retinal surface caused by the Dirac current in terms of operators and then by assuming a definite state of the photon–electron–positron field, we shall be able to compute the mean and covariance of the e.m. field on the retinal screen surface in this state. The second quantized Dirac fermionic field can be expressed on the other hand in terms of fermionic creation and annihilation operators. In the photon case, the antiparticle of a photon is again a photon and so its creation operator at a given momentum helicity is the adjoint of the corresponding annihilation operator. On the other hand, the antiparticle of the electron is another particle, the positron, and hence the Dirac field should be expressed as a superposition of electron annihilation operators and positron creation operators. The creation of a positron of positive energy is according to Dirac equivalent to annihilating an electron of negative energy. Thus, the Dirac field is expressed as the superposition, given a quantum matter field, either bosonic or fermionic, it satisfies a classical wave equation of the Klein–Gordon type, or the 3D wave type or the Dirac type with certain boundary conditions. This second quantized field can therefore be expressed as a superposition of eigen functions corresponding to the boundary conditions with coefficients being particle creation and annihilation operators. If the field is bosonic, these creation and annihilation operators will satisfy CCRs while if the field is fermionic, they will satisfy CCRs. If H corresponds to the Hamiltonian of the first quantized theory, then we can write the second quantized Hamiltonian of the field as when the electron–positron field interacts with the bosonic radiation field, the Dirac current density acquires extra terms involving coupling between the fermionic and bosonic components. We analyze this interaction in what follows. The Dirac equation in the presence of the radiation field is given by the free Dirac current, i.e. in the absence of interactions with the photon field. We have already indicated how to compute the far-field radiation pattern produced by this field and how to evaluate the moments of this field. Specifically, if $G(x, y)$ denotes causal Green's function for the wave operator, then the e.m. four potentials produced by the Dirac current are given. We can consider a fermionic coherent state rather than a fermionic number state. Such a state is parameterized by a Grassmannian vector variable.

By imposing such restrictions, **we can calculate easily the moments of the current density field and hence of the radiated field is a state that is jointly coherent for the bosons (i.e. photons) and for the fermions. We observe that the perturbation to the current density of the Dirac field caused by the interactions between the electron–positron field and the photon field is given up to first order in the photon field and second order in the fermionic field by an expression of the form. This expression is manifestly trilinear in the operators**. Specifically, it is quadratic in the electron–positron field and linear in the photon field, totally yielding a trilinear term. It can be expressed as entanglement of the fermionic modes caused by interaction with the photon radiation field. When the electron–positron field interacts with the bosonic radiation field, the **Dirac current density acquires extra terms involving coupling between the fermionic and bosonic components**. We analyze this interaction in what follows.

The Dirac equation is used in the presence of the radiation field. From this expression, it is clear that the **photon operators tend to couple the other modes of the electron–positron field and hence produce additional terms in the far-field radiation**, a homogeneous polynomial of degree three in the electron–positron–photon operators with the photon operators appearing with a total degree of m and the electron–positron operators appearing with a total degree of 2 m. We can also evaluate the previous moment in a joint coherent state.

7.4 Theory of terahertz cylindrical DRA and mathematical formulations

Magnetic and electric dipole and dipole moment formations are an important phenomenon before radiations are achieved in photonic antennas.

At a nanometer scale or optical frequency the device has modeled as in the following in comparison with the microwave antenna using required mathematical derivations to establish the fact.

Let d be the height of the cylindrical cavity and a its radius as shown terahertz cylindrical DRA (TCDRA) design. Assume that all its walls, including the top and bottom surfaces, are perfect conductors. At the classical scale, using the fundamental identities resulting from the Maxwell curl equations relating the transverse components of the e.m. field to the longitudinal component is given as follows:

$$E_{\perp} = \partial_z \left(\frac{1}{h^2}\right) \nabla_{\perp} E_z - \left(\frac{j\omega\mu}{h^2}\right) \nabla_{\perp} H_z \times \hat{z} \tag{7.1}$$

$$H_{\perp} = \partial_z \left(\frac{1}{h^2}\right) \nabla_{\perp} H_z - \left(\frac{j\omega\mu}{h^2}\right) \nabla_{\perp} E_z \times \hat{z} \tag{7.2}$$

Also the longitudinal components of the Maxwell curl equation give us

$$\nabla_{\perp} \times E_{\perp} = -j\omega\mu\, H_z \hat{z} \tag{7.3}$$

$$\nabla_{\perp} \times H_{\perp} = -j\omega\epsilon\, E_z \hat{z} \tag{7.4}$$

In terms of the cylindrical coordinate components,

$$E_{\rho} = \left(-\frac{\gamma}{h^2}\right) \partial_{\rho} E_z - \left(\frac{j\omega\mu}{h^2}\right) \partial_{\phi} H_z \tag{7.5}$$

$$E_{\phi} = \left(-\frac{\gamma}{h^2}\rho\right) \partial_{\phi} E_z + \left(\frac{j\omega\mu}{h^2}\right) \partial_{\rho} H_z \tag{7.6}$$

$$H_{\rho} = \left(-\frac{\gamma}{h^2}\right) \partial_{\rho} H_z - \left(\frac{j\omega\epsilon}{h^2}\rho\right) \partial_{\phi} E_z \tag{7.7}$$

$$H_{\phi} = \left(-\frac{\gamma}{h^2}\right) \partial_{\rho} H_z - \left(\frac{j\omega\epsilon}{h^2}\rho\right) \partial_{\phi} E_z \tag{7.8}$$

where γ is to be replaced by the operator $-\partial/\partial_z$; it should be noted that the values assumed by γ are $\pi p/d$, where p is an integer in order that the tangential components of the electric field and the normal component of the magnetic field vanish on the top and bottom surfaces of the CDRA. We note that

$$E_{\perp}(\rho, \phi, z) = E_{\rho}\hat{\rho} + E_{\phi}\hat{\phi} \tag{7.9}$$

And likewise for $H_{\perp}$ also

$$\nabla_{\perp} = \hat{\rho} \cdot \partial_{\rho} + \left(\frac{\hat{\phi}}{\rho}\right) \partial_{\phi} \tag{7.10}$$

We are using the abbreviation

$$\partial_{\xi} = \frac{\partial}{\partial \xi},\ xi = \rho, \phi, z \tag{7.11}$$

Then using (7.1)–(7.4), we get

$$(\nabla_\perp^2 + h^2)(E_z,\ H_z) = 0 \tag{7.12}$$

We note that application of the boundary conditions, namely E_z, H_ϕ, H_ρ vanish at ρ, while E_ρ, E_ϕ, H_z vanish at $\rho = 0, d$, i.e. the tangential components of the electric field and the normal components of the magnetic field vanish on all the boundaries gives us for the TM modes,

$$h^2 = h^2(E_{mn}) = \frac{\alpha_m[n]^2}{a^2} = \omega(E, mnp)^2\epsilon_\mu - \left(\frac{\pi p}{d}\right)^2 \tag{7.13}$$

where m, n, p are integers and $\alpha_m[n], n = 1, 2, \ldots$ are the roots of the Bessel function $J_m(x)$, and for the TE modes,

$$h^2 = h^2(H, mn) = \frac{\beta_m[n]^2}{a^2} = \omega(H, mnp)^2\epsilon_\mu - \left(\frac{\pi p}{d}\right)^2 \tag{7.14}$$

where $\beta_m[n]^2 = 1, 2\ldots$ are the roots of $J_m'(x) = 0$; thus, we obtain the following expansions:

$$\begin{aligned} E_z(t,\rho,\phi,z) &= \sum_{mnp} J_m\left(\frac{\alpha_m[n]\rho}{a}\right)\cos\left(\frac{p\pi z}{d}\right) Re(c(E,mnp)\exp(j(m\phi - \omega(E_{mnp})t))) \\ &\sum_{mnp} R_e\left(c(E,mnp)\exp(-j\omega(E_{mnp})t)\mu_{mnp}(r)\right) \end{aligned} \tag{7.15}$$

$$\begin{aligned} H_z(t,\rho,\phi,z) &= \sum_{mnp} J_m\left(\frac{\beta_m[n]\rho}{a}\right)\sin\left(\frac{p\pi z}{d}\right) Re(c(H,mnp)\exp(j(m\phi - \omega(H,mnp)t))) \\ &\sum_{mnp} R_e(c(H,mnp)\exp(-j\omega(H_{mnp})t)v_{mnp}(r)) \end{aligned} \tag{7.16}$$

From these expressions, we can easily derive the corresponding expansions for the tangential components of the e.m. field in the time domain:

$$\begin{aligned} E_\perp = &\sum_{mnp}\left(\frac{1}{h}(E_{mn})^2\right) Re(c(E_{mnp})\exp(-j\omega(E_{mnp})t)\partial_z u_{mnp}(r)) \\ &- \sum\nolimits_{mnp}\left(\frac{\mu}{h}(H_{mn})^2\right) Re((j\omega(H_{mnp})c(H_{mnp})\exp(-j\omega(H_{mnp})t))\nabla_\perp v_{mnp}(r)\times\hat{z}) \end{aligned} \tag{7.17}$$

And likewise

$$\begin{aligned} H_\perp = &\sum_{mnp}\left(\frac{1}{h}(H_{mn})^2\right) Re(c(H_{mnp})\exp(-j\omega(H_{mnp})t)\partial_z v_{mnp}(r)) \\ &+ \sum\nolimits_{mnp}\left(\frac{\epsilon}{h}(H_{mn})^2\right) Re((j\omega(E_{mnp})c(E_{mnp})\exp(-j\omega(E_{mnp})t))\nabla_\perp u_{mnp}(r)\times\hat{z}) \end{aligned} \tag{7.18}$$

Note that the characteristic frequencies of oscillation of the TM modes are

$$\omega(E_{mnp}) = (\mu\epsilon)^{-1/2}\left(\frac{\alpha_m[n]^2}{a^2} + \left(\frac{\pi p}{d}\right)^2\right)\frac{1}{2} \tag{7.19}$$

And those for the TE mode are

$$\omega(H_{mnp}) = (\mu\epsilon)^{-1/2}\left(\frac{\beta_m[n]^2}{a^2} + \left(\frac{\pi p}{d}\right)^2\right)\frac{1}{2} \tag{7.20}$$

Thus, we can write

$$\begin{aligned} E(t,\rho,\phi,z) = E(t,r) = &\sum_{mnp} Re\left(c(E_{mnp})\exp(-j\omega(E_{mnp})t)\psi_{E_{mnp}}(r)\right) \\ &+\sum_{mnp} Re\left(c(H_{mnp})\exp(-j\omega(H_{mnp})t)\chi_{E_{mnp}}(r)\right) \end{aligned} \tag{7.21}$$

Equations (7.1)–(7.21) have been used to mathematically formulate E and H fields of TDRA.

Where, now $\psi_{E_{mnp}}(r)$ and $\chi_{E_{mnp}}(r)$ are $\mathbb{C}^3$—vector-valued complex function of the position variable. The first summation before is the contribution to the total electric field coming from the TM components, while the second summation is the contribution to the total electric field coming from the TE components.

Likewise,

$$\begin{aligned} H(t,\rho,\phi,z) = H(t,r) = &\sum_{mnp} Re\left(c(E_{mnp})\exp(-j\omega(E_{mnp})t)\psi_{H_{mnp}}(r)\right) \\ &+\sum_{mnp} Re\left(c(H_{mnp})\exp(-j\omega(H_{mnp})t)\chi_{H_{mnp}}(r)\right) \end{aligned} \tag{7.22}$$

Note that

$$u_{mnp}(r) = \cos\left(\frac{p\pi z}{d}\right)\cdot\exp(jm\phi)J_m\left(\frac{\alpha_m[n]\rho}{a}\right) \tag{7.23}$$

$$v_{mnp}(r) = \sin\left(\frac{p\pi z}{d}\right)\cdot\exp(jm\phi)J_m\left(\frac{\beta_m[n]\rho}{a}\right) \tag{7.24}$$

Thus,

$$\partial_z u_{mnp}(r) = -\left(\frac{p\pi}{d}\right)\sin\left(\frac{p\pi z}{d}\right)J_m\left(\frac{\alpha_m[n]\rho}{a}\right)\exp(jm\phi) \tag{7.25}$$

The vector-valued function $\psi_{E_{mnp}}, \chi_{E_{mnp}}, \psi_H(mnp), \chi_H(mnp)$ possess the usual orthogonality properties. After appropriate normalization of these function, using the orthonormality of the vector-valued functions, we can express the total energy in the e.m. field within the cavity C as

$$H_F = \left(\frac{\epsilon}{2}\right) \int_C \left| E(t,r)\right|^2 d^3r + \left(\frac{\mu}{2}\right) \int_{cavity} H(t,r)|^2 d^3r$$
$$\sum_{mnp} \left[\omega(E_{mnp})c(E_{mnp})^* c(E_{mnp}) + \omega(H_{mnp})c(H_{mnp})^* c(H_{mnp})\right] \tag{7.26}$$

Hence it is clear that after quantization, in order to obtain the correct time dependence of the coefficients, $iec(E_{mnp}, t)$ it should vary with time as $\exp(-i\omega(E_{mnp})twhilec(H_{mnp}, t)$ should vary as $\exp(-i\omega(H_{mnp})t)$, we must enforce the bosonic commutation relations

$$\begin{aligned} \left[c(E_{mnp}), c\left(E'_{m'n'p'})^*\right)\right] &= \delta[m-m']\delta[n-n']\delta[p-p'] \\ \left[c(H_{mnp}), c\left(H'_{m'n'p'})^*\right)\right] &= \delta[m-m']\delta[n-n']\delta[p-p'] \end{aligned} \tag{7.27}$$

with all the other commutators vanishing. We then easily obtain using these commutation relations that

$$\frac{dc(E_{mnp}, t)}{dt} = i\left[H_F, c(E_{mnp}, t)\right] = -i\omega(E_{mnp})c(E_{mnp}, t) \tag{7.28}$$

$$\frac{dc(H_{mnp}, t)}{dt} = i\left[H_F, c(H_{mnp}, t)\right] = -i\omega(H_{mnp}, t)c(H_{mnp}, t) \tag{7.29}$$

On solving which, the desired time dependence of these coefficients is obtained. These commutation relations can alternatively be obtained by requiring that the Maxwell equations follow from the Heisenberg equations for the e.m. field operators. We shall now briefly discuss the computation of the far-field radiation pattern. The surface current density induced on the sidewalls of the cavity is given by

$$\begin{aligned} J_s^s(t,\phi,z) &= -\hat{\rho} \times H(t,a,\phi,z) \\ &= -H_\phi(t,a,\phi,z)\hat{z} + H_z(t,a,\phi,z)\hat{\phi} \end{aligned} \tag{7.30}$$

That induced on the bottom surface is

$$\begin{aligned} J_s^b(t,\rho,\phi) &= \hat{z} \times H(t,\rho,\phi,0) \\ &= H_\rho(t,\rho,\phi,0)\hat{\phi} + H_\phi(t,\rho,\phi,0)\hat{\rho} \end{aligned} \tag{7.31}$$

And that on the top surface is

$$J_s^t(t,\rho,\phi,d) = -H_\rho(t,\rho,\phi,d)\hat{\phi} + H_\phi(t,\rho,\phi,d)\hat{\rho} \tag{7.32}$$

By using the formulas

$$\hat{\rho} = \hat{x} \cdot \cos(\phi) + \hat{y} \cdot \sin(\phi), \hat{\phi} = -\hat{x} \cdot \sin(\phi) + \hat{y} \cdot \cos(\phi) \tag{7.33}$$

We can evaluate the far-field-radiated magnetic vector potential as

$$A(\omega, r) = \left(\frac{\mu}{4\pi}\right) \exp\left(-\frac{jkr}{r}\right) \int_{SC} J_s(\omega, r') \exp(jk\hat{r} \cdot r') dS(r') \tag{7.34}$$

where SC denotes the boundary surface of the CDRA. The far-field quantum e.m. field radiated by the CDRA is easily seen to be a linear function of $c(E_{mnp}), c(E_{mnp})^*, c(H_{mnp}), c(H_{mnp})^*, m, n, p \in Z$. This is because the magnetic field is linear in these observables and hence the surface current density on the CDRA is also linear observables. Thus, we can express the radiated e.m. field in the form

$$\begin{aligned} E(t,r) &= \sum_{mnp} [c(E_{mnp}) F_1(mnp, r) \exp(-i\omega(E_{mnp})t) \\ &\quad + c(E_{mnp})^* \bar{F}_1(mnp, r) \exp(i\omega(E_{mnp})t)] \\ &\quad + \sum_{mnp} [c(H_{mnp}) F_2(mnp, r) \cdot \exp(-i\omega(H_{mnp})t] \\ &\quad + c(H_{mnp})^* \bar{F}_2(mnp, r) \cdot \exp(i\omega(H_{mnp})t)) \end{aligned} \tag{7.35}$$

And likewise for the radiated magnetic field:

$$\begin{aligned} H(t,r) = \sum_{mnp} &[c(E_{mnp}) G_1(mnp, r) \exp(-i\omega(E_{mnp})t) \\ &+ c(E_{mnp})^* \bar{G}_1(mnp, r) \exp(i\omega(E_{mnp})t)) \\ &+ \sum_{mnp} [c(H_{mnp}) G_2(mnp, r) \cdot \exp(-i\omega(H_{mnp})t) \\ &+ c(H_{mnp})^* \bar{G}_2(mnp, r) \cdot \exp(i\omega(H_{mnp})t)) \end{aligned} \tag{7.36}$$

where $F_k, G_k, k = 1, 2$ are complex 3×1 vector-valued functions of position only. The far-field electric field pattern has only the $1/r$ dependence and hence we can express it in the from

$$
\begin{aligned}
E(t,r) = r^{-1}\sum_{mnp}\Big[&c(E_{mnp})\exp\Big(-i\omega(E_{mnp})\Big(t-\frac{r}{c}\Big)Q_1(mnp,\hat{r})\Big)\\
&+c(E_{mnp})^*\cdot\exp\Big(i\omega(E_{mnp})\Big(t-\frac{r}{c}\Big)\bar{Q}_1(mnp,\hat{r})\Big]\\
&+r^{-1}\sum_{mnp}\Big[c(H_{mnp})\exp\Big(-i\omega(H_{mnp})\Big(t-\frac{r}{c}\Big)Q_2(mnp,\hat{r})\Big)\\
&+c(H_{mnp})^*\cdot\exp\Big(i\omega(E_{mnp})\Big(t-\frac{r}{c}\Big)\bar{Q}_2(mnp,\hat{r})\Big)\Big]
\end{aligned}
\tag{7.37}
$$

with a similar expression for the far-field magnetic field pattern. In fact, using the Maxwell equation

$$curlE = -j\omega\mu H$$

It easily follows that the far-field magnetic field pattern is given by

$$
\begin{aligned}
H(t,r) = r^{-1}\eta^{-1}\sum_{mnp}\Big[&-c(E_{mnp})\exp\Big(-i\omega(E_{mnp})\Big(t-\frac{r}{c}\Big)\hat{r}\times Q_1(mnp,\hat{r})\Big)\\
&+c\Big((E_{mnp})^*\cdot\exp\Big(i\omega(E_{mnp})\Big(t-\frac{r}{c}\Big)\Big)\hat{r}\times\bar{Q}_1(mnp,\hat{r})\Big)\Big]\\
&r^{-1}\eta^{-1}\sum_{mnp}\Big[-c(H_{mnp})\exp\Big(-i\omega(H_{mnp})\Big(t-\frac{r}{c}\Big)\hat{r}\times Q_2(mnp,\hat{r})\Big)\\
&+c\Big((H_{mnp})^*\cdot\exp\Big(i\omega(E_{mnp})\Big(t-\frac{r}{c}\Big)\Big)\hat{r}\times\bar{Q}_2(mnp,\hat{r})\Big)\Big]
\end{aligned}
\tag{7.38}
$$

where

$$\eta = \sqrt{\frac{\mu}{\epsilon}}$$

From these expressions, it is immediate that the far-field time-averaged quantum pointing vector field pattern is given by

$$
\begin{aligned}
P(r) = \Big(\frac{2\hat{r}}{\eta}\Big)r^{-2}\Bigg[&\sum_{mnp}\Big[c\Big(E_{mnp})^*c(E_{mnp})|Q_1(mnp,\hat{r})|^2\\
&+c(H_{mnp})^*c(H_{mnp})Q_2(mnp,\hat{r})|^2\Bigg]
\end{aligned}
\tag{7.39}
$$

We can calculate the quantum average of the fields and Poynting vector in any state of the photons, for example in a coherent state of the CDRA. In such a state

$|\phi(u)\rangle$, we have

$$\langle\phi(u)|c(E_{mnp})|\phi(u)\rangle = u(E_{mnp}), \tag{7.40}$$

$$\langle\phi(u)|c(E_{mnp})^*|\phi(u)\rangle = \bar{u}(E_{mnp}), \tag{7.41}$$

where $u = ((u(E_{mnp}), u)(H_{mnp}))$ is an infinite dimensional complex vector that parameterizes the state of the photons in the quantum e.m. field within the CDRA. In order to calculate the covariance of quantum fluctuations in the e.m. field in a coherent state, or the average value the Poynting vector in a coherent state or more generally, the higher order moments of the field in a coherent state, we require identities as

$$\langle\phi(u)|c(E_{mnp})^* c(E_{mnp})|\phi(u)\rangle = (E_{mnp})|^2$$

$$\langle\phi(u)|\Pi_{k=1}^{r} c(Em_k n_k p_k)^* q_k \Pi_{l=1}^{s}\left(E_{m_l' n_l' p_l'}\right)^{tl}|\phi(u)\rangle$$

$$= \Pi_{k=1}^{r}\bar{u}(Em_k n_k p_k)^{qk}\Pi_{l=1}^{s} u\left(E_{m_l' n_l' p_l'}\right)^{tl} \tag{7.42}$$

These identities should be combined with the commutation rules to ensure that the moments of these field creation and annihilation operators taken in any order can be expressed as linear combinations of the moments of these operators in the normal order, i.e., in each turn all the creation operators appear to the left of all the annihilation operators. It should be noted that these moment can also be computed in a state of the field corresponding to a finite number of photons in prescribed number states. For example, the state.

$$|\psi\rangle = |N_1(m_1 n_1 p_1), \ldots, N_k(m_k n_k p_k) = \underset{j=1}{\overset{k}{\otimes}} |N_j(m_j n_j p_j)\rangle \tag{7.43}$$

Represents a state in which the $(m_j n_j p_j)$th-mode photon is in the N_j-th number state. Thus we have

$$c(mnp)|N_j(m_j n_j p_j)\rangle = \delta[m-m_j]\delta[n-n_j]\ \delta[p-p_j]\sqrt{N_j(m_j n_j p_j)}|N_j(m_j n_j p_j)-1\rangle \tag{7.44}$$

And

$$c(mnp)^*|N_j(m_j n_j p_j)\rangle = |1(mnp),\ N_j(m_j n_j p_j)\rangle \tag{7.45}$$

Provided that $(mnp) \neq (m_j n_j p_j)$ and otherwise

$$c(m_j n_j p_j)^*|N_j(m_j n_j p_j)\rangle = \sqrt{N_j(m_j n_j p_j)+1}|N_j(m_j n_j p_j)+1\rangle \tag{7.46}$$

Equations (7.23)–(7.46) have been used to formulate state of quantized fields of TCDRA.

7.5 Optical CDRA description

A coarse approximation of this matter is to regard as a **sea of electrons and positrons** and then apply the formalism of **quantum electrodynamics** to describe its interaction with the **incident photons**, both before the light hits the cylindrical cavity matter directed based on boundary conditions. **The e.m. response is then dictated by collective electron oscillations (plasmons) characteristic of a strongly coupled plasma** [1–6]. In CAR and CCR, the momentum of the current density field and hence of the radiated field is a state that is jointly coherent for the bosons (photons) and for the fermions (electrons and positrons). Hence, e.m. fields are realized by quantum wave operator fields.

Here, the electrons **and positrons interaction with photons or the quantum light–matter field in a fermionic bosonic coherent state (scattering, distortions, SPP and Hamiltonian as the amplitudes for scattering). The parameters of** the **fermionic and bosonic annihilation operators define fermionic and bosonic as Dirac field** (momentum–spin, photonic wave field). The Dirac current (displacement type, tangential magnetic field), combined effect of electrons positrons and photons, eigen solution to the Dirac wave equation using 4×4 Dirac α-matrices satisfying the anti-commutation relation express the **Dirac current density** (e.m. wave field pattern), find **far-field magnetic vector potential** (retardation potentials), determine **far-field radiation pattern** (quantum wave operator field) and Berezin weight function parameterized by $\overline{\theta}$ (Grassmann variables). Minimizing the **quantum fluctuation correlation** in the pattern is used for possible matching the **average antenna pattern to the desired pattern** (boson Fock space) [7]. The mean and mean square fluctuations of quantum e.m. fields (antenna and free photons) are analyzed. Radiations are super directive (emitted vector potential, photons with different momenta and helicities, e.m. four potential fields), and **radiated field is a state of jointly coherent for the bosons (i.e. photons) and for the fermions (positrons–electrons)** [8].

Using the previous formulae, we calculate the expected value of

$$\Pi_{j=1}^{r} c\left(Ep_j q_j r_j\right)^{*sj} \Pi_{k=1}^{m}\left(Ep'_k q'_k r'_k\right) \text{ in the previous state } |\psi\rangle. \tag{7.47}$$

The photon field inside the CDRA interacts with matter in the form of electrons and positrons. The free Dirac field as

$$i\partial_t \psi(t,r) = ((\alpha, -i\nabla) + \beta m)\psi(t,r) \tag{7.48}$$

where $\alpha_1, \alpha_2, \alpha_3, \beta$ are the Dirac 4×4 matrices mutually anticommuting with each having the identity as its square. $\psi(t,r)$ is a four-component operator wave field. We must solve this equation within the cavity with the boundary condition that ψ vanishes on the boundary surface of the cavity. Thus, we take our basis functions as

$$\chi_{mnp}(r) = N(mnp) J_m\left(\frac{\alpha_m[n]\rho}{a}\right)\exp(im\phi)\sin\left(\frac{p\pi z}{d}\right), \quad (mnp)\epsilon\mathbb{Z}^3 \tag{7.49}$$

where $N(mnp)$ are appropriate normalizing constants. We expand the wave field as

$$\psi(t,r) = \sum_{mnp\sigma} [a(mnp\sigma)\chi_{mnp}(r)u(mnp\sigma)\exp(-i\omega(1mnp)t)$$
$$+b(mnp\sigma)^{*}\bar{\chi}_{mnp}(r)v(mnp\sigma)\exp(i\omega 2mnp)t \quad (7.50)$$

Using the orthonormality of the functions χ_{mnp} and their gradients, we find that the condition that ψ satisfy the Dirac equation is that

$$\omega(1mnp)u(mnp\sigma) = \left[\int_c \bar{\chi}_{mnp}(r)\left(\alpha, -i\nabla\chi_{mnp}(r)\right)d^3r + \beta m\right] u\,(mnp\sigma) \quad (7.51)$$

And

$$-\omega(2mnp)v(mnp\sigma) = \left[\int_c \chi_{mnp}(r)\left(\alpha, -i\nabla\bar{\chi}_{mnp}(r)\right)d^3r + \beta m\right] v\,(mnp\sigma) \quad (7.52)$$

This implies that on defining

$$P(mnp) = \int_c \bar{\chi}_{mnp}(r)\left(-i\nabla\bar{\chi}_{mnp}(r)d^3r \quad (7.53)$$

which implies by integration by parts that $P(mnp)$ is a real vector that

$$\omega(1mnp)\mu(mnp\sigma) = [(\alpha, P(mnp)) + \beta m]\; u(mnp\sigma) \quad (7.54)$$
$$-\omega(2mnp)v(mnp\sigma) = [(-(\alpha, P(mnp)) + \beta m]\; v(mnp\sigma) \quad (7.55)$$

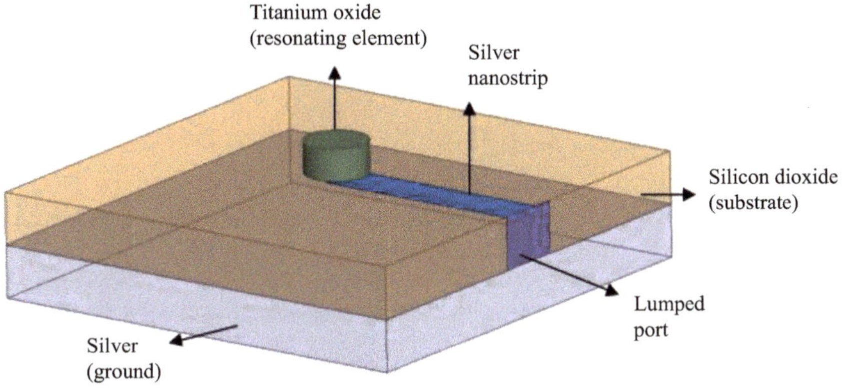

Figure 7.1 Terahertz CDRA design with silver nano waveguide

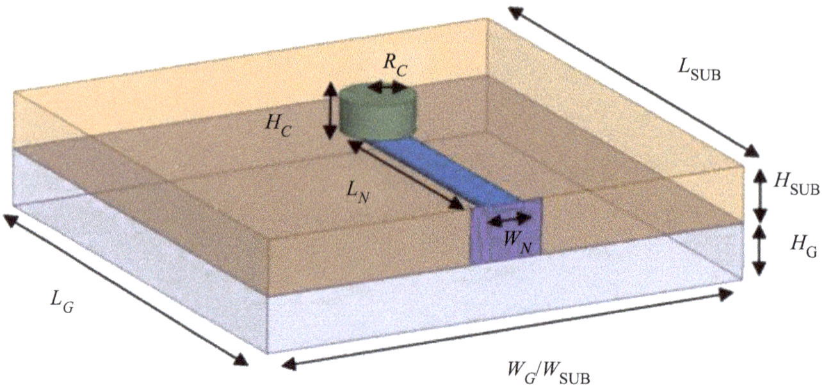

Figure 7.2 Terahertz CDRA with dimensions

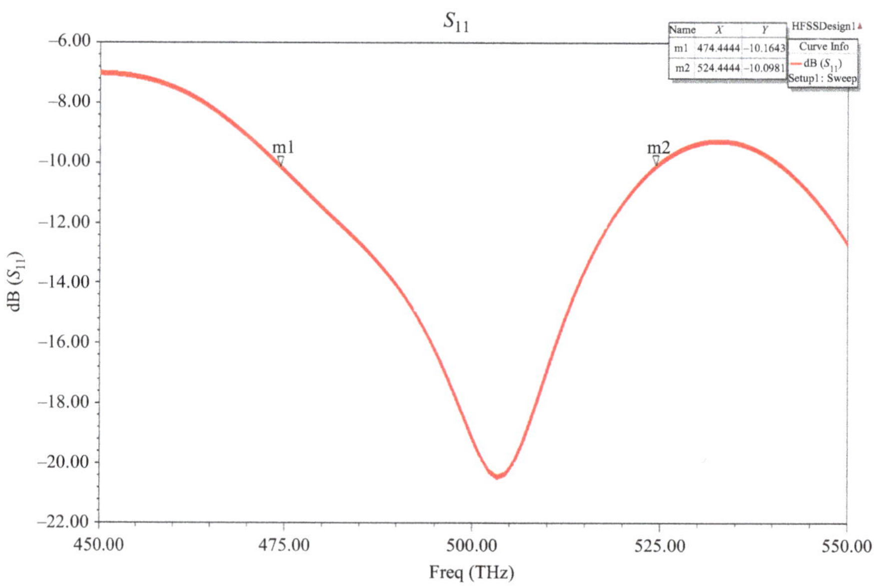

Figure 7.3 S_{11} (reflection coefficient of TCDRA)

From which it follows that $\omega(2mnp) = \omega(1mnp) = \omega(mnp)$ is the positive eigenvalue of the 4×4 matrix$(\alpha, P(mnp)) + \beta m$, i.e.

$$\omega(mnp) = \sqrt{m^2 + P(mnp)^2} \tag{7.56}$$

Then, we have that $\mu(mnp\sigma)$, $\sigma = 1, 2$ are two orthnormal eigenvectors of $(\alpha, P(mnp)) + \beta m$ corresponding to the eigenvalue $\omega(mnp)$ while $v(mnp\sigma)$ are two

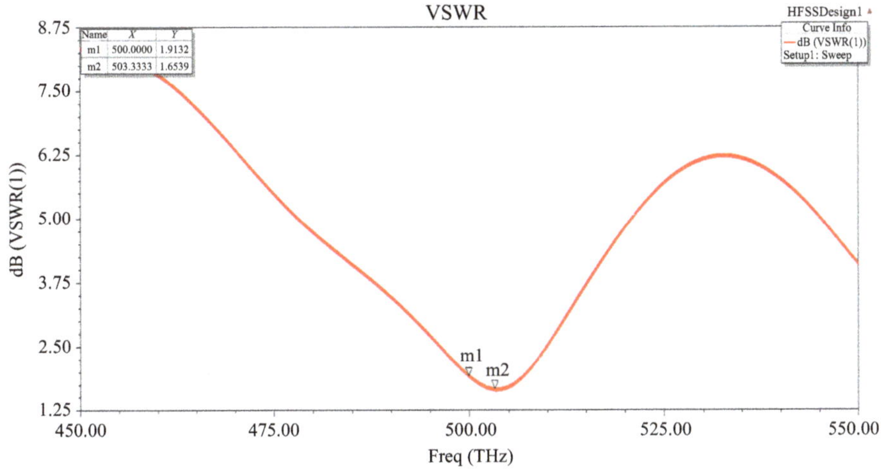

Figure 7.4 VSWR of TCDRA

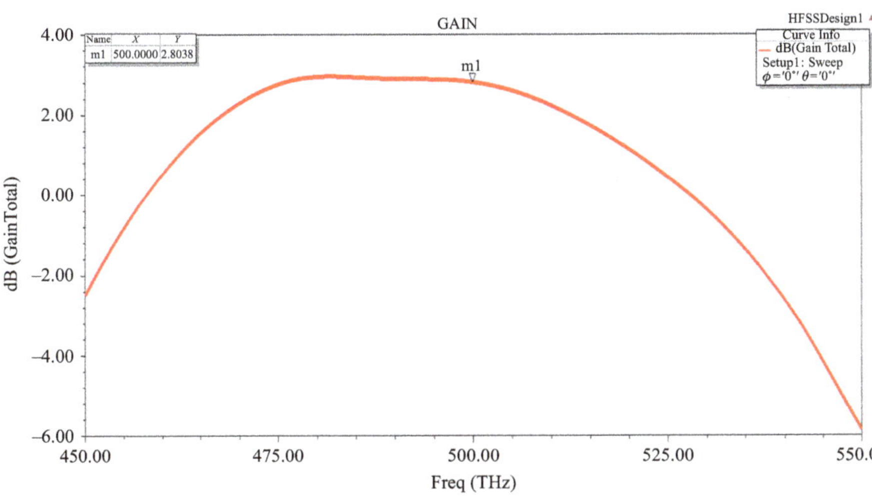

Figure 7.5 Gain vs frequency of TCDRA

orthonormal eigenvectors of $(\alpha, P(mnp)) - \beta m$ corresponding to the same eigenvalue.

When the quantum e.m. field within the resonator cavity interacts with the electron–positron field, the total second quantized Hamiltonian is given by

$$H(t) = H_F + H_D + H_I(t) \tag{7.57}$$

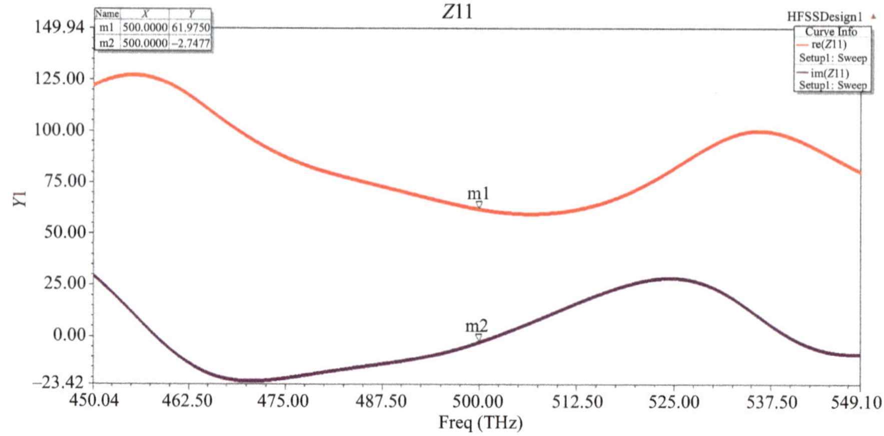

Figure 7.6 Impedance vs frequency of TCDRA

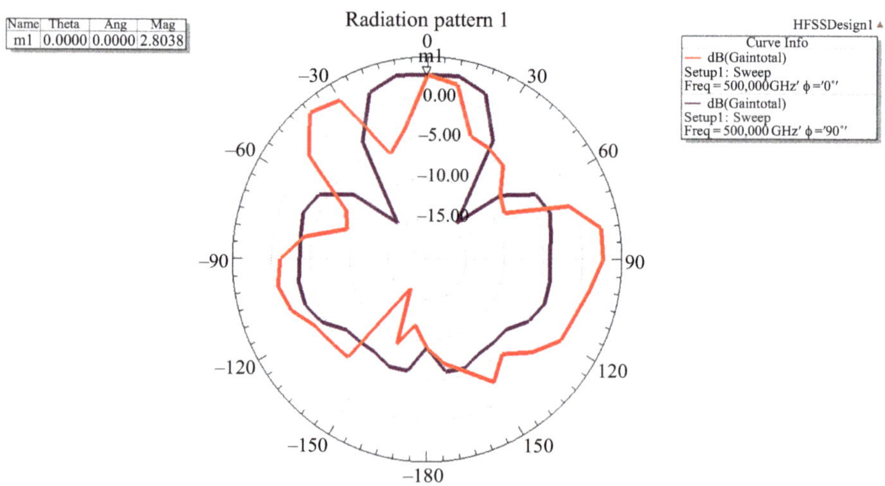

Figure 7.7 Radiation pattern (ϕ—0° and 90°) of TCDRA

where

$$H_F = \sum\nolimits_{mnp} \left[\omega(E_{mnp}) c(E_{mnp})^* c(E_{mnp}) + \omega(H_{mnp}) c(H_{mnp})^* c(H_{mnp}) \right]$$

is the Hamiltonian of the free constrained e.m. field,

$$\begin{aligned} H_D &= \int \psi(r)^* ((\alpha, -i\nabla) + \beta m)\psi(r) d^3 r \\ &= \sum\nolimits_{mnp} \omega(mnp)(a(mnp\sigma)^* a(mnp\sigma) + b(mnp\sigma)^* b(mnp\sigma)) \end{aligned} \tag{7.58}$$

is the Hamiltonian of the free constrained Dirac field and

$$H_I = e \int \psi(r)^* (\alpha, A(t, r)\psi(r) d^3 r \tag{7.59}$$

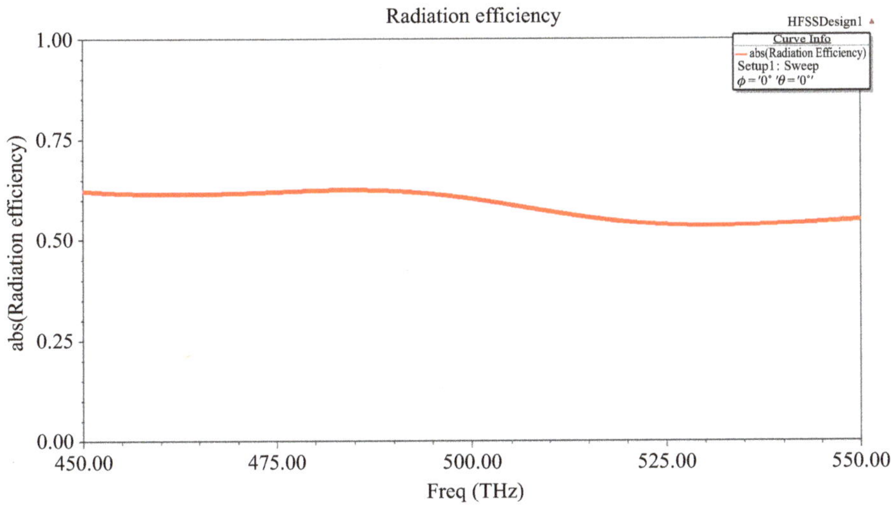

Figure 7.8 Radiation efficiency vs frequency of TCDRA

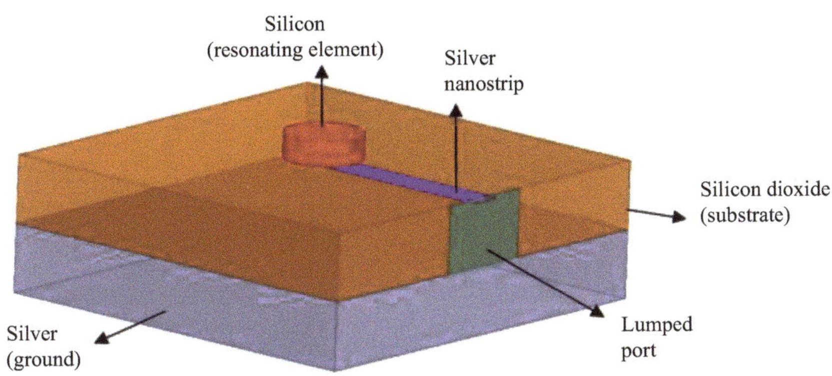

Figure 7.9 TCDRA 10-THz design

is the interaction Hamiltonian between the Dirac field of electrons and positrons. We can express it in terms of the photon and electron–position creation and annihilation operators as follows. First note that since $\text{div}E = 0$ within the resonator and we are adopting the coulomb gauge for which $\text{div}A = 0$, it follows for the Maxwell theory that $\nabla^2\Phi = 0$ and hence $\Phi = 0$ where Φ is the electric scalar potential. This is because the equation $\text{div}E = 0$ implies that there is zero-charge density inside. Hence the electric field and magnetic vector potential are related by

$$E(t, r) = -\partial_t A(t, r) \tag{7.60}$$

Equations (7.47)–(7.60) have been used to define Dirac fields.

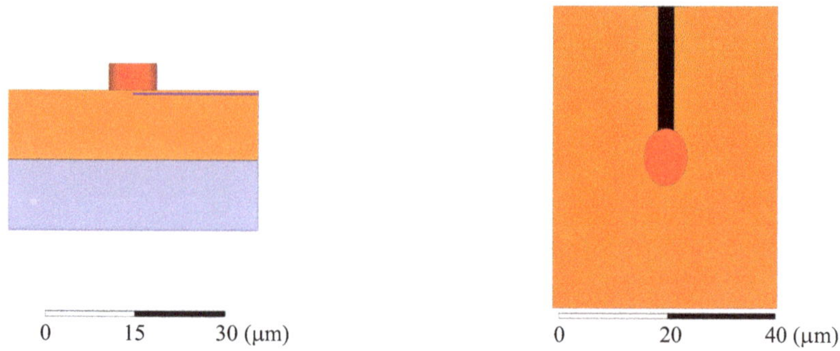

Figure 7.10 *TCDRA (cylindrical DRA) TiO_2 at 10 THz (radius = 2 μm and height = 3 μm)*

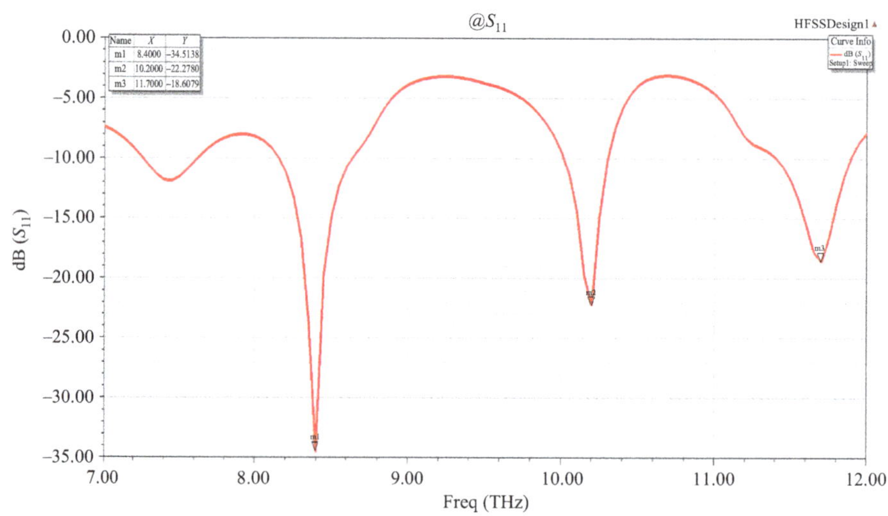

Figure 7.11 *S_{11} at 10-THz TCDRA*

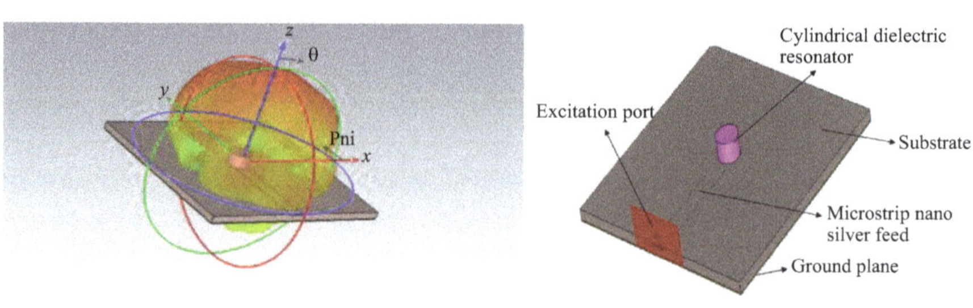

Figure 7.12 *TCDRA super directive nature*

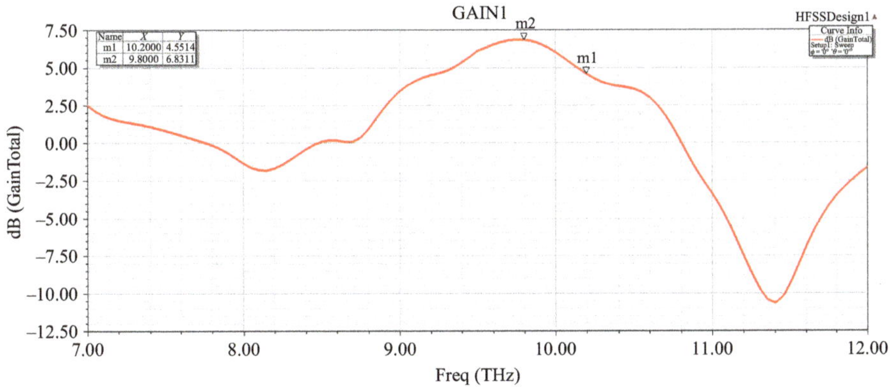

Figure 7.13 Gain vs frequency 6.8 dBi of TCDRA

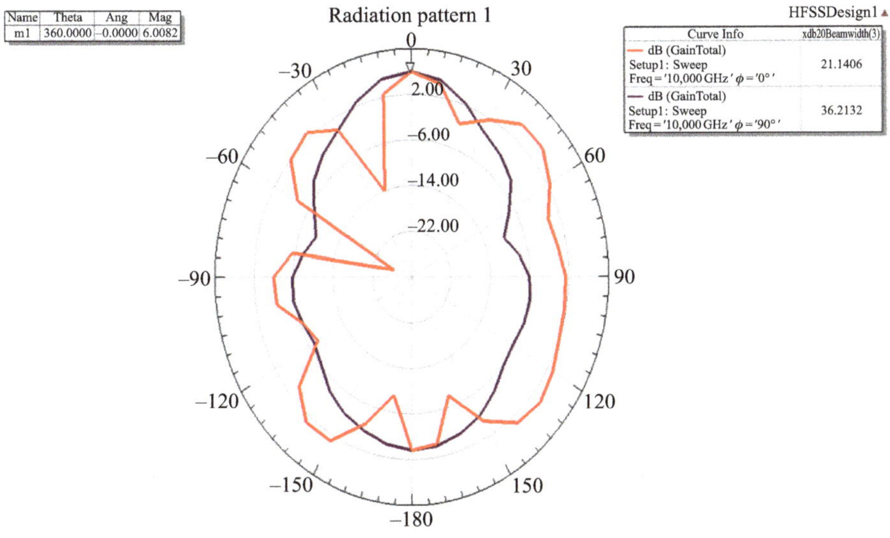

Figure 7.14 Radiation pattern co- and cross-polarization at 10 THz of TCDRA

Design formulation of resonant frequency of CDRA is given in the following:

$$f_r = \frac{c}{2\pi r}\frac{6.324}{\sqrt{\epsilon_r + 2}}\left[0.27 + 0.36\left(\frac{r}{2h}\right) + 0.02\left(\frac{r}{2h}\right)^2\right] \tag{7.61}$$

where h is height of CDRA and r is the radius of DRA to get resonant frequency f_r, c is the velocity of light and permittivity of material used is defined by ε_r in the previous formulation. Equation (7.61) is used as formulation to compute resonant frequency of TCDRA at dominant mode.

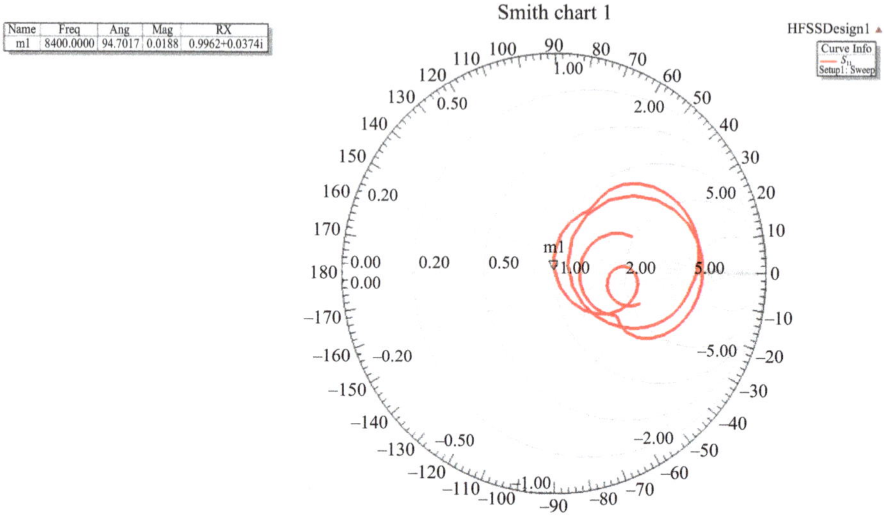

Figure 7.15 Smith chart of TCDRA

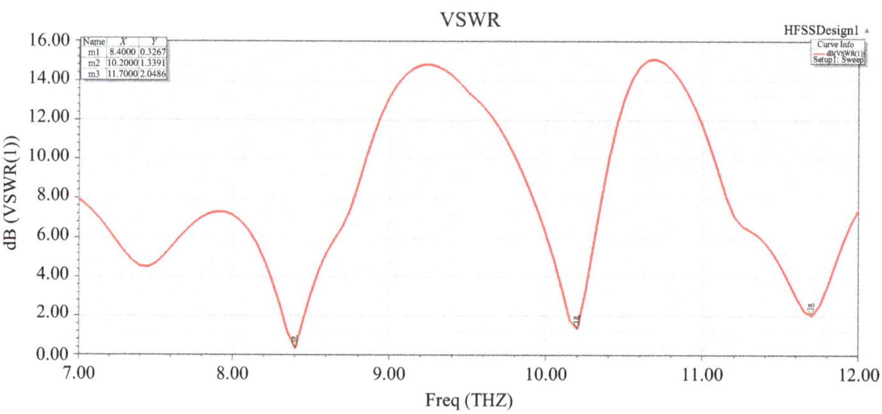

Figure 7.16 VSWR of TCDRA

The equations developed for defining the Poynting vector of CDRA are given in the following.

The vector-valued functions possess the usual orthogonality properties. After an appropriate normalization of these functions, using the orthonormality of the vector-valued functions, we can express the total energy in the e.m. field within the cavity *C* as

CDRA optical has been developed using nano-DRA of TiO_2 (titanate) materials at terahertz frequency with silver nano feed, which have been developed. Proximity feed has been used to excite higher order modes in a photonic antenna.

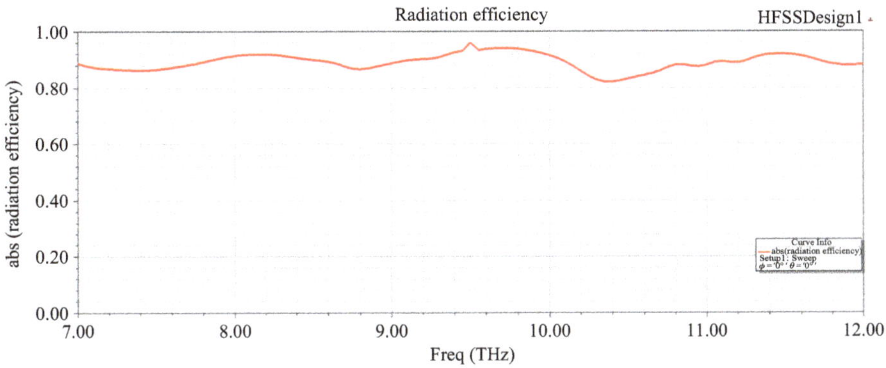

Figure 7.17 Efficiency vs frequency of TCDRA

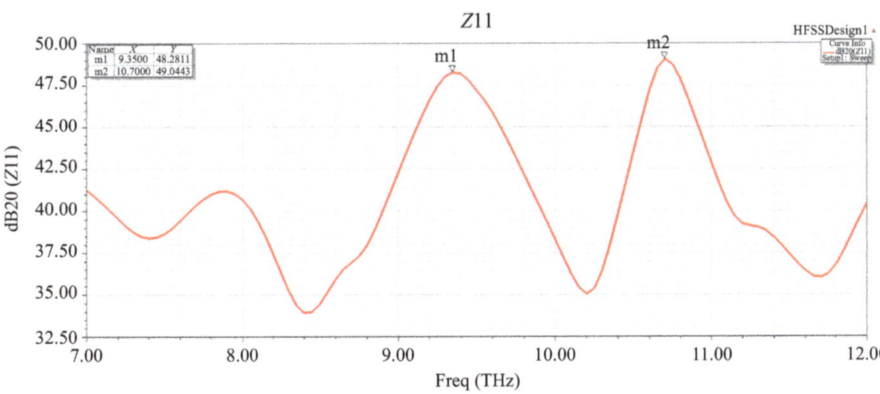

Figure 7.18 Z11 of TCDRA

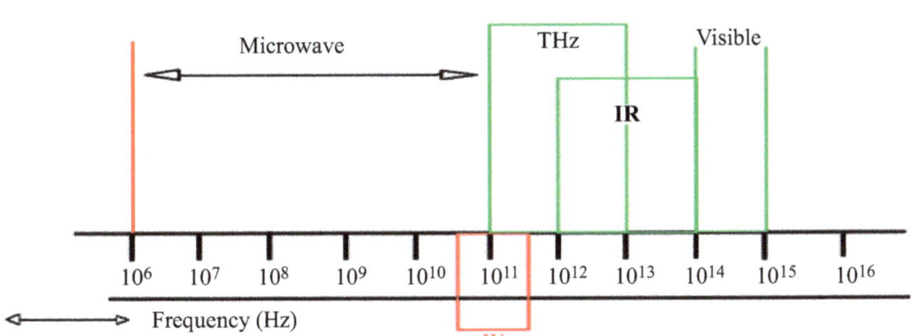

Figure 7.19 Frequency spectrum

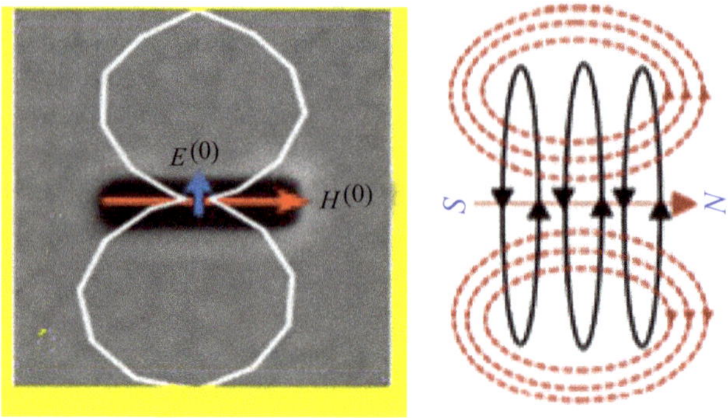

Figure 7.20 Origin of magnetic and electric dipoles, which turns into antenna radiation pattern

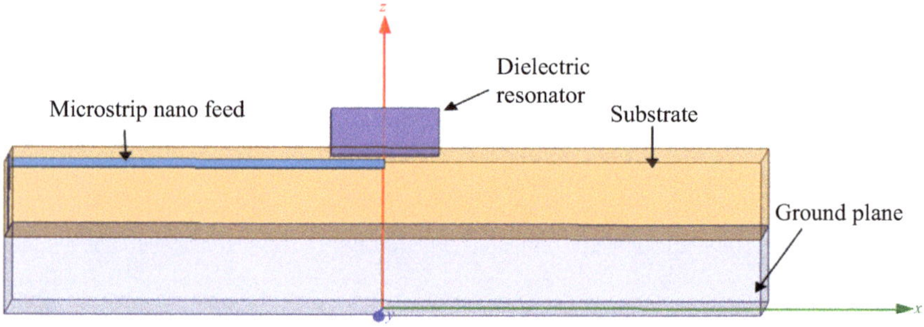

Figure 7.21 Optical CDRA at 521 THz with proximity feed and LASER input

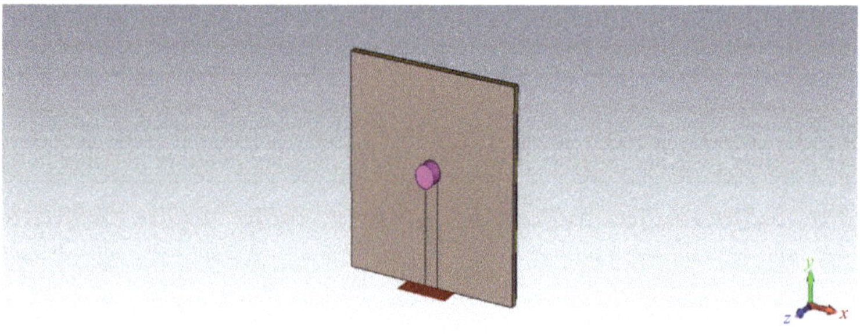

Figure 7.22 Optical CDRA

Input to silver nano waveguide is fed through laser as Gaussian beam to excite SPP into an SiO_2 (silicon oxide) substrate. Drude's nonlinear model has been used to define SPP. The electrons and positrons interaction solution is solved by Dirac second quantized field equations. This is based on quantum electrodynamics to

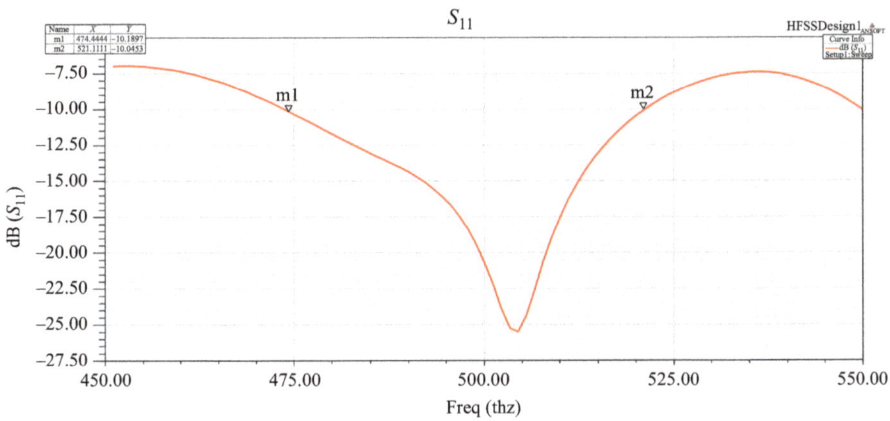

Figure 7.23 S_{11} of optical CDRA

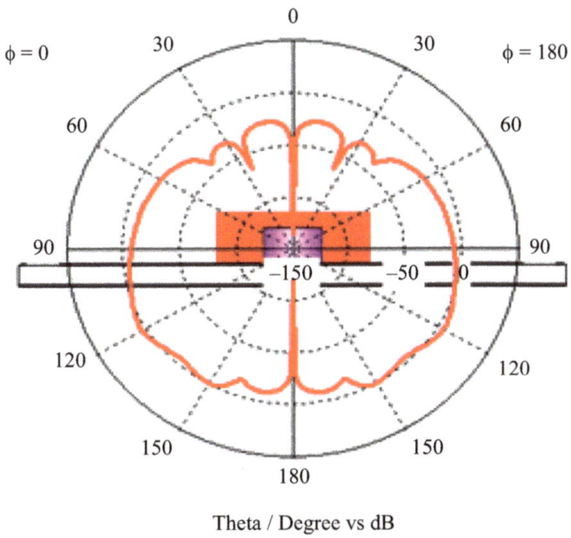

Figure 7.24 Radiation pattern of a photonic antenna at 521 THz, gain 4.04 dBi

describe photon spin. Here, e.m. field is produced by linear superposition of creation and annihilation operators of photon fields. The electron positron creation and annihilation operator fields plus photons creation and annihilation operators fields are total fields produced in a photonic antenna. The current densities of fields are obtained by quadratic functions of the Dirac field operators. The current density produced quantum e.m. field is described by the retardation potentials as Dirac wave function. Hence computations of mean and mean square fluctuations of quantum e.m. fields produced by antenna plus free photon e.m. fields in any state

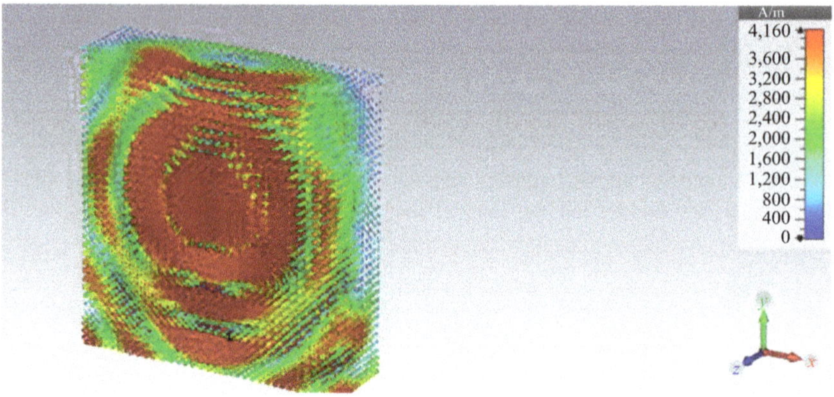

Figure 7.25 H-Field

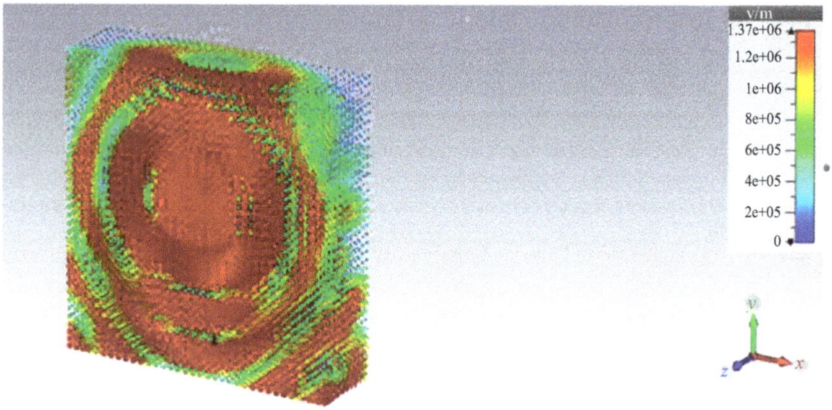

Figure 7.26 E-Field

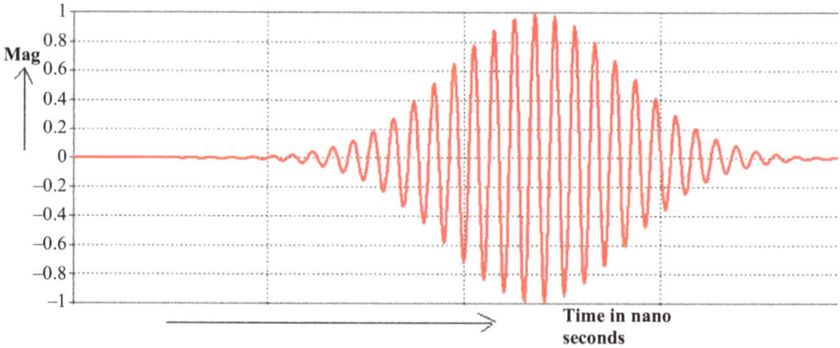

Figure 7.27 Excitation used in optical CDRA (laser output)

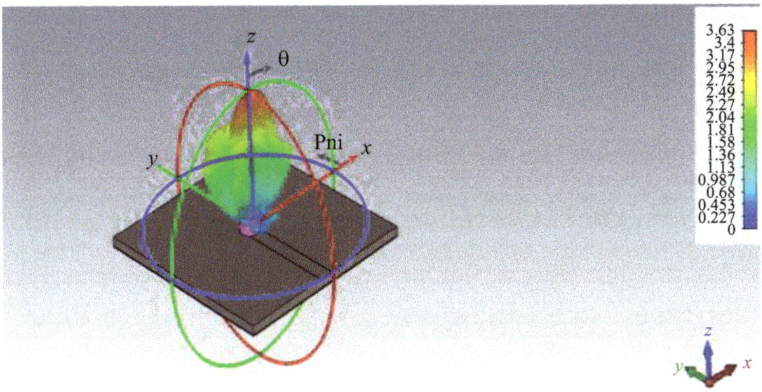

Figure 7.28 3D super directive radiations in optical CDRA (highly directive) at 521 THz

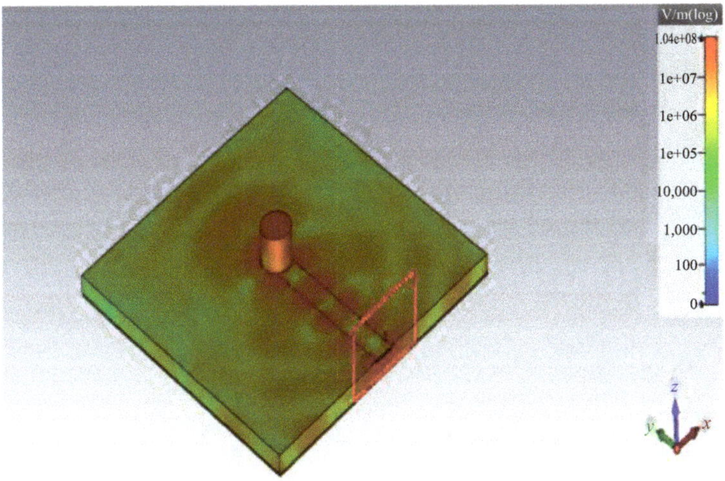

Figure 7.29 Current (electric) in optical CDRA at 521 THz

will provide the complete solution. Dirac current moments are combined state of electrons positrons and photons fields. The interaction of free-photon fields and radiated fields of electrons positrons takes place to introduce spin fields known as current moments. Drude's modeling can provide solution at optical frequencies. At optical frequency, it can penetrate through metals and give rise to excitations of free elections. At optical frequencies scattering, absorption and local field enhancement takes place. Higher order resonant modes can be excited in optical antennas, which can result into super directivity. The characteristics of photonic antennas are listed in the following:

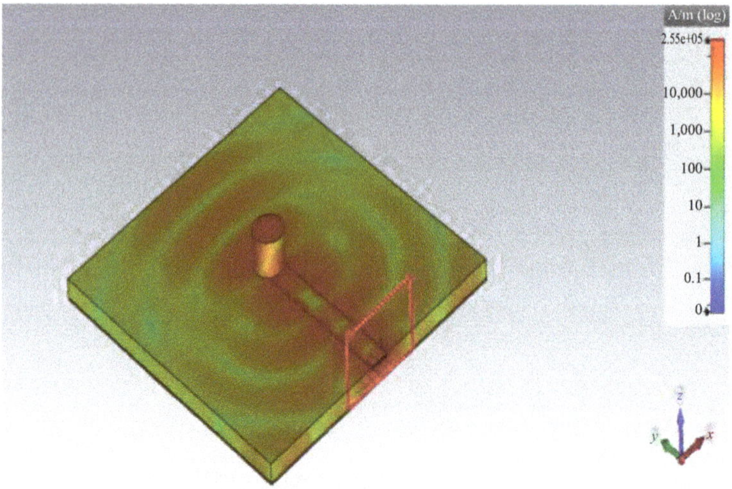

Figure 7.30 Current (magnetic) in photonic CDRA

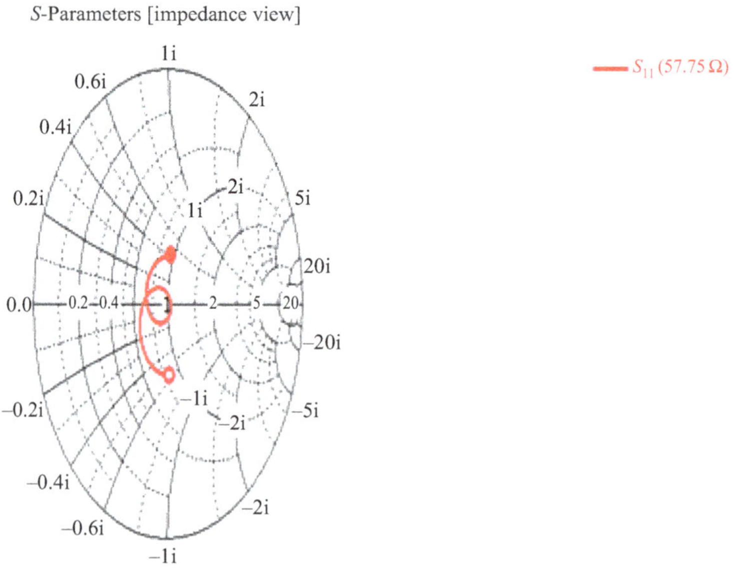

Figure 7.31 Smith chart showing the impedance of photonic antenna at 521 THz

1. Light in light out is principle of working.
2. Input excitation is provided by semiconductor LASER.
3. Surface plasmonic resonance phenomenon takes place into substrate and metal interface.

Table 7.1 Terahertz CDRA dimensions table

Parameters	Materials	Dimensions (nm)
L_{SUB}	Silicon dioxide ($\varepsilon_r = 4$)	900
W_{SUB}		900
H_{SUB}		100
L_G	Silver ($\varepsilon_r = 1$)	900
W_G		900
H_G		100
L_N	Silver ($\varepsilon_r = 1$)	450
W_N		67
R_C	Titanium oxide ($\varepsilon_r = 8.29$)	64
H_C		60

Table 7.2 TCDRA 10 THz dimensions table

Parameters	Materials	Dimensions (μm)
L_{SUB}	Silicon dioxide ($\varepsilon_r = 4$, $\mu_r = 1$ and mass density = 2,220)	45
W_{SUB}		45
H_{SUB}		10
L_G	Silver ($\varepsilon_r = 1$, $\mu_r = 0.99998$ and mass density = 10,500)	45
W_G		45
H_G		10
L_N	Silver ($\varepsilon_r = 1$, $\mu_r = 0.99998$ and mass density = 10,500)	22.5
W_N		3
R_C	Silicon ($\varepsilon_r = 11.9$, $\mu_r = 1$, loss tangent = 0.001 and mass density = 2,330)	4
H_C		3

Table 7.3 Dimensions of optical CDRA at 521 THz

S. no	Name	Materials	Permittivity	Length (μm)	Width (μm)	Height (μm)
1	Ground plane	Silver	1	4.6	4.5	0.175
2	Substrate	Arlon	2.17	4.5	4.6	0.175
3	Microstrip nano feed	Silver	1	2.3	0.34	0.02
4	Cylindrical DRA	Silicon	11.9	$R = 0.255$	–	0.325

4. e.m. Coupling takes place to QRDRA through plasmonic resonance.
5. Capacitive coupling takes place due to proximity feed mechanism.
6. Screening of electronic–hole takes place near band gap due to accelerated charge carriers.

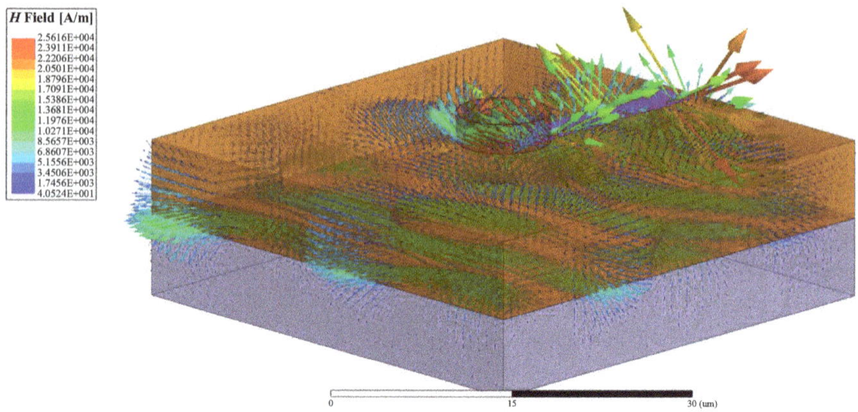

Figure 7.32 TDRA fields H

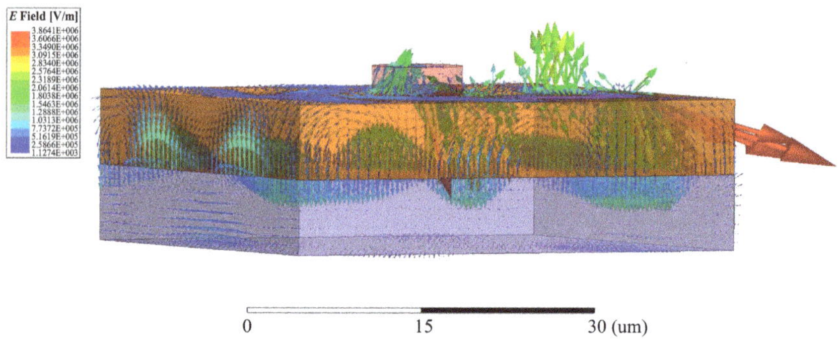

Figure 7.33 TDRA fields E

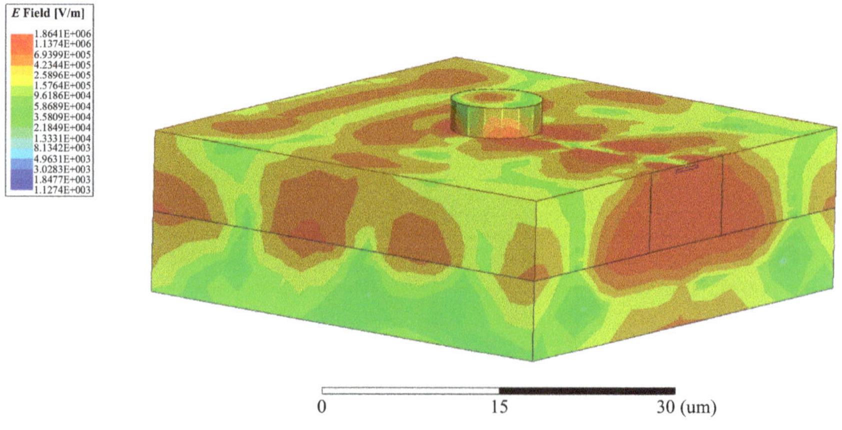

Figure 7.34 TDRA fields scalar

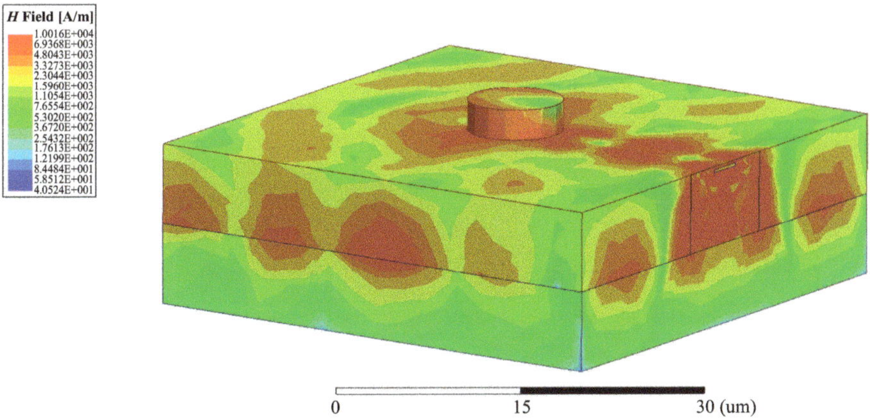

Figure 7.35 TDRA fields scalar H

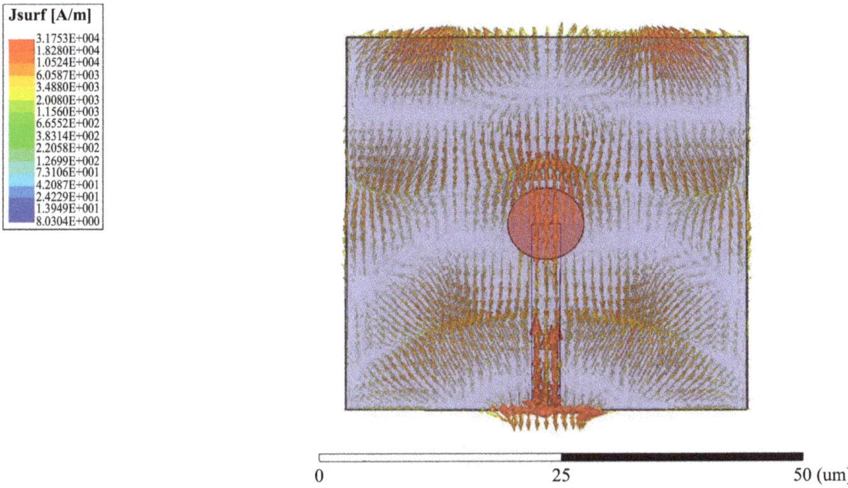

Figure 7.36 TDRA surface current J

7. Initially very small band-gap energy was observed.
8. LASER provided higher photon energy then initial band-gap energy.
9. LASER accelerated photons due to excitation.
10. LASER created clouds of charge carriers due to which screening of fields takes place.
11. Quick change in fields takes place to generate transient currents.

Skin effect role is important at optical frequency and it is given as follows:

$$\delta = \sqrt{\frac{2\rho}{\omega\mu}}\sqrt{\sqrt{1 + (\rho\omega \in)^2} + \rho\omega \in} \tag{7.62}$$

In these previous formulations, (7.62) shows that resistance becomes the function of frequency. Drude's model can provide characterization of noble metals at optical frequencies based on (7.63) formulation:

$$\varepsilon(\omega) = \varepsilon_\infty - \frac{\omega_p^2}{\omega^2 + j\gamma\omega} \tag{7.63}$$

Light–matter interaction is presented in the previous equation, where skin effect is presented and it is frequency dependent. Metal shall behave as a transparent material to optical frequency, as it does not reflect e.m. wave at an optical spectrum similar to microwave regime. Light gets inserted into metals. Plasmon resonance takes place due to light–matter interaction and hence it is clear that after quantization, in order to obtain the correct time dependence of the coefficients, i.e. $c(E_{mnp}; t)$ should vary with time as $\exp(-i\omega(E_{mnp})t)$ while $c(H_{mnp}; t)$ should vary as $\exp(-i\omega(H_{mnp})t)$, we must enforce the bosonic commutation relations on solving which the desired time dependences of these coefficients are obtained. These commutation relations can alternatively be obtained by requiring that the Maxwell equations follow from the Heisenberg equations for the e.m. field operators. We shall now briefly discuss the computation of the far-field radiation pattern. The surface current density is induced on the sidewalls of the cavity. Here, surface current denotes the boundary surface of the CDRA. The far-field photonic or quantum e.m. field radiated by the CDRA is easily seen to be a linear function of $c(E_{mnp})$; $c(E_{mnp})$; $c(H_{mnp})$; $c(H_{mnp})$. This is because the magnetic field is linear in these observables and hence the surface current density on the CDRA is also linear in these observables.

The optical CDRA and TCDRA parameters have been obtained by simulations on HFSS software and these results are shown in Figures 7.1–7.31. Tables 7.1–7.3 are design dimensions. Figures 7.32–7.36 are fields as shown later for E and H and current density J.

7.6 Conclusion

High efficiency, super directivity and large bandwidth with compact size are important parameter of an optical CDRA. Mathematical formulations at terahertz frequency have been broadly developed to make the subject more interesting. They are best in communication having razor-sharp beam with highly effective communication even in lossy environments. Comprehensive investigations on optical CDRA have been made. This chapter will be very useful to develop retinal photoreceptors and rectennas if integrated with photo diodes. They can be used as an implant antenna as photoreceptors in retinal photoreceptors applications.

References

[1] J. W. Choong, N. Nefedkin, and A. Krasnok, "Collectively driven optical nanoantennas," *Phys. Rev. A*, vol. 103, 043714, 2021.

[2] P. Bharadwaj, B. Deutsch, and L. Novotny, "Optical antennas," *Adv. Opt. Photon*, vol. 1, pp. 438–483, 2009.

[3] G. N. Malheiros-Silveira, G. S. Wiederhecker, and H. E. Hernández-Figueroa, "Dielectric resonator antenna for applications in nanophotonics," *Opt. Express*, vol. 21, no. 1, 2013.

[4] W. T. Sethi, H. Vettikaladi, H. Fathallah, and M. Himdi, "Dielectric resonator nantennas for optical communication," Optical Communication Technology, Pedro Pinho, IntechOpen, doi: 10.5772/intechopen.69064.

[5] G. N. Malheiros-Silveira and H. E. Hernández-Figueroa, "Wireless optical coupling evaluation in a dielectric resonator nanoantenna," *OSA Continuum*, vol. 1, no. 3, pp. 805–811, 2018, https://doi.org/10.1364/OSAC.1.000805.

[6] A. Alù and N. Engheta, "Wireless at the nanoscale: Optical interconnects using matched nanoantennas," *Phys. Rev. Lett.*, vol. 104, no. 21, p. 213902, 2010.

[7] D. M. Solís, J. M. Taboada, F. Obelleiro, and L. Landesa, "Optimization of an optical wireless nanolink using directive nanoantennas," *Opt. Express*, vol. 21, no. 2, pp. 2369–2377, 2013.

[8] S. Weinberg, *The Quantum Theory of Fields*, vol. 1, Cambridge University Press, New York, NY, 1995.

Chapter 8
Spherical terahertz and optical DRA—design and implementations

Abstract

In this chapter, terahertz spherical DRA (dielectric resonator antenna) has been simulated for radiation parameters and its mathematically modeling is carried out. The terahertz DRA has been simulated at 511 THz. This frequency corresponds to a visible spectrum, and suitable for LiDAR and retinal photoreceptor applications. One more model has been developed at 10 THz. The spherical DRA is built with silicon ($\varepsilon_r = 11.9$) materials, and analyses on exciting higher order modes have been carried out. It has a gain of 18 dBi, in which multipole generation has been observed. Two separate models (silicon and titanate) at optical frequencies have also been developed and validated with 2.4-dB gain using computer simulation technology. Multi-input–multi-output spherical DRA has also been realized at 10 THz to get excellent directivity. Mathematical formulation on super directivity is also formulated. The proximity coupled feed has been used with laser input, and Gaussian beam excitation. **The silver nano waveguides** have been used in the modeling to provide **feed to terahertz DRAs** of different geometries, i.e. spherical DRA, etc. **LASER input interacts with silver metal and dielectric SiO_2** substrate. Hence, in terahertz DRA, **SPP waves are generated** due to light–matter interaction, which will give rise to plasmon frequency generation. **This plasmon frequency is always kept lesser than LASER input frequency for forward propagation, i.e.** so as to enable it to propagate in forward direction. The **radiated field is a state of jointly coherent for the bosons (i.e. photons) and for the fermions (positrons–electrons).**

8.1 Introduction

Dielectric resonator antennas (DRAs) can efficiently operate right from microwave to an optical frequency spectrum [1,2]. DRAs have high efficiency, low-loss, high permittivity, design simplicity, ease of fabrication, design flexibility and high stability in a wide frequency spectrum. They can operate at higher frequencies such as terahertz and optical frequencies because of low conductor losses as compared to metallic antennas. Quantum electromagnetic (e.m.) fields can be described as understanding of light since Sir Isaac Newton started with first assuming light to be corpuscles, namely small particles of different colors corresponding to the different frequencies. This idea was first propounded by Newton and came to be known as the corpuscular theory of light [3–5]. Several years later, based on experiments involving interference of light, Huygens came to understand that light is to be regarded as being composed not of particles but rather is a wave and that any light field can be regarded as a superposition of plane waves of different frequencies and wave numbers. This meant that during the process of interference involving computation of the total intensity of light coming from different sources, one should not add the respective intensities, but rather the amplitudes, then square the resultant amplitude and then form its time average [6–8]. This meant that in the resultant total intensity, there would appear cross terms that are oscillatory, which lead us to the phenomenon of interference. Later on, Fermat propounded the ray theory of light in order to explain Snell's theorems on reflection and refraction of light, according to which a light ray always follows the path of minimum time [9–11]. The wave nature of light as propounded by Christian Huygens gained more support when Maxwell unified electromagnetism with light by showing that the basic equations of electromagnetism imply that the electric and magnetic fields in space satisfy the wave equation with a velocity of propagation being equal to that of light. This fundamental discovery gave firm support to the wave theory of light [12,13]. However, with the birth of the quantum theory in the early part of the nineteenth century, it became clear that light comes in discrete packets called quanta and each quantum of radiation carries an energy proportional to its frequency and that the intensity of light is proportional to the number of quanta [14,15]. These quanta came to be known as photons and it appeared therefore that with the advent of the quantum theory, physicists had reverted back to Newton's corpuscular theory of light. However, while Planck was busy creating his quantum theory of light, simultaneously Einstein wrote some beautiful papers proving that light can behave sometimes as a particle and sometimes as a wave, specifically showing that the variance of fluctuations in the energy of black-body radiation has two components: a particle component in agreement with **Planck's quantum hypothesis** and a wave component in agreement with **Rayleigh's theory on the numbe**r of "modes" of a wave field within a given energy shell. The particle theory of light got further support from Einstein's special theory of relativity giving the relationship between the energy and momentum of a particle of any given mass and in particular for light particles having zero rest mass. However, the crucial breakthrough into the wave–particle duality matter and in particular of light came with de Broglie's discovery of the fundamental relationship between the wavelength of a wave and the momentum of the associated particle. The de Broglie theory was applicable to all of matter not just light. It gave conclusive evidence that even particles like electrons exhibit wave–particle duality, i.e. they sometimes behaved like particles and sometimes like waves. The final crunch describing the wave–particle duality came with the wave equation for particles discovered by Erwin Schrödinger using which he was able to calculate the energy levels of an electron bound to a nucleus in terms of the eigenvalues of a partial differential equation. The corresponding eigen functions were complex wave fields that Schrödinger could not immediately interpret. These eigen functions approximately corresponded to the kind of "matter–waves" predicted by Louis de Broglie having frequency determined by Planck's quantum hypothesis relating frequency of waves to the energy of the quanta and having wavelength determined by de Broglie's theory relating the wavelength of waves to the momenta of the particles. Schrödinger's wave equation was a nonrelativistic wave equation in that it could be "derived" by assuming the nonrelativistic relationship between energy and momentum and then assigning operators to energy and momentum. The correct interpretation of the eigen functions of Schrödinger's equations was provided about a year later by Max Born with the suggestion that the modulus square of the wave function gave the probability density of the particle to be present at a given point in space at a given time. Equivalently, this modulus square could be interpreted as the number density of particles or the intensity of the wave at that point in space at that time. In fact, Schrödinger's equation could also be generalized to determine the wave functions and energy levels for any particle with a prescribed energy–momentum relationship by replacing the energy and momentum with appropriate operators and applying both sides of this relationship to a wave function. Thus, Schrödinger's formalism could be applied to determine the kind of de Broglie waves associated with a particle having any given energy–momentum relation. The de Broglie wave solutions then have the interpretation that their modulus square represents the intensity/probability density of particles in space at a given time. In special relativity, the energy momentum relationship for a relativistic particle is a quadratic form in the energy and momenta unlike the Newtonian-nonrelativistic case where it is linear in the energy and quadratic in the momenta. Consequently the Schrödinger equation is a linear PDE that is of the first order in time and second order in spatial derivatives. The nonrelativistic Schrödinger equation as a consequence then leads to a unitary evolution of the wave function with time and with the additional pleasing property that evolution from time zero to time $t_1 + t_2$ can be expressed as the composition of evolution from time zero to time t_1 followed by evolution from time t_1 to time $t_1 + t_2$. In other words, Schrödinger evolution follows the semigroup property. This is in contrast to the case of special relativity where in the wave equation is quadratic in time leading to the wave function at a given time being a superposition of a forward-propagating solution and a backward-propagating solution. Thus, the total probability will be conserved in Schrödinger's nonrelativistic theory but not so in the relativistic theory. This difficulty was finally resolved by Paul Dirac who factorized the energy–momentum relationship of special relativity into linear factors in energy and momentum using 4×4 complex anticommuting matrices and it led to Dirac's relativistic theory of the electron which along with Maxwell's equations is at the heart of quantum electrodynamics describing the interaction of light with matter. Dirac's equation being linear in the space and time derivatives is a truly relativistic wave equation for quantum theory because first it respects Lorentz invariance in that space and time are treated on an equal footing and that we can find a spin or representation of the Lorentz group under which Dirac's equation remains invariant and second it is first order in time and therefore preserves the unitary semigroup property of the evolution guaranteeing, therefore, conservation of total probability. Light as we understand today is a second quantized e.m. field in which the vector potential components in the spatial frequency domain represent creation and annihilation operators of photons with different momenta and helicities. This is owing to the fact that the total energy of the e.m. field can in the Coulomb gauge be expressed as a quadratic functional of the spatial frequency components of the vector potential and each spatial frequency component of the vector potential, according to Maxwell's equations, evolves harmonically with time with a frequency being ω, where k is the wave vector. Such a picture is in perfect accord with the Hamiltonian theory of a collection of independent quantum harmonic oscillators. However, this model of light can actually be traced back to the work of Satyendranath Bose who proposed a statistical method for deriving Planck's law of black-body radiation wherein we are interested in distributing a total amount of energy E having N quanta at frequency ν so that $E = Nh\nu$, amongst p oscillators with the quanta being regarded as indistinguishable particles. When an oscillator has n such quanta, we say in the modern language of quantum mechanics that it has been excited to the n energy level by an application of n creation operators to the vacuum. In this way, the entire field of black-body radiation is simply a collection of quantum harmonic oscillators and if we maximize the entropy, i.e. the total number of ways of distributing this energy then we end up with the famous Bose–Einstein statistics that gives us the relative number of photons at each frequency. This in turn enables us to determine the intensity of black-body radiation as a function of frequency. Summarizing the modern point of view, the entire photon field described as an operator e.m. four potential field is just a superposition of plane waves of different wave vectors whose coefficients are creation and annihilation operators at the different wave vectors and a given state of the photon field is actually a linear combination of number states wherein a number state is specified by specifying the number of photons that occupy each state of definite wave-vector/momentum and definitive helicity/spin. A particular kind of photon state called a coherent state is a state constructed by an appropriate linear combination of this kind that turns out to be an eigen state for all the photon annihilation operators. This model for the quantum e.m. field implicitly contains both the particle nature and the wave nature of light. The particle nature is contained in the presence of the creation and annihilation operators while the wave nature is contained in the plane waves that act as carrier signal fields for the creation and annihilation operators.

Second quantization means a quantization of a classical field theory. It is called second quantization for the following reasons:

The first quantization is simply a wave equation like the three-dimensional wave equation, the Schrödinger wave equation for a single or an infinite number of quantum particles, the Klein Gordon equation or the Dirac equation. If such a classical wave equation is quantized, then it describes an infinite number of quantum particles.

This can be seen clearly from the following example:

Take the Klein–Gordon equation that is the wave equation described before corresponding to Einstein's energy momentum relation with the energy and momenta replaced by appropriate operators. We expand the solution wave field as a three-dimensional Fourier series within a cube of side length L. The coefficient of this Fourier series then becomes quantum operators each one and its adjoint, i.e. Hermitian conjugate describes a single quantum particle. Thus, a second quantized field can equivalently be described by a countably infinite number of quantum particles. For bosonic quantum fields like the Maxwell e.m. photon field, the state of the field can be described, for example by specifying how many particles are occupying each state of definite momentum and helicity or equivalently by specifying the momentum and helicity of each of the particles. In this case, there can be zero, one or more than one particle having a specified momentum and helicity. In the case of fermionic fields like the second quantized Dirac field, we cannot have more than one particle having a definitive value of momentum and spin.

In this chapter, terahertz spherical DRA characteristics equation has been developed for possible resonant frequency oscillations and compared with microwave working of spherical DRAs [16]. The far-field formulations have also been developed. Surface plasmon resonance has excited terahertz spherical DRA. *E* and *H* fields parameters have been obtained in terahertz spherical DRA. All relevant parameters are listed in Figures 8.1–8.8 and dimensions are given in Tables 8.1 and 8.2. The role of high-order multiples in the light scattering can be worked out for the development of meta-surfaces and metamaterials in terahertz range. The study

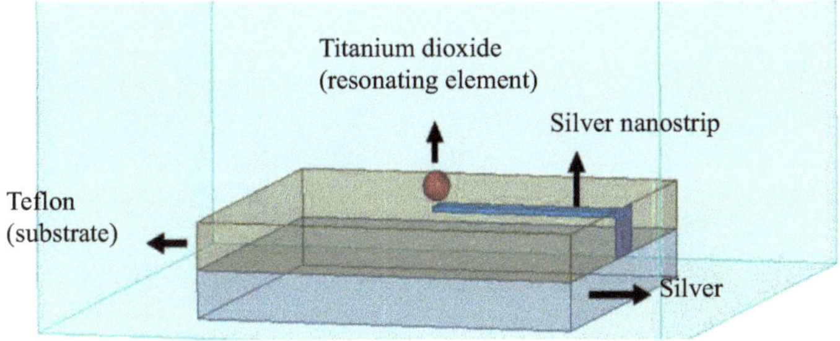

Figure 8.1 TiO_2 spherical DRA at 511 THz

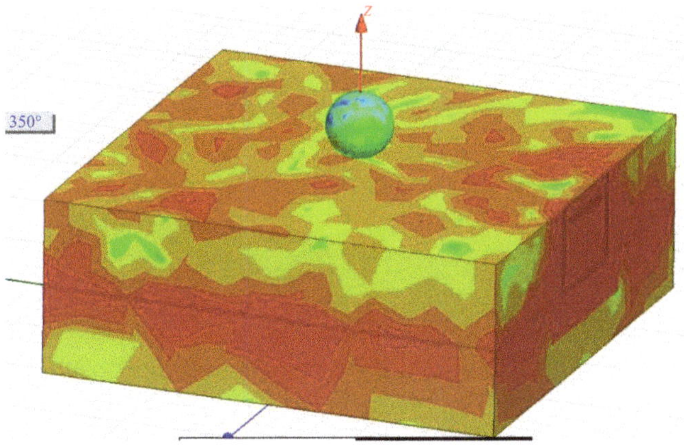

Figure 8.2 Silicon terahertz spherical DRA at 511-THz E fields

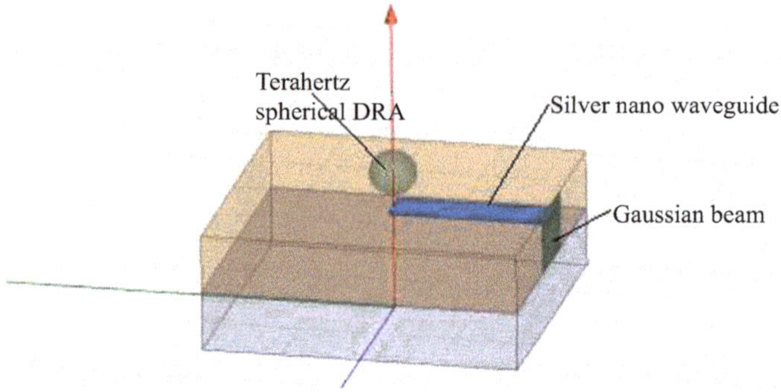

Figure 8.3 10-THz spherical DRA (radius = 3 μm and height = 6 μm)

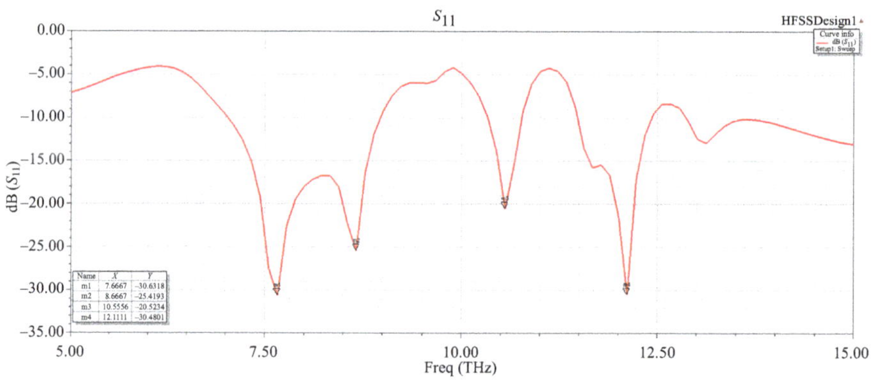

Figure 8.4 Reflection coefficient of spherical DRA at 10 THz

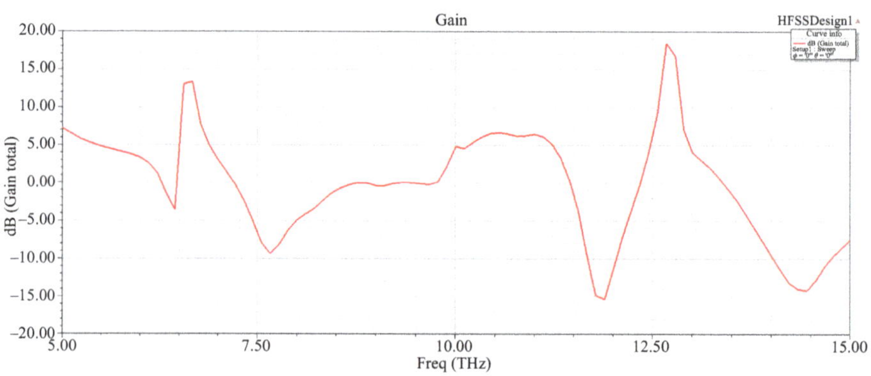

Figure 8.5 Gain of spherical DRA at higher order mode in high

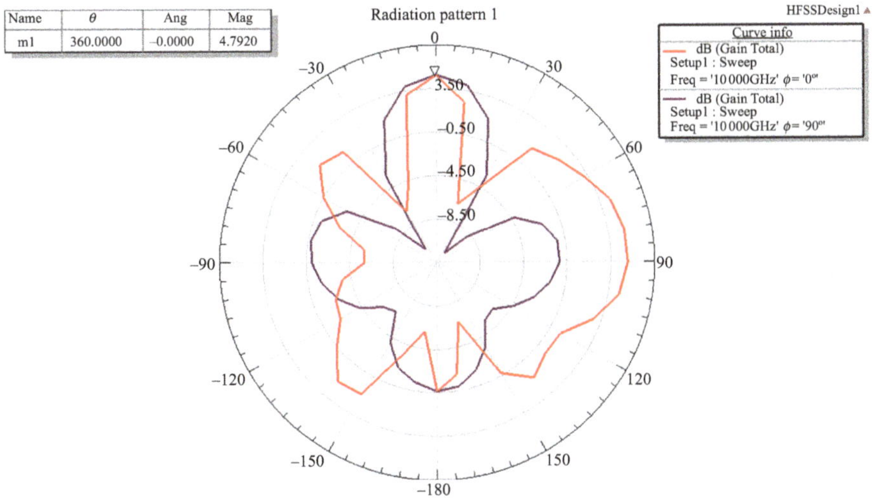

Figure 8.6 Radiation pattern of terahertz spherical DRA

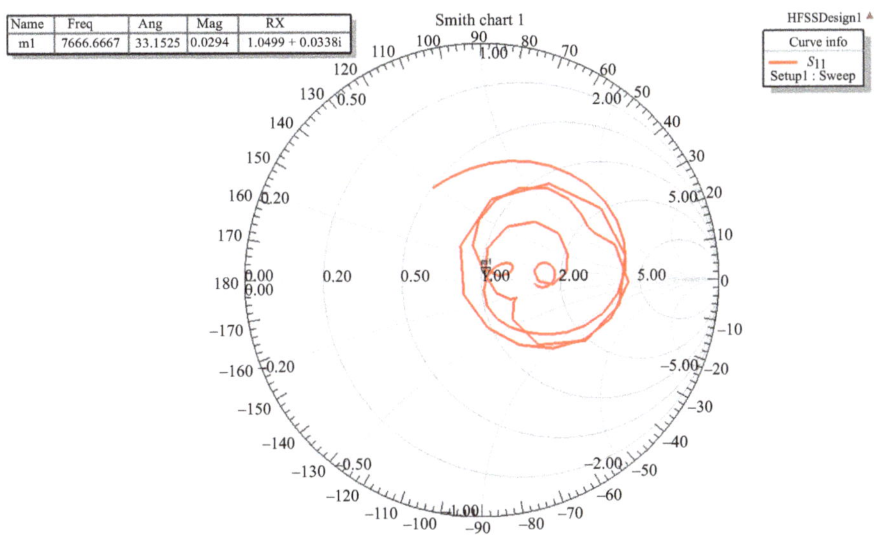

Figure 8.7 Smith chart of terahertz spherical DRA 10 THz

of excitation of both electric and magnetic multipole resonances at optical frequency is an important phenomenon, which has dependence on geometry, size and material characteristics. Scattering of light by nanosize particles can result into multi-resonance. Higher order resonances up to third-order multipole excitations can be investigated for useful applications. Total electric dipole moment in scattering can be computed using tensor analysis. Interference, polarization and

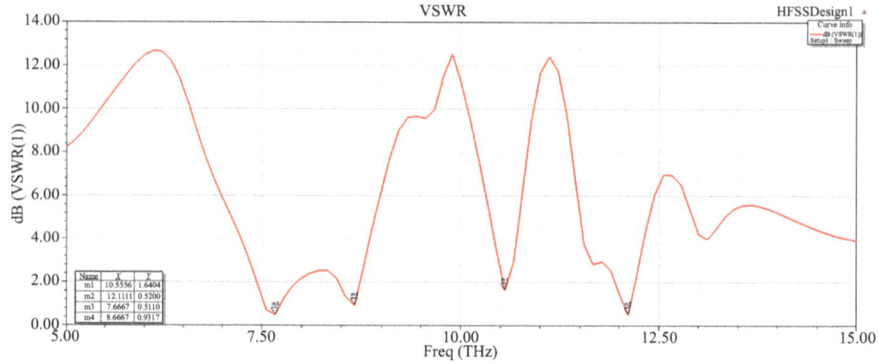

Figure 8.8 VSWR of terahertz spherical DRA

Table 8.1 TiO_2 terahertz spherical DRA dimensions

S. no	Dimensions	Materials	Value	Permittivity/ loss tangent
1	Ground plane	Silver	$900 \times 900 \times 100$ nm^3	1
2	Substrate	Teflon	$900 \times 900 \times 100$ nm^3	0.0006
3	Nanostrip waveguide	Silver	$450 \times 67 \times 10$ nm^3	1
4	Resonator radius	Titanium dioxide	32 nm	8.29
5	Resonator height	Titanium dioxide	64 nm	8.29

Table 8.2 Design dimensions of optical spherical resonator antenna

Name of components	Materials used	All dimensions (μm)
L_{SUB}	Substrate	42
W_{SUB}	Silicon dioxide ($\varepsilon_r = 4$)	42
H_{SUB}		8
L_G	Ground plane	42
W_G	Silver ($\varepsilon_r = 1$)	42
H_G		8
L_M	Nanostrip	6.5
W_M	Silver ($\varepsilon_r = 1$)	0.5
R_S	DRA used	3
H_S	Silicon ($\varepsilon_r = 11.9$)	6

collision integral are few important phenomena. This way, radiations of desired wave vector with polarization by scattering, suppression, forward and backward scattering can be obtained. These investigations can be applied to develop nanoantennas in optical spectrum. These DRAs have applications as sensors, chip-to-chip communication, retinal photo receptors, scanning and imaging devices, etc.

The mathematical modeling of terahertz spherical DRA is established using (8.1)–(8.16). Two DRAs and one multi-input–multi-output (MIMO) cases at terahertz frequency have been developed and validated along with super directivity derivations. No similar work exists for making its comparative study.

8.2 Design of terahertz spherical DRA at 511 THz

Case-1 TiO_2 spherical DRA at 511 THz

Case-2 10-THz spherical DRA

8.3 Mathematical formulations of terahertz spherical DRA

The multipole electric field can be expressed as

$$E(\omega, r) = \sum_{lm} \left[c(lm) h_l(kr) L Y_{lm}(\hat{r}) + d(lm) \frac{1}{j\omega\varepsilon} \mathrm{curl}(h_l(kr) L Y_{lm}(\hat{r})) \right]$$

And the multipole magnetic field as

$$H(\omega, r) = \sum_{lm} \left[-\frac{1}{j\omega\mu} c(lm) \mathrm{curl}(h_l(kr) L Y_{lm}(\hat{r})) + d(lm) h_l(kr) L Y_{lm}(\hat{r}) \right]$$

where

$$L = -ir \times \nabla \tag{8.1}$$

This is because $h_l(kr)LY_{lm}(\hat{r})$ satisfies the Helmholtz equation and is orthogonal to the radial direction $\hat{r}$. Further, it is easily verified that div $(h_l(kr)LY_{lm}(\hat{r})) = 0$ because $\hat{r} \cdot L = 0$ and $\mathrm{div}(LY_{lm}(\hat{r})) = 0$. So all the Maxwell equations in free space are satisfied by the multipole fields. For a detailed analysis of multipole field, see JD Jackson, "Classical electrodynamics," Wiley. It should be noted that $h_l(kr)LY_{lm}(\hat{r})$ is purely transverse, i.e. its radial components vanish, while $\mathrm{curl}(h_l(kr)LY_{lm}(\hat{r}))$ satisfies the Helmholtz equation but has both nonvanishing radial and transverse components. Its divergence obviously vanishes because the divergence of a curl is zero. The Maxwell equations are satisfied by multipole fields because

$$\begin{aligned} \mathrm{curl}(\mathrm{curl}(h_l(kr)LY_{lm}(\hat{r}))) &= \nabla(\mathrm{div}(h_l(kr)LY_{lm}(\hat{r}))) - \nabla^2(h_l(kr)LY_{lm}(\hat{r})) \\ &= -k^2 h_l(kr) L Y_{lm}(\hat{r}) \end{aligned}$$

It should be noted that $h_l(x)$ are Hankel functions constructed as $h_l(x) = x^{-1/2} J_{l+1/2}(x)$ where $J_\nu(x)$ satisfies Bessel's equation:

$$x^2 J_\nu''(x) + x J_\nu'(x) + \left(x^2 - \nu^2\right) J_\nu(x) = 0$$

This can be verified by directly substituting
$\psi(h_l(kr)LY_{lm}(\hat{r}))$ into the following Helmholtz equation

$$(x^2 + k^2)\psi = 0$$

Expressed in spherical-polar coordinate and deriving from that the radial equation for $h_l(kr)$. It can also be obtained by the method of separation of variables.

The frequency of oscillation of the e.m. field within this spherical cavity gets quantized because of the requirement that the radial component of magnetic field and the tangential components, i.e. $\hat{\theta}, \hat{\varphi}$ components of the electric field, are required to vanish on the spherical surface, i.e. at $r = R$ now,

$$\begin{aligned} \hat{r} \cdot LY_{lm} &= 0, \\ r \cdot \mathrm{curl}(h_l(kr)LY_{lm}(\hat{r})) &= \mathrm{div}(h_l(kr)r \times LY_{lm}(\hat{r})) \end{aligned} \tag{8.2}$$

Since

$$\mathrm{curl}\, r = 0$$

Also,

$$\begin{aligned} \hat{r} \times LY_{lm} &= -i\hat{r} \times (r \times \nabla)Y_{lm} \\ &= -ir\partial_r Y_{lm} + ir\nabla Y_{lm} = ir\nabla Y_{lm} \end{aligned}$$

Thus,

$$r \cdot \mathrm{curl}(h_l(kr)LY_{lm}(\hat{r})) = -\mathrm{div}\left(ih_l(kr)r^2\nabla Y_{lm}(\hat{r})\right)$$

The boundary conditions imply that

$$\hat{r} \cdot H = 0,\ \hat{r} \times E = 0,\ r = R$$

Hence

$$\sum_{lm}\left[c(lm)h_l(kr)(\hat{r}) \times LY_{lm}(\hat{r}) + d(lm)\left(\frac{1}{j\omega\varepsilon}\right)\hat{r} \times (\mathrm{curl}(h_l(kr)LY_{lm}(\hat{r})))\right]\Bigg|_{r=R}$$
$$= 0$$

And multipole magnetic field as

$$\sum_{lm}[c(lm)\hat{r} \cdot \mathrm{curl}(h_l(kr)LY_{lm}(\hat{r}))]|_{r=R} = 0$$

Now

$$\hat{r} \times \mathrm{curl}(f(r)g(\hat{r})) = \hat{r} \times \left(f'(r)\hat{r} \times g(\hat{r})\right) + \hat{r} \times (f(r)\nabla \times g(\hat{r}))$$

Applied to our problem, this because

$$\begin{aligned}\hat{r}\times\mathrm{curl}(h_l(kr)LY_{lm}(\hat{r})) &= \hat{r}\times\left(kh_l'(kr)\hat{r}\times LY_{lm}(\hat{r})\right)\\ &\quad+\hat{r}\times(h_l(kr)\nabla\times(LY_{lm}(\hat{r})))\\ &= -kkh_l'(kr)LY_{lm}(\hat{r})+\hat{r}\times(h_l(kr)\nabla\times(LY_{lm}(\hat{r})))\end{aligned} \tag{8.3}$$

$$\begin{aligned}&i\nabla\times L\psi(\hat{r})=\\ &\nabla\times(r\times\nabla\psi(\hat{r}))=\\ &=\nabla^2\psi(\hat{r})r-(r,\nabla)\nabla\psi(\hat{r})+(\nabla\psi(\hat{r}),\nabla)r-3\nabla\psi(\hat{r})\\ &=r\nabla^2\psi(\hat{r})-r\partial_r\nabla\ \psi(\hat{r})-2\nabla\psi(\hat{r})\\ &=r\nabla^2\psi(\hat{r})-\nabla\psi(\hat{r})\end{aligned} \tag{8.4}$$

Thus,

$$\hat{r}\times\mathrm{curl}(h_l(kr)LY_{lm}(\hat{r}))=-\left(kh_l'(kr)+\frac{h_l(kr)}{r}\right)LY_{lm}(\hat{r})$$

Thus, our boundary conditions reduce to

$$\begin{aligned}&c(lm)h_l(kr)\hat{r}\times LY_{lm}(\hat{r})-\left(\frac{d(lm)}{j\omega\varepsilon}\right)\left(kh_l'(kr)+\frac{h_l(kr)}{r}\right)LY_{lm}(\hat{r})=0,\ r=R,\\ &c(lm)r\nabla^2Y_{lm}(\hat{r})=0,\ r=R\end{aligned}$$

These yield

$$c(lm)=0, kRh_l'(kR)+h_l(kR)=0$$

Thus, oscillation frequency $(ln), n=1,2,3,\ldots;$
the roots of the equation are

$$\omega Rh_l'(\omega Rc)+h_l\left(\frac{\omega R}{c}\right)=0$$

And the solution for the multipole field within the spherical DRA is given by the previous superpositions.

The far-field radiation pattern generated by the surface current density on the sphere:

The relevant multipole fields are

$$\begin{aligned}E(t,r)&=R_e\left[\sum_{lmn}\frac{d(lmn)}{j\omega(ln)\epsilon}\exp(j\omega(ln)t)\mathrm{curl}\left\{h_l\left(\frac{\omega(ln)r}{c}\right)LY_{lm}(\hat{r})\right\}\right]\\ H(t,r)&=R_e\left[\sum_{lmn}d(lmn)\exp(j\omega(ln)t)\mathrm{curl}\left\{h_l\left(\frac{\omega(ln)r}{c}\right)LY_{lm}(\hat{r})\right\}\right]\end{aligned}$$

where

$$r=(r,\theta,\varphi),\ 0\le r\le R,\ 0\le\theta<\pi,\ 0\le\varphi<2\pi \tag{8.5}$$

This e.m. field satisfies the Maxwell equations and all the relevant boundary conditions on the perfectly conducting spherical surface. We now evaluate the far-field radiation pattern produced by the surface currents: the surface current density is given by

$$J_S(t, R\hat{r}) = \hat{r} \times H(t, R\hat{r})$$
$$= R_e\left[\sum_{lmn} d(lmn)\exp(j\omega(ln)t)\left\{h_l\left(\frac{\omega(ln)R}{c}\right)\hat{r}LY_{lm}(\hat{r})\right\}\right]$$

Now, for a function $\psi(\hat{r})$ or $\hat{r} = (\theta, \varphi)$ only, we have that

$$\hat{r} \times LY_{lm}(\hat{r}) = -i\hat{r} \times (r \times \nabla\psi(\hat{r})) = ir\,\nabla\psi(\hat{r})$$

Since $\hat{r} \cdot \nabla\psi(\hat{r}) = 0$ because $\nabla\psi(\hat{r})$ has only θ and φ components. Then

$$\begin{aligned} J_S(t, r) &= R_e\left[\sum_{lmn} lRd(lmn)\exp(j\omega(ln)t)\left\{h_l\left(\frac{\omega(ln)R}{c}\right)\nabla Y_{lm}(\hat{r})\right\}\right] \\ &= -RIm\left[\sum_{lmn} d(lmn)\exp(j\omega(ln)t)\left\{h_l\left(\frac{\omega(ln)R}{c}\right)\nabla Y_{lm}(\hat{r})\right\}\right] \end{aligned} \tag{8.6}$$

And hence, the far-field magnetic vector potential in the time domain generated by this surface current density is given by

$$\begin{aligned} A(t, r) &= \frac{\mu}{4\pi}\int_S J_S\left(t - \frac{r}{c} + \frac{R\hat{r}\cdot\hat{r}'}{c}, R\hat{r}'\right)dS(\hat{r}') \\ &= Im\left[-\frac{\mu R^3}{4\pi}\sum_{lmn} d(lmn)h_l\left(\frac{\omega(ln)R}{c}\right)\exp\left(-\frac{j\omega(ln)r}{c}\right)\exp(j\omega(ln)t)\right. \end{aligned}$$

$$\left.\int_S \exp\left(j\omega(ln)\frac{R\hat{r}\cdot\hat{r}'}{c}\right)\nabla Y_{lm}(\hat{r})d\Omega(\hat{r}')\right]$$

Not that on the surface of the sphere

$$dS(\hat{r}') = R^2 d\Omega(\hat{r}') = \sum_{ln} R_e[A_{ln}(r)\exp(j\omega(ln)t]$$

Note that the characteristics oscillation frequencies $\omega(ln)$ do not depend upon m.

We now compute the energy of radiation field within the spherical cavity. This expression would enable us to quantize the field inside in terms of cavity creation and annihilation operators:

$$U = \int_B \left[\frac{\varepsilon}{2}|E(t, r)|^2 + \frac{\mu}{2}|H(t, r)|^2\right]d^3r \tag{8.7}$$

where B is the volume region of cavity:

$$B = \{(r, \theta, \varphi): 0 \le r \le R,\ 0 \le \theta < \pi, 0 \le \varphi < 2\pi\}$$

And S is its surface:

$$S = \{(R, \theta, \varphi): 0 \le \theta < \pi, 0 \le \varphi < 2\pi\}$$

$$= \frac{1}{2}\sum_{lmn} \frac{\varepsilon}{2}\left(\frac{|d(lmn)|^2}{\omega(ln)^2 \varepsilon^2}\right) \int_B \left|\text{curl}\left\{h_l\left(\frac{\omega(ln)r}{c}\right)LY_{lm}(\hat{r})\right\}\right|^2 drdS(\hat{r})$$

$$+ \frac{1}{2}\sum_{lmn} \frac{\mu}{2}\left(|d(lmn)|^2\right) \int_B \left|\left\{h_l\left(\frac{\omega(ln)r}{c}\right)LY_{lm}(\hat{r})\right\}\right|^2 drdS(\hat{r}) \tag{8.8}$$

Now using the orthonormality of the spherical harmonics as well as their eigen function property for operator L^2, we see that

$$\int_B \left|\left\{h_l\left(\frac{\omega(ln)r}{c}\right)LY_{lm}(\hat{r})\right\}\right|^2 drdS(\hat{r}) = l(l+1)\int_0^R \left|\left\{h_l\left(\frac{\omega(ln)r}{c}\right)LY_{lm}(\hat{r})\right\}\right|^2 r^2 dr$$

And

$$\int_B \left|\text{curl}\left\{h_l\left(\frac{\omega(ln)r}{c}\right)LY_{lm}(\hat{r})\right\}\right|^2 drdS(\hat{r}) = \int_B |\text{curl}\psi(\hat{r})|^2 d^3r$$

where

$$\psi(r) = h_l\left(\frac{\omega(ln)r}{c}\right)LY_{lm}(\hat{r})$$

Now,

$$|\text{curl}\psi|^2 = (\text{curl}\psi^*, \text{curl}\psi) = \text{div}(\psi^* \times \text{curl}\psi) + (\psi^*, \text{curlcurl}\psi)$$

The integral of the divergence term on the right over B by virtue of Gauss's integral there in once we note that $\hat{r} \times \text{curl}\psi = 0$ where $r = R$ in view of the vanishing boundary condition for tangential component of the electric field on surface of the surface. In fact, this condition was one of the defining conditions for the computing the characteristic oscillation frequency $\omega(ln)$. Further,

$$\text{curlcurl}\ \psi = -\nabla^2\psi = \left(\frac{\omega(ln)^2}{c}\right)\psi$$

Since as observed earlier, $\text{div}\psi = 0$ by the Maxwell equation that states that the magnetic field has a vanishing divergence. Thus, we get finally after combining all these equations that the total cavity field energy is given by

$$U = \sum_{lmn} \frac{\mu}{2} l(l+1) f(l,n) |d(lmn)|^2 \tag{8.9}$$

where

$$f(l,n) = \int_0^R \left| h_l \left(\frac{\omega(ln)r}{c} \right) \right|^2 r^2 dr \tag{8.10}$$

And this quadratic equation in the $d(lmn)$ can be immediately quantized using the harmonics oscillator method.

Equations (8.1)–(8.10) have been developed for the theoretical analysis of terahertz spherical DRA.

8.4 Results and discussions

Simulations results of terahertz spherical DRAs have been validated and results such as S_{11}, radiation pattern, VSWR, E and H fields are given in Figures 8.9–8.17.

8.4.1 Super directivity in spherical DRA

Equations (8.11)–(8.16) have been developed to define the concept of super directivity in spherical DRA at terahertz frequency.

Fermionic Dirac wave operator field can be defined as

$$\psi(t,r) = \sum_{mnp\sigma} [a(mnp\sigma)\exp(-j\omega(mnp)t) U(mnp\sigma, r) + b(mnp\sigma)^* \exp(j\omega(mnp)t) \tilde{X}(mnp\sigma, r) V(mnp\sigma)] \tag{8.11}$$

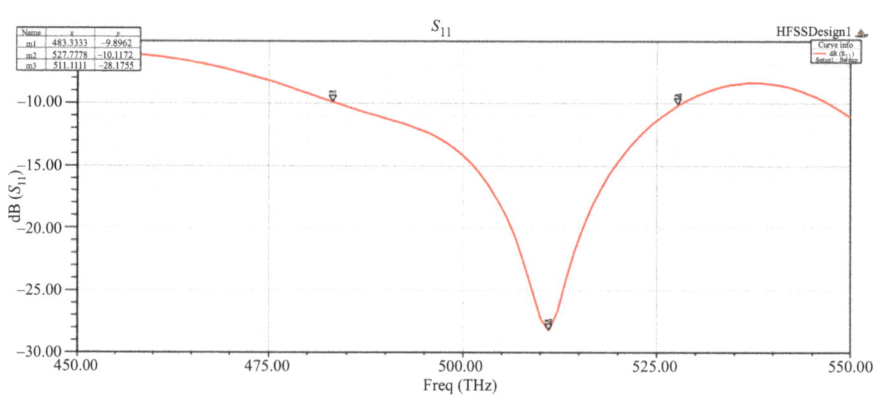

Figure 8.9 Reflection coefficient of silicon spherical optical DRA at 511 THz

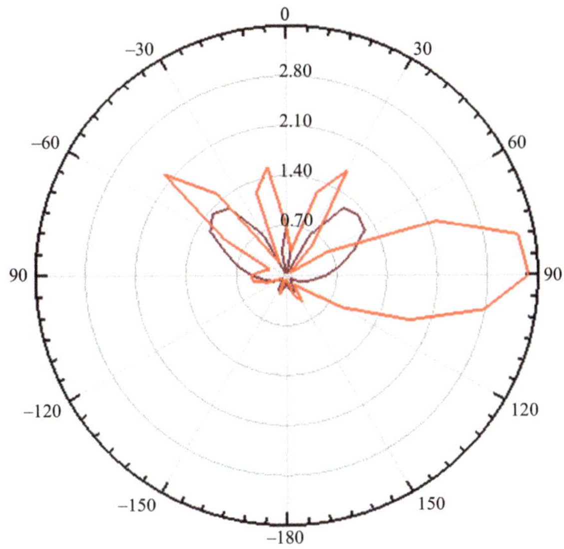

Figure 8.10 Radiation pattern with max gain 2.83 dBi of optical spherical DRA at 511 THz

The Dirac four-current density corresponding to this fermionic wave field is given by

$$J^{\mu}(t,r) = -e\psi(t,r)^{*}\alpha^{\mu}\psi(t,r),\ \alpha^{\mu} = \gamma^{0}\gamma^{\mu} \tag{8.12}$$

The e.m. four quantum potential of radiation generated by this current is given by the standard retarded potential formula given as follows:

$$A^{\mu}(t,r) = k\cdot\int_{C} J^{\mu}\left(t-\frac{|r-r'|}{c},r'\right)d^{3}r'/|r-r'|,$$

$$k = \frac{\mu}{4\pi}$$

The far-field magnetic vector potential is given by

$$(A^{r}(t,r)\colon r=1,2,3) = A(t,r) = \left(\frac{k}{r}\right)\cdot\int_{C} J\left(t-\frac{r}{c}+\hat{r}\cdot\frac{r'}{c},r'\right)d^{3}r' \tag{8.13}$$

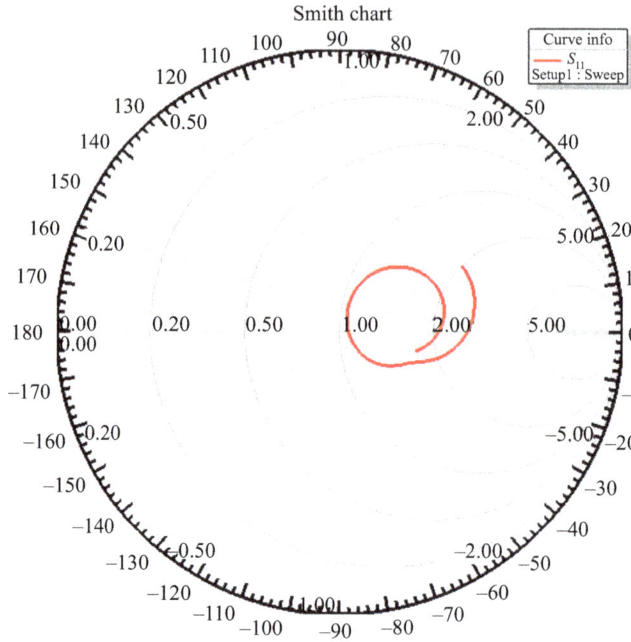

Figure 8.11 Smith chart showing an impedance of terahertz spherical DRA

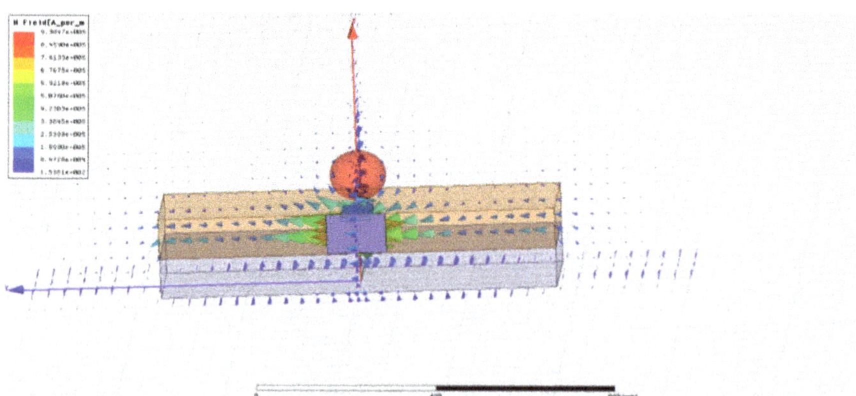

Figure 8.12 Laser beam input to terahertz spherical DRA at 511 THz

And in the frequency domain, this can be expressed as

$$A(\omega, r) = \left(\frac{k}{r}\right) \int J(\omega, r') \exp\left(j\left(\frac{\omega}{c}\right)\hat{r} \cdot r'\right) d^3 r'$$

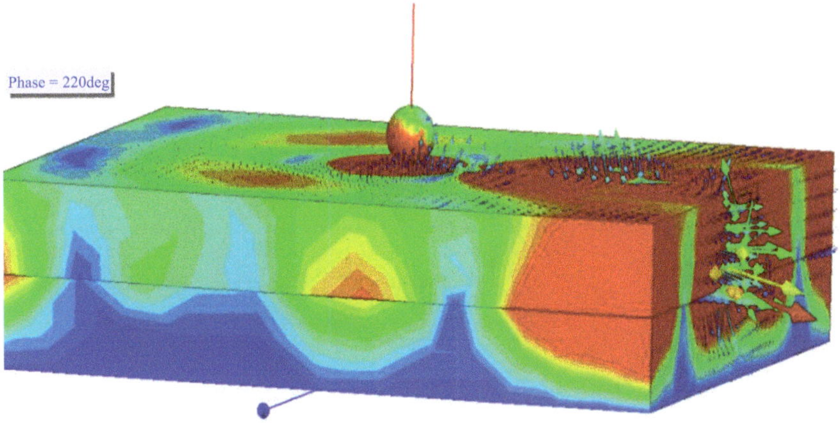

Figure 8.13 Silicon optical spherical DRA showing vector fields at 511 THz, E-field vector

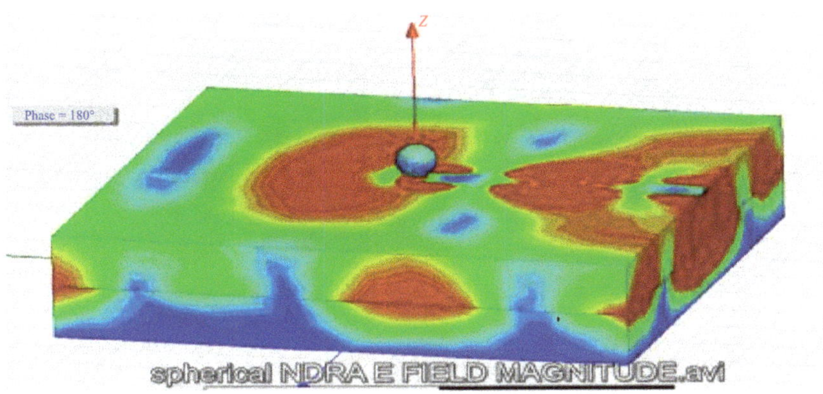

Figure 8.14 E-Field scalar terahertz spherical DRA

By the Fourier transform of the current density operator field, we get

$$J(\omega, r) = -e \int \psi(\omega' - \omega, r)^* \alpha \cdot \psi(\omega', r) d\omega' \tag{8.14}$$

By the convolution theorem for Fourier transform, we get

$$\begin{aligned}\psi(\omega, r) = \sum_{mnp\sigma} &[a(mnp\sigma)U(mnp\sigma)\delta(\omega - \omega(mnp))X(mnp, r) \\ &+ b(mnp\sigma)\delta\big(\omega + \omega(mnp)\tilde{X}(mnp, r)\big)]\end{aligned}$$

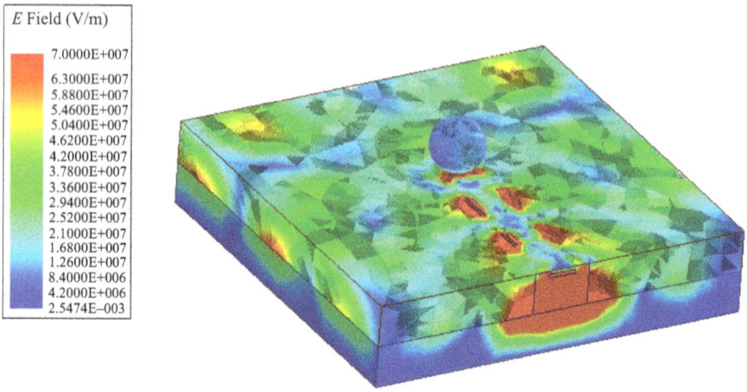

Figure 8.15 E Fields and input excitation of terahertz spherical DRA

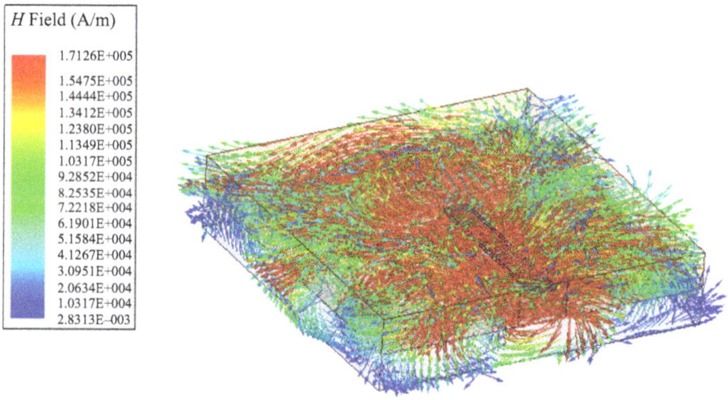

Figure 8.16 H-Field vector of terahertz spherical DRA

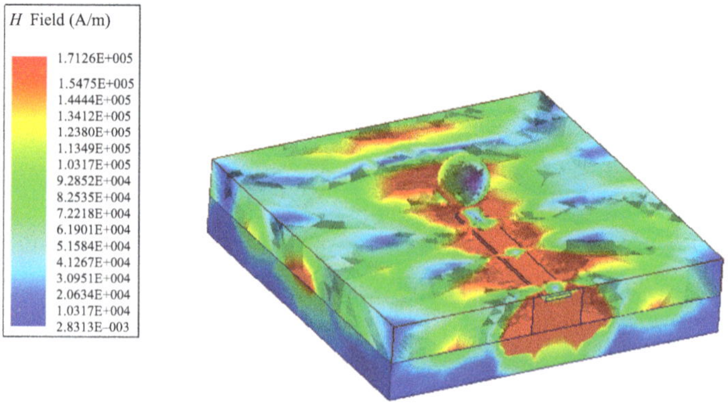

Figure 8.17 H-Field scalar of terahertz spherical DRA

Hence,

$$J(\omega, r) = -e \int d\omega' \Bigg[\sum_{mnp\sigma} \big[a(mnp\sigma)^* \tilde{U}(mnp\sigma) \tilde{X}_{mnp}(r) \delta(\omega' - \omega - \omega(mnp))$$

$$+ b(mnp\sigma) \tilde{V}(mnp\sigma) X_{mnp}(r) \delta(\omega' - \omega + \omega(mnp)) \big]$$

$$\times \Bigg[\sum_{mnp\sigma} \big[a(mnp\sigma) U(mnp\sigma) X_{mnp}(r) \delta(\omega'$$

$$-\omega(mnp)) b(mnp\sigma)^* V(mnp\sigma) \tilde{X}_{mnp}(r) \delta(\omega' + \omega(mnp)) \big]\big]\big]$$

$$= -e \sum_{mnp\sigma m'n'p'\sigma'} \big[a(mnp\sigma)^* a(m'n'p'\sigma') U(m'n'p'\sigma') \tilde{X}_{mnp}(r)$$

$$X_{m'n'p'}(r) \delta(\omega(m'n'p') - \omega(mnp) - \omega)$$

$$+ a(mnp\sigma)^* b(m'n'p'\sigma')^* \tilde{U}(mnp\sigma) V((m'n'p'\sigma') X_{m'n'p'}(r)$$

$$\tilde{X}_{mnp}(r) \delta(\omega(m'n'p') + \omega(mnp) + \omega))$$

$$+ b(mnp\sigma) a(m'n'p'\sigma') \tilde{V}(mnp\sigma) U((m'n'p'\sigma')$$

$$X_{mnp}(r) \tilde{X}_{m'n'p'}(r) \delta(\omega(m'n'p') + \omega(mnp) - \omega))$$

$$+ b(mnp\sigma), b(m'n'p'\sigma')^* \tilde{V}, (mnp\sigma), V, ((m'n'p'\sigma')$$

$$\tilde{X}_{mnp}(r) \tilde{X}_{m'n'p'}(r) \delta(\omega(mnp) - \omega(m'n'p') - \omega))] \tag{8.15}$$

It is now a simple matter to evaluate the far-field vector potential and hence the far-field e.m. field and hence also the far-field Poynting vector and determine a state $|n\rangle$ of the electron–positron field such that if $S(\omega, r) = P(\omega, \hat{r})\hat{r}/r^2$ denote the far-field Poynting vector, then for a given frequency (i.e. one among the set $\omega(mnp) \pm \omega(m'n'p')$ such that $\langle n|P(\omega, \hat{r})|n\rangle$ is a maximum for a given direction, etc. To compute this quantum average, all we require all the matrix elements

$$\begin{aligned} &\langle \eta | a(mnp\sigma)^* a(m'n'p'\sigma') | \eta \rangle, \langle \eta | a(mnp\sigma)^* b(m'n'p'\sigma') | \eta \rangle, \\ &\langle \eta | b(mnp\sigma) a(m'n'p'\sigma') | \eta \rangle, \langle \eta | b(mnp\sigma) b(m'n'p'\sigma')^* | \eta \rangle \end{aligned} \tag{8.16}$$

8.5 MIMO (multi-input–multi-output) spherical DRA

MIMO DRAs are directive antennas for enhanced gain and diversity. The aspect ratio of spherical DRA is **(*h*/*r*) = 7/3.5 = 2. Figures 8.18 and 8.19 have shown two spherical DRA connected with independent feed but common ground plane. The isolation has been measured between two. Results simulated using HFSS are shown in Figures 8.20–8.25.**

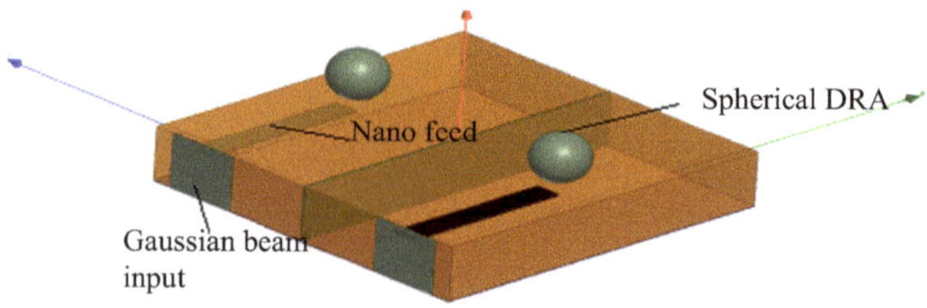

Figure 8.18 MIMO spherical DRA top view

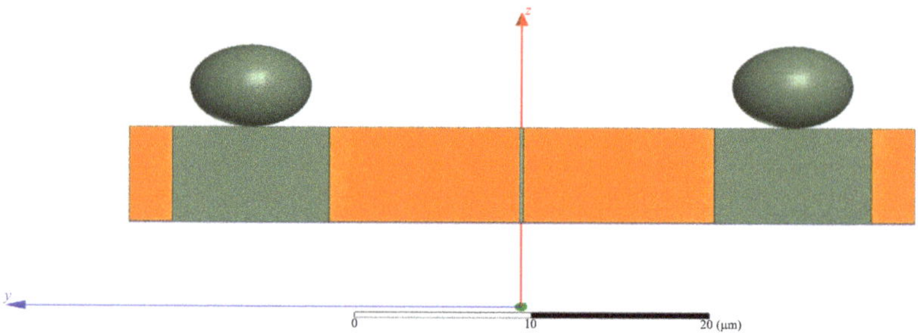

Figure 8.19 MIMO spherical DRA side view

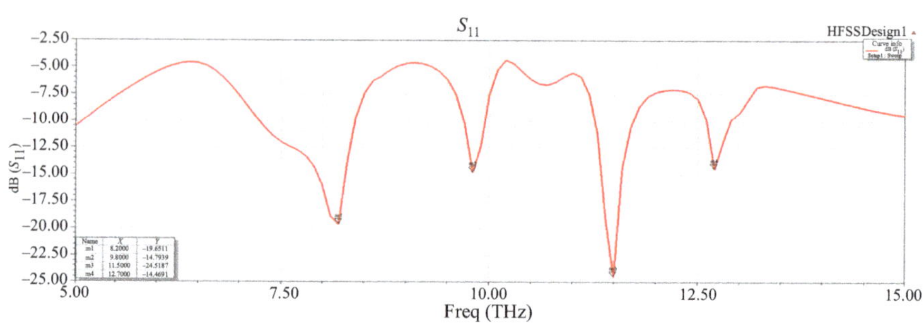

Figure 8.20 S_{11} of MIMO spherical DRA

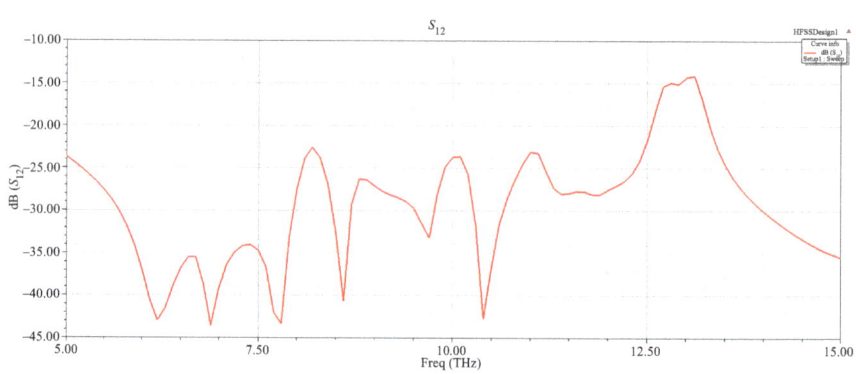

Figure 8.21 S_{12} MIMO spherical DRA

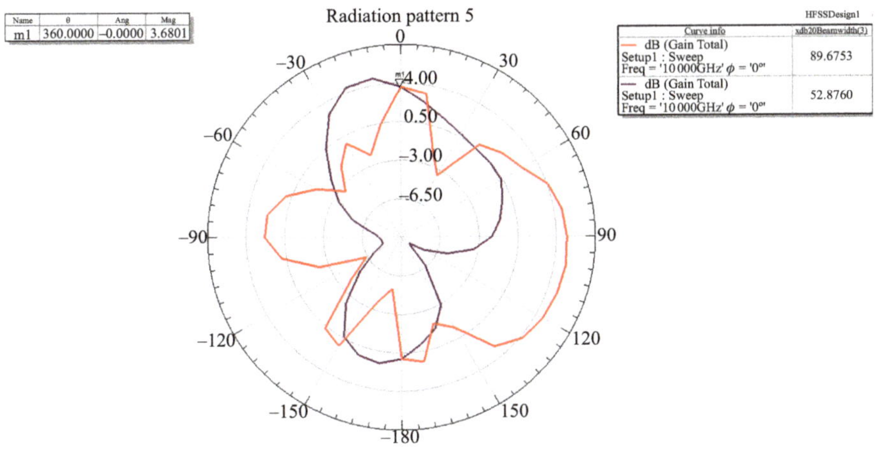

Figure 8.22 MIMO spherical DRA radiation pattern

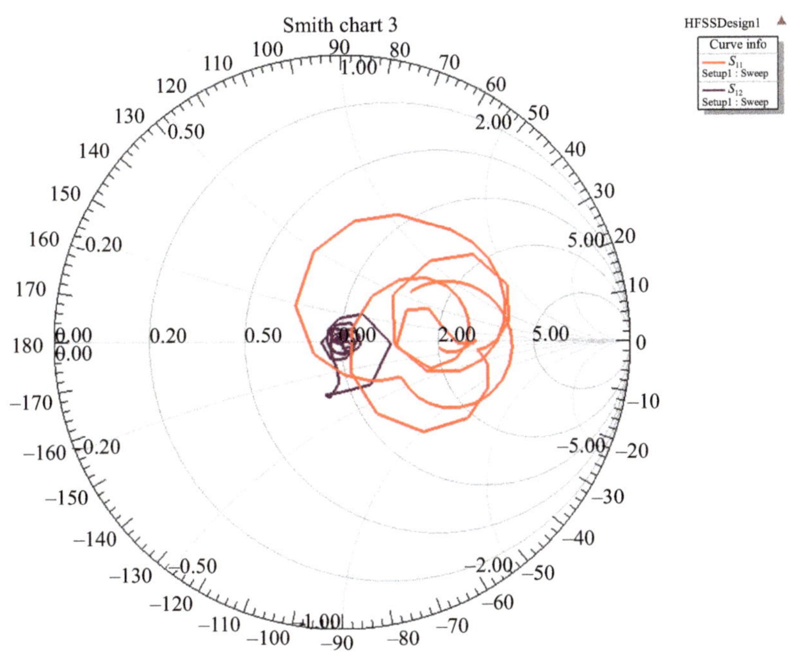

Figure 8.23 MIMO spherical DRA smith chart

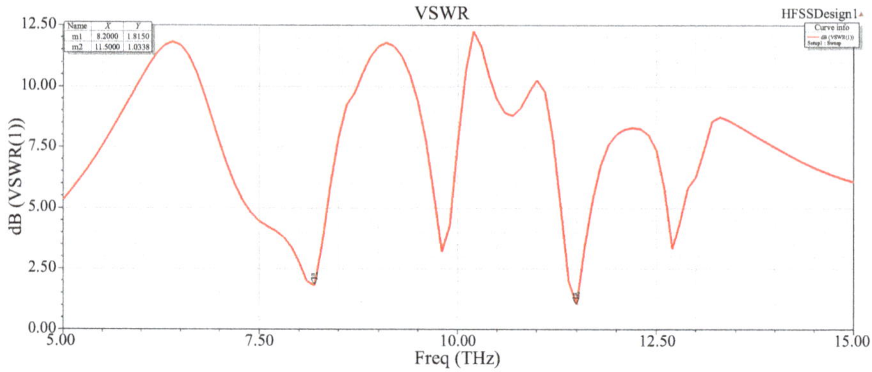

Figure 8.24 MIMO spherical DRA VSWR

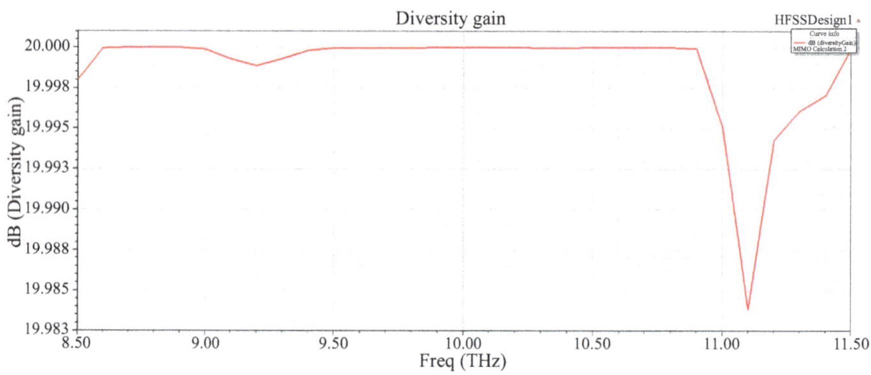

Figure 8.25 MIMO spherical DRA diversity gain

Model-3 terahertz spherical DRA designed at 8.5 THz

8.6 Conclusion

Microwave optical spherical DRA at 511 THz with 2.45-dB gain has been simulated using HFSS for optical applications such as imaging, sensor and scanning. Theoretical modeling of radiation pattern has been formulated in (8.1)–(8.10). THz DRA with a high gain of 18 dBi and a frequency of 10 THz has been developed for communication applications. It is having super directivity. These spherical DRAs have applications such as biomedical sensors, scanning, imaging, chip-to-chip communications and retinal photoreceptors at optical frequencies. MIMO DRA has advantages of diversity reception and super directivity. Results at 8.5 THz frequency have also been seen in Figures 8.26–8.28.

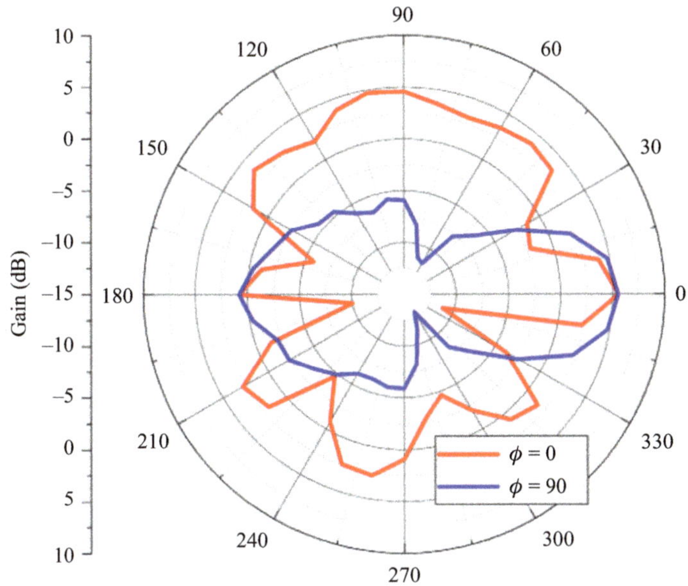

Figure 8.26 Radiation pattern at 8.5 THz

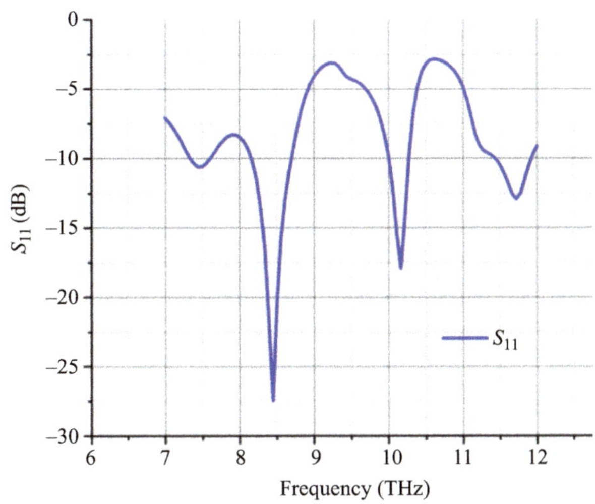

Figure 8.27 S_{11} at 8.5 THz

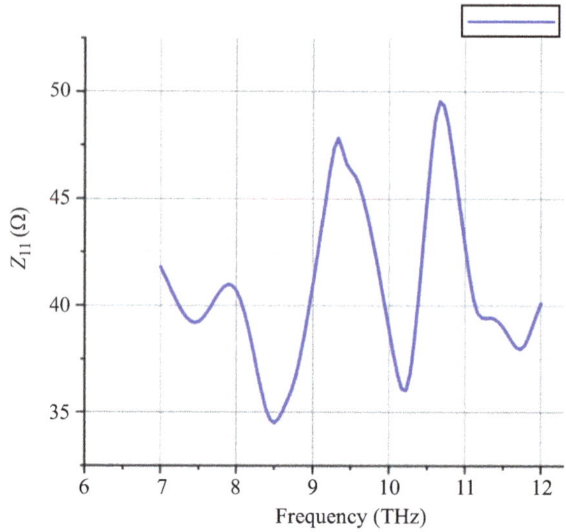

Figure 8.28 Z11 at 8.5 THz

References

[1] R. S. Yaduvanshi and H. Parthasarathy, *Rectangular Dielectric Resonator Antennas: Theory and Design*, Springer, New Delhi, 2016.

[2] R. S. Yaduvanshi and V. Gaurav, *Nano Dielectric Resonator for 5G Applications*, CRC Press, Boca Raton, FL, 2020.

[3] L. Novotny and N. van Hulst, "Antennas for light," *Nat. Photonics*, vol. 5, no. 2, pp. 83–90, 2011.

[4] P. Biagioni, J.-S. Huang, and B. Hecht, "Nanoantennas for visible and infrared radiation," *Rep. Prog. Phys.*, vol. 75, p. 024402, 2012.

[5] L. Zou, W. Withayachumnankul, C. Shah, *et al.*, "Dielectric resonator nanoantennas at visible frequencies," *Opt. Express*, vol. 21, no. 1, pp. 1344–1352, 2013.

[6] G. N. Malheiros-Silveira, G. S. Wiederhecker, and H. E. Hernandez Figueroa, "Dielectric resonator antenna for applications in nanophotonics," *Opt. Express*, vol. 21, no. 1, pp. 1234–1239, 2013.

[7] M. W. Knight, H. Sobhani, P. Nordlander, and N. J. Halas, "Photodetection with active optical antennas," *Science*, vol. 332, no. 6030, pp. 702–704, 2011.

[8] P. Mühlschlegel, H.-J. Eisler, O. J. F. Martin, B. Hecht, and D. W. Pohl, "Resonant optical antennas," *Science*, vol. 308, no. 5728, pp. 1607–1609, 2005.

[9] Y. Zhao, N. Engheta, and A. Alù, "Effects of shape and loading of optical nano antennas on their sensitivity and radiation properties," *J. Opt. Soc. Am.*, vol. B28, no. 5, pp. 1266–1274, 2011.

[10] S. A. Maier, *Plasmonic—Fundamentals and Applications*, Springer, New York, NY, 2007.
[11] E. Ozbay, "Plasmonics: merging photonics and electronics at nano scale dimensions," *Science*, vol. 311, p. 189, 2006.
[12] L. Zou, W. Withayachumnankul, C. Shah, *et al.*, "Efficiency and scale ability of dielectric resonator antennas at optical frequencies," *IEEE Photon. J.*, vol. 6, p. 110, 2014.
[13] J. D. Jackson, *Classical Electrodynamics*, Wiley, New York, NY, 1962.
[14] A. Bonakdar and H. Mohseni, "Impact of optical antennas on active optoelectronic devices," *Nanoscale*, vol. 6, pp. 10961–10974, 2014.
[15] P. D. Terekhova, K. V. Baryshnikovab, Y. A. Artemyevb, A. Karabchevskya, A. S. Shalinb, and A. B. Evlyukhinb, "Optical multipole resonances of non-spherical silicon nanoparticles and the influence of illumination direction," Proc. of SPIE Vol. 10528 1052802-1, SPIE OPTO, San Francisco, CA, United States, 2018.
[16] D. Storozhenko, A Nepomnyaschiy, and A. V. Deev, "High directivity in the narrow band of spherical dielectric antennas in GHz and THz ranges," *Semiconductors*, vol. 53, no. 14, pp. 1967–1969, 2019.

Chapter 9

Rectangular terahertz DRA—design, simulation and implementations

Abstract

In this, design and implantations of a terahertz rectangular dielectric resonator antenna (DRA) using an HFSS simulator has been presented. The aspect-ratio has played a deterministic role for exciting desired multiple resonant modes at terahertz frequencies. The design has large bandwidth hence can be used for high-speed communications and terahertz sensors. It is simple due to rectangular geometry, efficient due to use of DR, compact due to terahertz frequency and unique using silicon as a radiator. Mathematical modeling of 10-THz frequency using a terahertz DRA is presented along with simulations results of S_{11} below -10 dB, gain of 4.89 dBi with good radiation pattern, which have been validated. DRAs at photonic wavelength have also been developed for 511-THz frequency with all desired parameters and simple geometry. These DRAs fall within visible frequency; hence they can be used as artificial retinal photoreceptors. A DR material has advantages over metallic antennas as they can operate with stability from microwave to optical regime. These optical antennas can be suitable for wireless transmission of TBPS data rate. Also, coherent states and control at optical frequencies have been formulated mathematically with unique characteristics at terahertz frequency.

9.1 Introduction

It is well known that a dielectric resonator antenna (DRA) is an efficient radiator within microwave frequency to optical frequency regime. Dielectric materials were first suggested by Richtmyer in 1939 for use as an antenna-radiating element, later SA Long experimentally used DRA for practical implementations [1]. Since then, RK Mongia, A Kisk, YMM Antaar, Guha, Van Bladel, Pan, Petosa, Leung, Yaduvanshi and many other researchers haves done research work on DRAs [1–11]. Yaduvanshi and Parthasarathy have published on rectangular DRAs (RDRAs) in Springer 2016 on RDRA. A nano-DRA for 5G applications is published in CRC Press 2020 by Yaduvanshi, Varshney. Terahertz is a new field of applications for high-speed communications due to its large bandwidth characteristics, i.e., 300 GHz to 10 THz. This band is still lying vacant for telecommunication application [1–5]. Terahertz technology has effectively miniaturized the size of DRA. They consume less power due to small size. Hence, terahertz DRAs (TDRAs) are in much demand nowadays for its high-speed communications and sensing applications to be used in Internet on Things networks. A frequency spectrum of microwave, terahertz and photonics is given in Figure 9.1. The design of TDRA is shown in Figure 9.2 and dimensions are shown in Table 9.1 for a 10-THz DRA. Figures 9.3–9.10 show simulated results. Here, most of our work shall be in the range of 10^{11}–10^{15} Hz (THz, IR and optical spectrum). The DR materials used were Al_2O_3, TMM, sapphire, $BaSrTiO_2$ at microwave frequencies and silicon, graphene, TiO_2, arlon and teflon at terahertz and optical

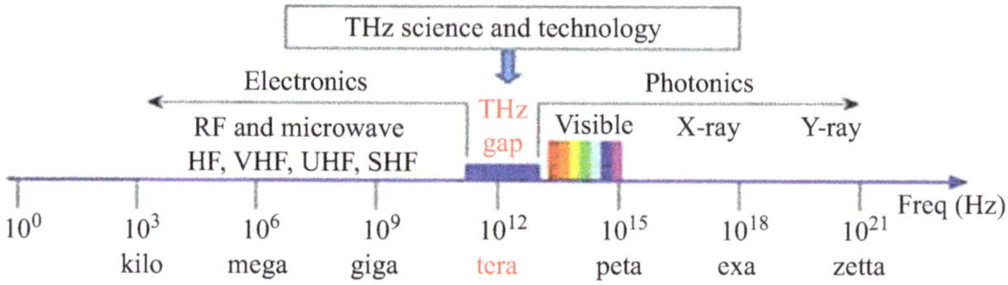

Figure 9.1 Microwave, terahertz and photonic science (spectrum)

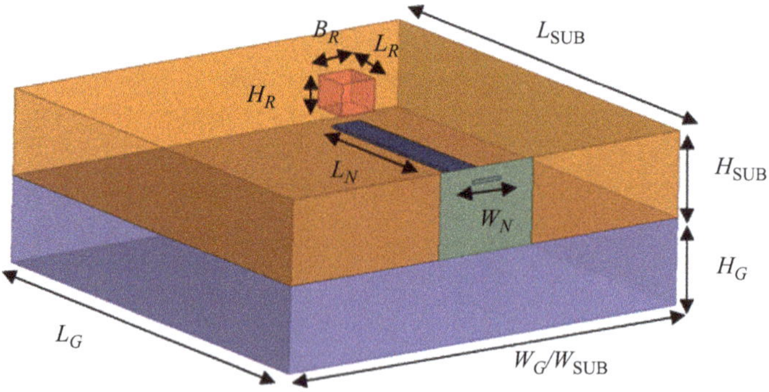

Figure 9.2 Design of TRDRA (CST model)

Table 9.1 Dimension table

Name	Material	Permittivity and loss tangent	Dimension (nm)
Substrate top	Silicon dioxide	4 and 0	900*900*100
Substrate bottom	Silver	1 and 0	900*900*100
Nano waveguide	Silver	1 and 0	450*67*10
CDRA	Titanium dioxide	8.29 and 0.01	$L=64$, $B=64$, $L=60$
Gaussian beam	LASER pulses	500 THz	133.5*100

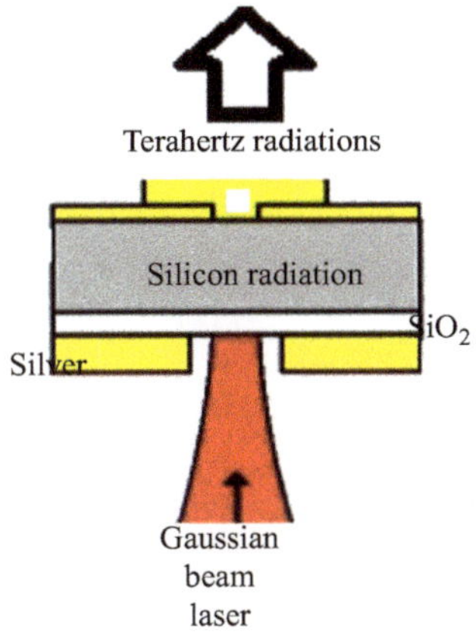

Figure 9.3 Terahertz DRA side view with flow (TRDRA)

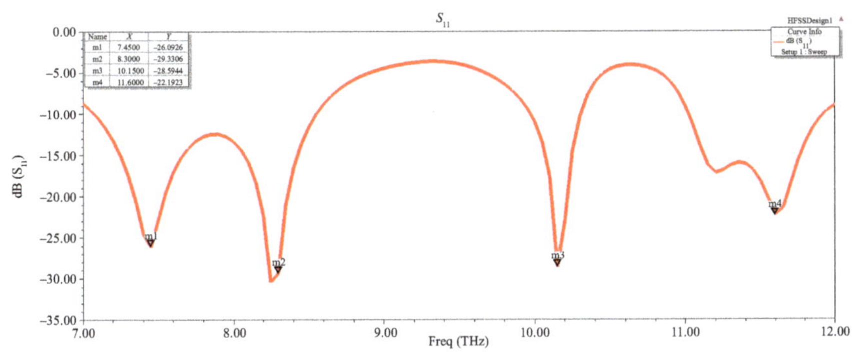

Figure 9.4 Terahertz RDRA gain plot with max gain 7.38 dBi (TRDRA)

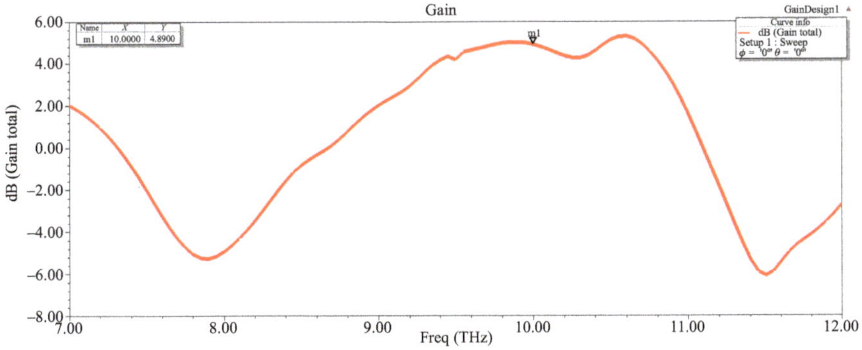

Figure 9.5 Gain of rectangular TDRA

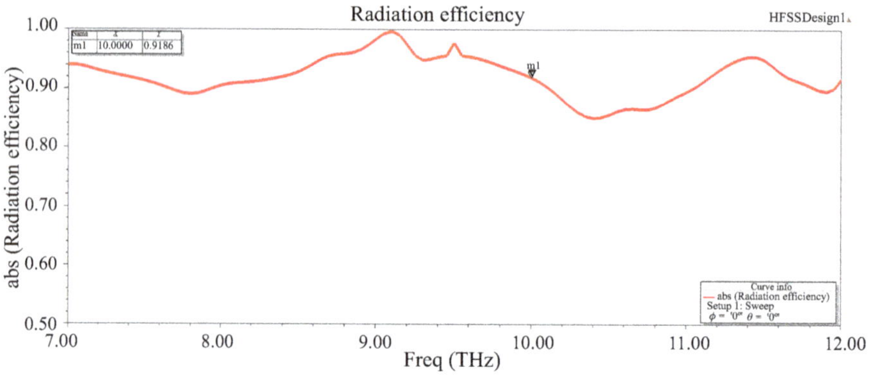

Figure 9.6 Efficiency of rectangular TDRA

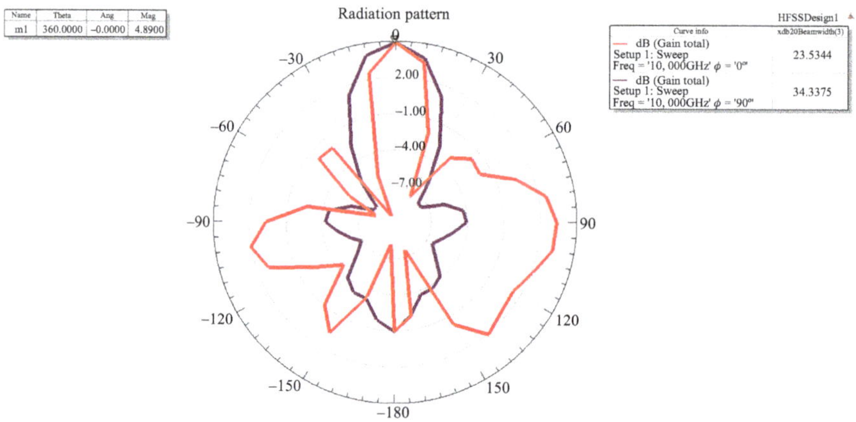

Figure 9.7 Radiation pattern of rectangular DRA

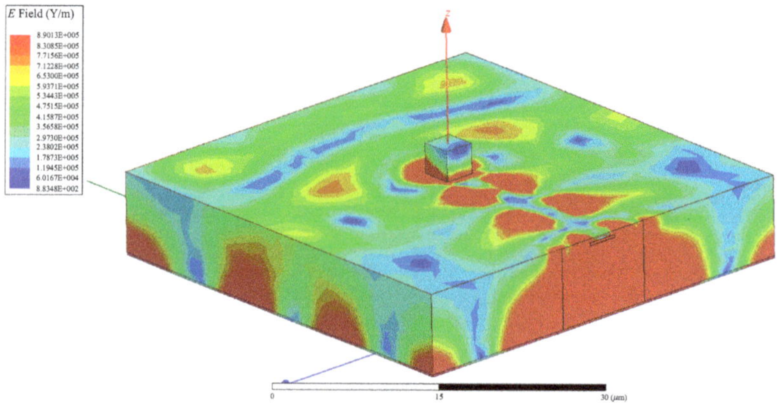

Figure 9.8 E field of rectangular DRA

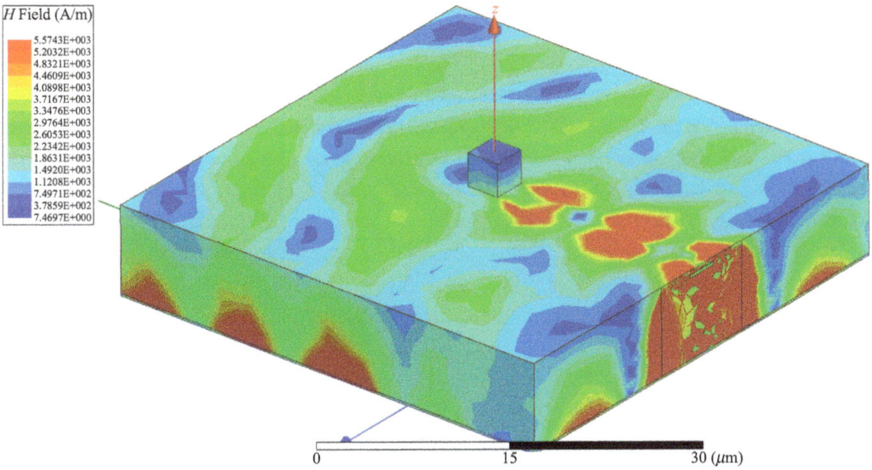

Figure 9.9 H Fields at 10 THz

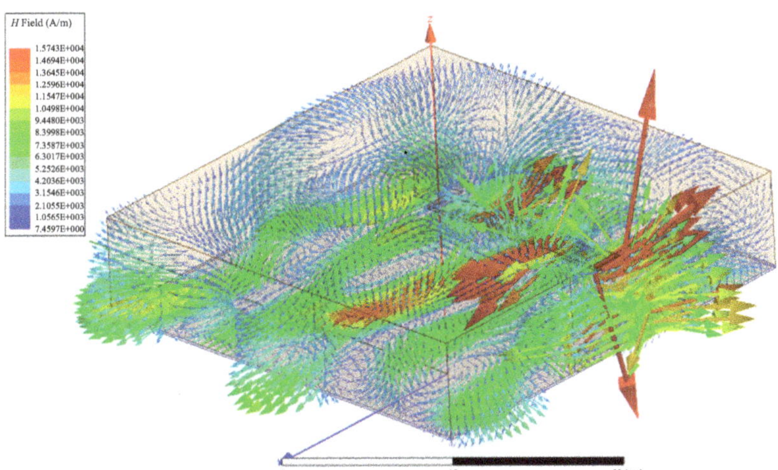

Figure 9.10 H-Field vector

spectrum. DRAs are efficient in radiating elements throughout the entire frequency spectrum. The permittivity of these materials is available from 10 to 1,600. Hence, due to high permittivity, the materials help to reduce device sizes as it is inversely proportional to size, i.e. high permittivity corresponds to small size. They have design flexibility as DRAs command two different aspect-ratio height/base-length or height/base-width. They have capability of generating higher order modes along with fundamental modes depending upon aspect-ratios. It is generally chosen between 0.5 and 2.5 for generating resonant modes of our choice. Higher order modes can have higher gain and better directivity as compared to fundamental mode. Terahertz regime can offer higher bandwidth. Hence terahertz is suitable for higher speed communication. This terahertz band is lying vacant thus can be used for high-speed communication. Due to skin effect, metal behaves like plasma and Drude's analysis will be required to assess scattering properties. Wave-port analysis to a terahertz device may not give accurate results; however, Gaussian pulse along with sliver nano waveguide shall give accurate results. The Gaussian beam at terahertz is used for phase matching, and proximity feed is used to accommodate electrical effects of surface plasmon polariton (SPP) length. Plasma frequency of each metal is different and for propagation it is always kept lower than laser input excitation frequency. The Dirac equation is used to compute radiated fields at optical frequencies. The development of quantum antenna has been a slow process and operators are used for mathematical analysis of quantum antennas.

There is great contribution of Bhardwaj, Deutsch and Novotny in the development of light antenna or optical antennas [12–17]. Propagation loss is the main disadvantage at higher frequencies. These losses can be minimized by making use of DRAs at terahertz and optical frequencies. The quantum cascade lasers can be used at 1–5 THz, 90 mW, narrow width line and single-mode operation source to feed. Noble metal silver is used as waveguide at higher frequency in place of microstrip line feed. With this antenna, TBPS of data rate can be achieved. This high data rate is good for 5G and beyond communication systems. Microwave antennas have limited capability to deliver data speed only up to few GBPS. These terahertz antennas can be placed in chip (CMOS). Of course, they will be costly to develop, but cost can be reduced if they are to be produced at large industrial scale. Initially light antennas were to be operated only near field. By making use of DR, their gain can be enhanced and they are able to produce far fields and most suitable for long distance wireless communications at THz and optical frequencies. TDRAs of cylindrical, conical and spherical shape at optical frequency and dimension can be very useful for the development of artificial retinal photoreceptors as these DRA photonic antennas are efficient [18,19].

9.2 Propagation of light

The understanding of light since Sir Isaac Newton started with first assuming light to be corpuscles, namely small particles of different color corresponding to the different frequencies. This idea was first propounded by Newton and came to be known as the corpuscular theory of light. Several years later, based on experiments involving interference of light, Huygens came to understand that light is to be regarded as being composed not of particles but rather is a wave and that any light field can be regarded as a superposition of plane waves of different frequencies and wave numbers. This meant that during the process of interference involving computation of the total intensity of light coming from different sources, one should not add the respective intensities, but rather the amplitudes, then square the resultant amplitude and then form its time average. This meant that in the resultant total intensity, there would appear cross terms that are oscillatory, which lead us to the phenomenon of interference. Later, Fermat propounded the ray theory of light in order to explain Snell's theorems on reflection and refraction of light, according to which a light ray always follows the path of minimum time. The wave nature of light as propounded by Christian Huygens gained more support when Maxwell unified electromagnetism with light by showing that the basic equations of electromagnetism imply that that the electric and magnetic fields in space satisfy the wave equation with the velocity of propagation being equal to that of light. This fundamental discovery gave firm support to the wave theory of light. However, with the birth of the quantum theory in the early part of the nineteenth century, it became clear that light comes in discrete packets called quanta and each quantum of radiation carries an energy proportional to its frequency and that the intensity of light is proportional to the number of quanta. These quanta came to be known as photons and it appeared therefore that with the advent of the quantum theory, physicists had reverted back to Newton's corpuscular theory of light. However, while Planck was busy creating his quantum theory of light, simultaneously Einstein wrote some beautiful papers proving that light can behave sometimes as a particle and sometimes as a wave, specifically showing that the variance of fluctuations in the energy of black-body radiation has two components, a particle component in agreement with **Planck's quantum hypothesis** and a wave component in agreement with **Rayleigh's theory** on the number of "modes" of a wave field within a given energy shell. The particle theory of light got further support from Einstein's special theory of relativity giving the relationship between the energy and momentum of a particle of any given mass and in particular for light particles having zero rest mass. However, the crucial breakthrough into the wave–particle duality matter and in particular of light came with de Broglie's discovery of the fundamental relationship between the wavelength of a wave and the momentum of the associated particle. The de Broglie theory was applicable to all of matter not just light. It gave conclusive evidence that even particles like electrons exhibit wave–particle duality, i.e., they sometimes behaved like particles and sometimes like waves. The final crunch describing the wave–particle duality came with the wave equation for particles discovered by Schrödinger using which he was able to calculate the energy levels of an electron bound to a nucleus in terms of the eigen values of a partial differential equation. The corresponding eigen functions were complex wave fields that Schrödinger could not immediately interpret. These eigen functions approximately corresponded to the kind of "matter-waves" predicted by Louis de Broglie having frequency determined by Planck's quantum hypothesis relating frequency of waves to the energy of the quanta and having wavelength determined by de Broglie's theory relating the wavelength of waves to the momenta of the particles. Schrödinger's wave equation was a nonrelativistic wave equation in that it could be "derived" by assuming the nonrelativistic relationship between energy and momentum and then assigning operators to energy and momentum. The correct interpretation of the eigen functions of Schrödinger's equations was provided about a year later by Max Born with the suggestion that the modulus square of the wave function gave the probability density of the particle to be present at a given point in space at a given time. Equivalently, this modulus square could be interpreted as the number density of particles or the intensity of the wave at that point in space at that time. In fact, Schrödinger's equation could also be generalized to determine the wave functions and energy levels for any particle with a prescribed energy–momentum relationship by replacing the energy and momentum with appropriate operators and applying both sides of this relationship to a wave function. Thus, Schrödinger's formalism could be applied to determine the kind of de Broglie waves associated with a particle having any given energy–momentum relation. The de Broglie wave solutions then have the interpretation that their modulus square represents the intensity/probability density of particles in space at a given time. In special relativity, the energy–momentum relationship for a relativistic particle is a quadratic form in the energy and momenta unlike the Newtonian-nonrelativistic case where it is linear in the energy and quadratic in the momenta. Consequently, the Schrödinger equation is a linear PDE that is of the first order in time and second order in spatial derivatives. The nonrelativistic Schrödinger equation as a consequence then leads to a unitary evolution of the wave function with time and the additional pleasing property that evolution from time zero to time $t_1 + t_2$ can be expressed as the composition of evolution from time zero to time t_1 followed by evolution from time t_1 to time $t_1 + t_2$. In other words, Schrödinger evolution follows the semigroup property. This is in contrast to the case of special relativity wherein the wave equation is quadratic in time leading to the wave function at a given time being a superposition of a forward-propagating solution and a backward-propagating solution and the loss of unitarily of the evolution. Thus, the total probability will be conserved in Schrödinger's nonrelativistic theory but not so in the relativistic theory. This difficulty was finally resolved by Paul Dirac who factorized the energy–momentum relationship of special relativity into linear factors in energy and momentum using 4×4 complex anticommuting matrices and it led to Dirac's relativistic theory of the electron which along with Maxwell's equations is at the heart of quantum electrodynamics describing the interaction of light with matter. Dirac's equation being linear in the space and time derivatives is a truly relativistic wave equation for quantum theory because first it respects Lorentz invariance in that space and time are treated on an equal footing and that we can find a spin or representation of the Lorentz group under which Dirac's equation remains invariant and second it is first order in time and therefore preserves the unitary semigroup property of the evolution guaranteeing therefore conservation of total probability. Light as we understand today is a second quantized electromagnetic field in which the vector potential components in the spatial frequency domain represent creation and annihilation operators of photons with different momenta and helicities. This is owing to the fact that the total energy of the electromagnetic field can in the Coulomb gauge be expressed as a quadratic functional of the spatial frequency components of the vector potential and each spatial frequency component of the vector potential, according to Maxwell's equations, evolves harmonically with time with a frequency being ω, where k is the wave vector. Such a picture is in perfect accord with the Hamiltonian theory of a collection of independent quantum harmonic oscillators. However, this model of light can actually be traced back to the work of

Satyendranath Bose who proposed a statistical method for deriving Planck's law of black-body radiation wherein we are interested in distributing a total amount of energy E having N quanta at frequency ν so that $E = Nh\nu$, amongst p oscillators with the quanta being regarded as indistinguishable particles. When an oscillator has n such quanta, we say in the modern language of quantum mechanics that it has been excited to the nth energy level by an application of n creation operators to the vacuum. In this way, the entire field of black-body radiation is simply a collection of quantum harmonic oscillators and if we maximize the entropy, i.e., the total number of ways of distributing this energy then we end up with the famous Bose–Einstein statistics that gives us the relative number of photons at each frequency. This in turn enables us to determine the intensity of black-body radiation as a function of frequency. Summarizing the modern point of view, the entire photon field described as an operator electromagnetic four-potential field is just a superposition of plane waves of different wave vectors whose coefficients are creation and annihilation operators at the different wave vectors and a given state of the photon field is actually a linear combination of number states wherein a number state is specified by the number of photons that occupy each state of definite wave-vector/momentum and definitive helicity/spin.

A particular kind of photon state called a coherent state is a state constructed by an appropriate linear combination of this kind that turns out to be an eigen state for all the photon annihilation operators. This model for the quantum electromagnetic field implicitly contains both the particle nature and the wave nature of light. The particle nature is contained in the presence of the creation and annihilation operators, while the wave nature is contained in the plane waves that act as carrier signal fields for the creation and annihilation operators.

Second quantization means a quantization of a classical field theory. It is called second quantization for the following reasons: the first quantization is simply a wave equation like the three-dimensional wave equation, the Schrödinger wave equation for a single or an infinite number of quantum particles, the Klein Gordon equation or the Dirac equation. If such a classical wave equation is quantized, then it describes an infinite number of quantum particles. This can be seen clearly from the following example.

Take the Klein–Gordon equation that is the wave equation described before corresponding to Einstein's energy momentum relation with the energy and momenta replaced by appropriate operators. We expand the solution wave field as a three-dimensional Fourier series within a cube of side length L. The coefficient of this Fourier series then becomes quantum operators, each one of which and its adjoint, i.e., Hermitian conjugate describes a single quantum particle. Thus, a second quantized field can equivalently be described by a countable infinite number of quantum particles. For bosonic quantum fields like the Maxwell electromagnetic photon field, the state of the field can be described, for example by specifying how many particles are occupying each state of definite momentum and helicity or equivalently by specifying the momentum and helicity of each of the particles. In this case, there can be zero, one or more than one particle having a specified momentum and helicity. In the case of fermionic fields like the second quantized Dirac field, we cannot have more than one particle having a definitive value of momentum and spin [20,21].

9.3 Design and simulation of terahertz dielectric resonator antenna

Figure 9.1 shows the frequency spectrum and Figure 9.2 shows the design of terahertz rectangular antenna based on dimensions available in Table 9.1, respectively. An SPP phenomenon takes place at light–matter interaction and Gaussian beam is given as an input to feed. Silver nano waveguide has been used as feed. The resonant frequency of TRDRA can be computed based on the following formulations given, f_{mnl} is resonant frequency, where a, b, d are dimensions and m, n, l are indices in x, y, z directions, c is the velocity of light and ε_r is the permittivity of material used for radiating element:

$$f_{mn\ell} = \frac{ck_{mn\ell}}{2\pi\sqrt{\mu_r \epsilon_r}} = \frac{c}{2\pi\sqrt{\mu_r \epsilon_r}}\sqrt{\left(\frac{m\pi}{a}\right)^2 + \left(\frac{n\pi}{b}\right)^2 + \left(\frac{\ell\pi}{d}\right)^2}.$$

9.4 Synthesis of a terahertz rectangular DRA at optical frequency and its radiation theory

1. $[q_a, p_b] = \frac{ih}{2\pi}\delta(a,b)$

   where q is the position vector; p is the time-derivative moment vector; $b = 1/kT$ and T is temperature.
2. $\rho = \frac{\exp(-\beta H)}{T_r(\exp(-\beta H))}$; mean field energy in Gibbs state $= \rho$ is given by $T_r(\rho H)$ and the $T_r(\rho p_\epsilon)$ classically we would get

$$\frac{\int \exp(-\beta H) p_a d^N q d^N p}{\int \exp(-\beta H) d_q^N d_p^N} = 0$$

$a_k = \frac{a_k + ip_k}{\sqrt{2}}$ are annihilation and creation operators

$$a_k^* = \frac{a_k - ip_k}{\sqrt{2}}, \; k = 1, 2, 3$$

where $\left[a_k, a_m^*\right] = (h/2\pi)\delta(k, m)$ is the Heisenberg commutation relation.

Electromagneticfield $= H = \frac{1}{2}\sum_{k=1}^{N} a_k^* a_k + \frac{Nh}{4\pi}$; field energy.

Quantam states or eigen states of H

$$|\overline{n}\rangle = |n_1, n_2, n_3, \ldots, n_N\rangle, n_1, n_2, n_3, \ldots, \; n_N = 0, 1, 2, 3, \ldots$$

$$\mathrm{H}|\overline{n}\rangle = \frac{h}{2\pi}\left(\frac{N}{2} + n_1 + n_2 + \ldots + n_N\right)$$

Thus, the quantum a_k in the Gth state is

$$\exp\frac{(-\beta Nh)}{4\pi}\langle n|a_k\cdot\exp\left(\frac{(-\beta h)}{2\pi}\sum_m a_m^* a_m\right)$$

$$(n\rangle=\exp\frac{(-\beta Nh)}{4\pi}\sum\nolimits_{\overline{n}}\exp\left(\frac{(-\beta h)}{2\pi}\sum_m n_m\right)\delta(n,m-e_k)=0$$

$$e_k=[0,0,0,1,0]$$

Here

$$T_r\left(p_k^2\exp(-\beta h)\right)$$

$$=-\frac{\frac{1}{2}T_r\left[\left(p_k^2+p_k^{*2}-a_k a_k^*-a_k^* a_k\right)e^{-\beta h}\right]}{z_q(\beta)}$$

$$z_q(\beta)=\text{quantam partition function}$$

$$=\frac{1}{2z_q(\beta)}T_r\left[\left(a_k^* a_k+\frac{h}{4\pi}\right)\exp(-\beta h)\right]$$

$$=\frac{1}{2z_q(\beta)}\exp\left(\frac{(-\beta Nh)}{4\pi}\right)\sum\nolimits_{\overline{n}}\left(n_k+\frac{1}{2}\right)\frac{h}{2\pi}\exp\left(-\frac{\beta h}{4\pi}\sum_m n_m\right)$$

$$\int\exp\left(-\beta\sum_k\frac{(a_k^2+p_k^2)}{2}\right)a_k^2 d_q^N d_p^N/z_c(\beta)$$

where $z_c(\beta)$ is quantum part (Figures 9.11–9.14).

9.5 Mathematical analysis of resonant modes excited into a terahertz rectangular DRA

Consider a cavity resonator of one angstrom size, i.e., a cube with each side of length $a=10^{-10}$ m. The Maxwell equations in such a cube have solutions of the

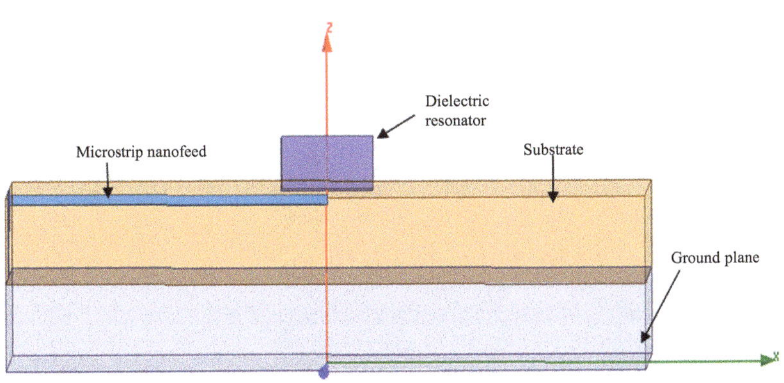

Figure 9.11 Optical RDRA at 484 THz

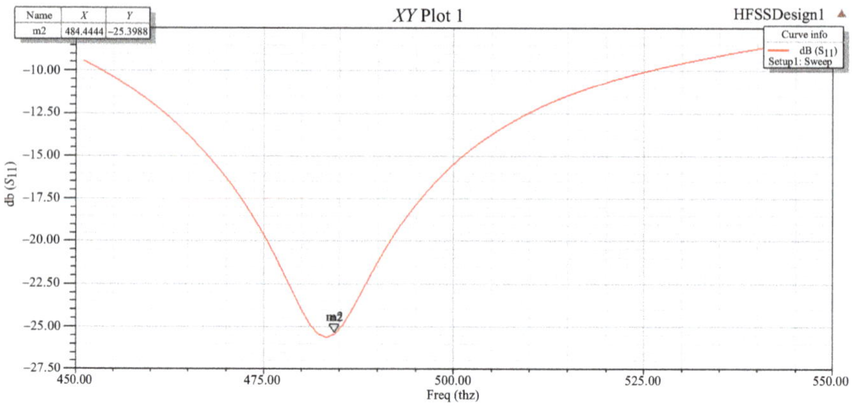

Figure 9.12 S_{11} at 484 THz

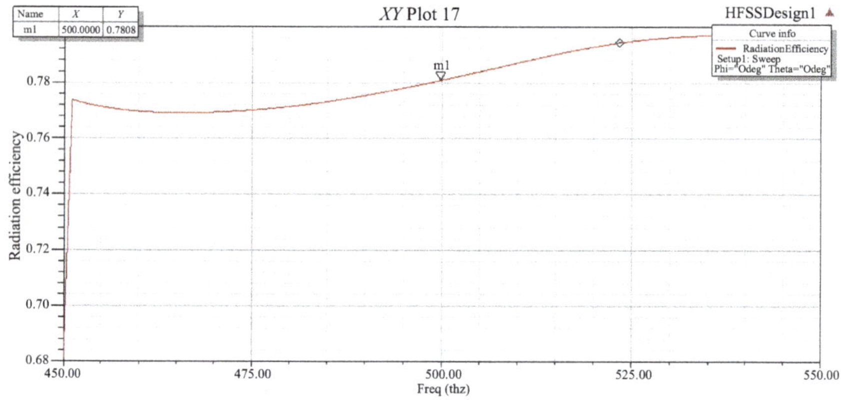

Figure 9.13 Radiation efficiency at 484 THz

from

$$A_r(t,x,y,z) = \sum_{mnp} c(mnp,t)u_r, mnp(x,y,z), r = 1,2,3 \tag{9.1}$$

where u_r, mnp are spatial functions obtained by integrating the electric field w.r.t.

$$\left\{\cos\left(\frac{m\pi x}{a}\right), \sin\left(\frac{m\pi x}{a}\right)\right\} \otimes \left\{\cos\left(\frac{n\pi y}{a}\right), \ \sin\left(\frac{n\pi y}{a}\right) \otimes \left\{\cos\left(\frac{p\pi y}{a}\right), \sin\left(\frac{p\pi z}{a}\right)\right\}\right.$$

Multiplied by some constants depending on the indies (m,n,p). We may, without loss of generality, assume those as normalized so that

$$\int_C u_r, mnp(r)\bar{u}_s, m'n'p'(r)d^3r = \delta_{rs}\delta_{mm'}\delta_{nn'}\delta_{pp'}$$

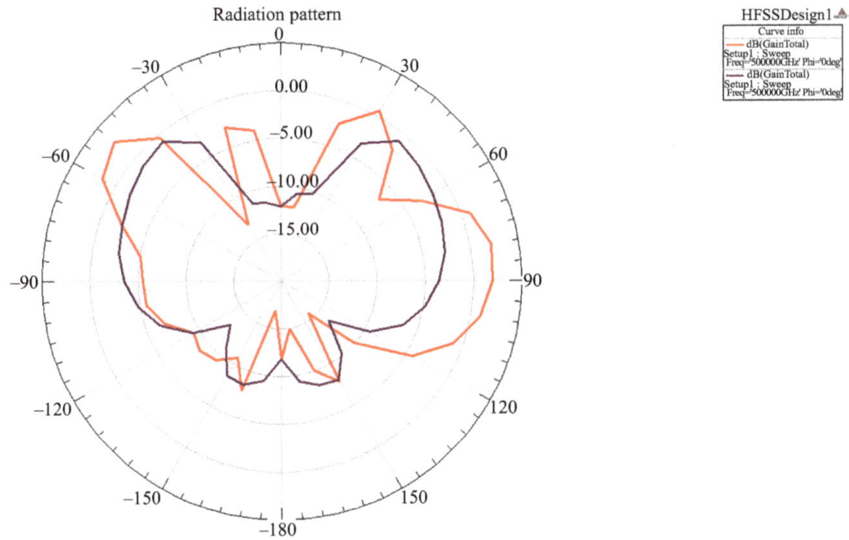

Figure 9.14 Radiation pattern of TRDRA at 484 THz

The dependence of $c(mnp,t)$ on t is $\exp(i\omega(mnp)t)$ where $\omega(mnp)$ are the characteristic frequencies of oscillation:

$$\omega(mnp) = \left(\frac{\pi c}{a}\right)\sqrt{m^2+n^2+p^2},\, m,n,p = 1,2 \tag{9.2}$$

which are of the order of magnitude:

$$\omega = \frac{\pi c}{a}$$

The electric field is

$$E_r = \partial_t A_r = \sum_{mnp} c(mnp,t)i\omega(mnp)u_r, mnp(r)$$

The magnetic field is $B = \text{curl}A$

Which is of the order of magnitude $|c(mnp,t)|/a$ where by $c(mnp,t)$ we actually mean that it is an average in a coherent state. The total electric field energy within the cavity C is

$$U_E = \left(\frac{\epsilon_0}{2}\right)\int_C |E|^2 d^3r$$

which has components of the order of magnitude

$$\frac{\epsilon_0}{\omega}(mnp)c(mnp,t)|^2 a^3 = \epsilon_0\omega(mnp)^2 a^3 |c(mnp,t)|^2$$

The total magnetic field energy within the cavity is

$$U_B = (2\mu_0)^{-1}\int_c |B|^2 d^3r$$

which have components of the order of magnitude

$$\left|\frac{c(mnp,t)}{a}\right|^2 \frac{a^3}{\mu_0} = |c(mnp,t)|^2 \frac{a}{\mu_0}$$

The ratio of the orders of magnitude of the electric field energy and the magnetic field energy within the cavity therefore has the order of magnitude:

$$\frac{U_E}{U_B} \approx \mu_0\epsilon_0\omega(mnp)^2 a^2 \approx \frac{\omega^2 a^2}{c^2} \approx 1$$

As expected, the canonical commutation relations are

$$\left[A_r(t,r), \partial_t A_s\left(t,r'\right)\right] = \left(\frac{ih}{2\pi}\right)\delta^3\left(r-r'\right) \tag{9.3}$$

These fields are

$$\left[c_r(mnp,t), \omega(mnp)c_s(m'n'p',t)^*\right] = \left(\frac{h}{2\pi}\right)\delta_{rs}\delta_{mm'}\delta_{nn'}\delta_{pp'}$$

So that the eigenvalues of $c_r(mnp,t)^* c_r(mnp,t)$ are positive integer multiples of $h/2\pi\omega(mnp)$. This means that the field energy within the cavity, when a finite number of modes are excited, assumes eigenvalues that are of the same order of magnitude as positive integer multiples of $h\omega/2\pi$ as expected by Planck's quantum theory of radiation. This fact also yields the result that $|c(mnp,t)|$ is of the order of magnitude of $\sqrt{h/(2\pi\omega)}$.

Now we come to the question of computing the order of the magnitude of Poynting vector power flux at a given radial distance R from the quantum cavity antenna caused by the surface current density induced by the magnetic field on the antenna surface. The magnetic field on the surface and hence the corresponding induced surface current density both have the order of magnitudes of $|c(mnp,t)|/a$ which is of the order of $a^{-1}\sqrt{h/\omega}$. Therefore, the far-field magnetic vector potential at a distance of R from the cavity is of the order of magnitude (use the retarded potential formula) $(a/R)\sqrt{h/\omega}$ and hence the corresponding far-field radiated magnetic field is of the order of magnitude $(\omega/c)(a/R)\sqrt{h/\omega}$ while the near-field magnetic field of the order of magnitude $(a/R^2)\sqrt{h/\omega}$. Actually, these expressions for the magnetic field must be multiplied by $\sqrt{N}$ where N is a positive integer corresponding to the largest modal eigenvalue of the operators $(2\pi\omega(mnp)/h)c(mnp,t)^* c(mnp,t)$.

The far-field Poynting vector has the order of magnitude of $B^2c/2\mu_0$ which is of the following order:

$$\left(\frac{c}{2\mu_0}\right)\left(\sqrt{N}\cdot\left(\frac{\omega}{c}\right)\left(\frac{a}{R}\right)\cdot\sqrt{\frac{h}{\omega}}\right)^2=\left(\frac{h}{2\mu_0}\right)N\cdot\left(\frac{\omega}{c}\right)\left(\frac{a^2}{R^2}\right)$$

And the total power radiated outward by this quantum antenna in the far-field zone is thus of the order of magnitude:

$$P=N=\left(\frac{h}{2\mu_0}\right)\left(\frac{a^2\omega}{c}\right)$$

Now we look at the order of magnitude of the power radiated in the far-field zone by the Dirac field of electrons and positrons within the cavity. The Dirac equation is

$$\left[i\gamma^\mu\partial_\mu - m\right]\psi(x)=0$$

Or more precisely in arbitrary units,

$$\left[\left(\frac{ih}{2\pi}\right)\partial_t - c\left(\alpha,\left(-\frac{ih}{2\pi}\right)\nabla\right)-\beta mc^2\right]\psi(x)=0$$

Here, the appearances of the constants h, m, c are explicitly shown. Now the $|\psi(x)|^2c$ is the probability density of the electron that must integrate to unity over the cavity volume. Thus $\psi(x)$ is of the order of magnitude $a^{-3/2}$. The Dirac current density $J^\mu = e\psi^*\gamma^0\gamma^\mu\psi$ has the same order of magnitude as $e|\psi(x)|^2c$ that is $ec/a^{3/2}$. Therefore, the far-field magnetic vector potential at a radial distance of R from the cavity is, in accordance with the retarded potential theory, of the order

$$\left(\frac{ec}{a^{3/2}}\right)\cdot\left(\frac{a^3}{R}\right)=\frac{eca^{3/2}}{R}$$

The electric field in the far-field zone is then of the order

$$E\approx\omega\cdot\frac{eca^{3/2}}{R} \tag{9.4}$$

where ω is the characteristic oscillation frequency of the Dirac current. The magnetic field is of the order

$$B\approx a^{-1}\cdot\frac{eca^{3/2}}{R}=ec\sqrt{\frac{a}{R}}$$

If P is the characteristic momentum of the electrons and positrons in a given state, for example P may be the average momentum of an electron in a given state, then according to de Broglie, P is of the order h/a since a is the order of the electron wavelength. Then the electron energy is of the order of

$$E_e=c\sqrt{m^2c^2+P^2}\approx c\sqrt{m^2c^2+\frac{h^2}{a^2}} \tag{9.5}$$

And the characteristic frequency of oscillation of the Dirac wave field is then given by

$$\omega = \frac{E_e}{h}$$

The Poynting vector corresponding to the power radiated by the Dirac field in the far-field zone then has the order of magnitude:

$$S \approx c\left(\epsilon_0 E^2 + \frac{B^2}{\mu_0}\right) = c^3 \epsilon_0 \omega^2 e a^3 + \frac{e^2 c^3 a}{\mu_0} \tag{9.6}$$

Equations (9.1)–(9.6) have been derived for formulating rectangular quantum or optical antennas.

9.6 Terahertz optical RDRA at 484 THz

The application of optical antennas in photodetectors is particularly promising. The main reason is that an optical antenna increases the absorption cross-section and hence the light flux that impinges on a detector, giving it a clear signal-to-noise advantage.

There is an upper limit to the performance of optical components. A minimum volume must be ensured to guarantee the necessary number of modes for information to be processed. The limit is valid for multiple scattering events and for linear interactions. The minimum volume scales with the dielectric constant of a material, and hence it can be reduced by using metallic structures such as optical antennas (Figure 9.15).

9.6.1 Approximate analysis of a rectangular terahertz DRA and its controlled electromagnetic fields

The TDRA is assumed to be the cuboid shape; let region be defined as $[0, a] \times [0, b] \times [0, d]$. In the rectangular cavity of a, b and d dimensions, it is assumed to be comprising photons initially due to feed and electrons and positrons

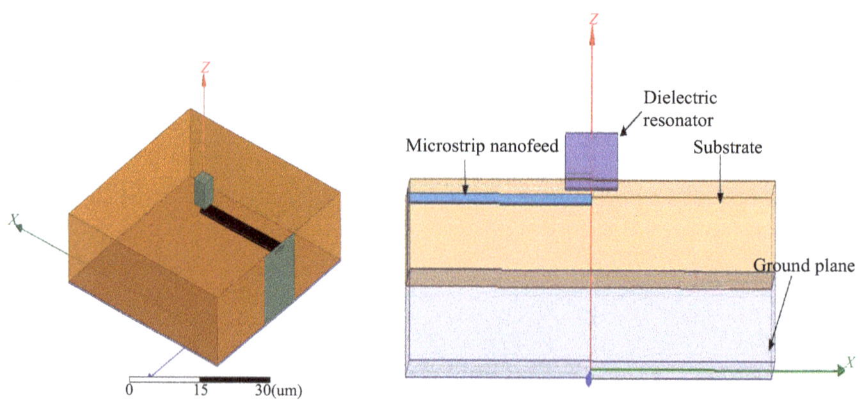

Figure 9.15 Optical rectangular DRA top and side view

due to substrate or DRA materials. Light–matter interaction takes place. The mathematical model is developed for this cavity under light–matter interaction. The exact equations governing the quantum fields corresponding to these particles are (a) the Maxwell equations for the four-vector potential driven by the Dirac field current and (b) the Dirac field equations driven by an interaction between the Dirac field and the Maxwell field four vector potential. Value and correlations in a given quantum coherent state of the photons and electrons–positrons within the cavity are close to specified values.

These exact field equations are

$$A_\mu(x) = \mu_0 e\psi^*(x)\alpha^\mu\psi(x), x$$

$$= (t,r)((\alpha, -i\nabla) + \beta m_0)\psi(t,r) + eA_\mu(x)\alpha^\mu\psi(x) = i\partial_t\psi(t,r)$$

where

$$\alpha^\mu = \gamma^0\gamma^\mu, \beta = \gamma^0$$

And γ^μ are the Dirac-gamma matrices. Note that $(\gamma^0)^2 = I_4$ and hence $\alpha^0 = I_4$. The boundary conditions under which we need to solve these Maxwell–Dirac equations are that the Dirac operator wave field $\psi(x)$, the tangential components of the electric field $F_{or} = A_{r,0} - A_{0,r'} r = 1,2,3$ and the normal components of the magnetic field $F_{rs} = A_{s,r} - A_{r,s}, 1 \leq \bar{r} < s \leq 3$ must vanish on the boundaries of the cavity. In particular, the free Dirac field must have an expansion as

$$\psi^{(0)}(t,r) = \sum_{mnp} c(mnp,t)\mu_{mnp}(r)$$

where m,n,p run over positive integers and

$$\mu_{mnp}(r) = \left(\frac{2\sqrt{2}}{\sqrt{abd}}\right)\sin\left(\frac{m\pi x}{a}\right)\sin\left(\frac{n\pi y}{b}\right)\sin\left(\frac{p\pi z}{d}\right)$$

Substituting this into the free Dirac equation, without any electromagnetic interactions, we get

$$\sum_{mnp}\left(i\partial_t c(mnp,t)\mu_{mnp}(r)\right) = ((\alpha, -i\nabla) + \beta m_0).\sum_{mnp} c(mnp,t)\mu_{mnp}(r)$$

From which we derive on taking the inner products on both sides with $u_{kls}(r)$ and using the orthonormality of this set of functions over the cavity volume:

$$\langle u_{kls}, u_{mnp}\rangle = \int_B u_{kls}(r)u_{mnp}(r)d^3r = \delta_{Km}\delta_{ln}\delta_{sp}$$

where B is the cavity volume

$$B = [0, a] \times [0, b] \times [0, d],$$

The following sequence of differential equations:

$$\begin{aligned} i\partial_t c(kls, t) &= \sum_{mnp} [\langle u_{kls} - i\partial_x u_{mnp}\rangle \alpha_1 c(mnp, t) + \langle u_{kls}, -i\partial_y u_{mnp}\rangle \alpha_2 c(mnp, t) \\ &\quad + \langle u_{kls}, -i\partial_z u_{mnp}\rangle \alpha_3 c(mnp, t)] + m_0 \beta c(kls, t) \end{aligned}$$

Now we evaluate

$$\begin{aligned} \langle u_{kls} - i\partial_x u_{mnp}\rangle - i\delta_{ln}\delta_{sp}\left(\frac{m\pi}{a}\right) \int_0^a \left(\frac{2}{a}\right) \sin\left(\frac{k\pi x}{a}\right) \cos\left(\frac{m\pi x}{a}\right) dx \\ = a_1(k, m)\delta_{ln}\delta_{sp} \end{aligned}$$

where

$$a_1(k, m) = \left(-\frac{2im\pi}{a^2}\right) \int_0^a \sin\left(\frac{k\pi x}{a}\right) \cos\left(\frac{m\pi x}{a}\right) dx$$

Likewise,

$$\langle u_{kls,} - i\partial_y u_{mnp}\rangle = a_2(l, n)\delta_{km}\delta_{sp},$$

And

$$\langle u_{kls,} - i\partial_z u_{mnp}\rangle = a_3(s, p)\delta_{km}\delta_{ln}$$

Combining all these equations gives us finally,

$$\begin{aligned} i\partial_t c(kls, t) &= \sum_m a_1(k, m)\alpha_1 c(mls, t) + \sum_n a_2(l, n)\alpha_2 c(kns, t) \\ &\quad + \sum_p a_3(s, p)\alpha_3 c(klp, t) + m_0\beta c(kls, t) \end{aligned}$$

Note that $\alpha_1, \alpha_2, \alpha_3, \beta$ are 4×4 Hermitian matrices while $c(mnp, t)$ is a 4×1 complex vector. Arranging the 4×1 vectors $c(mnp, t) m, n, p \geq 1$ in lexicographic order to give an infinite vector $c(t)$ and likewise defining a block structured infinite-dimensional Dirac Hamiltonian matrix H_0 by

$$H_0 = \sum_{klsm} a_1(k,m)(I_4 \otimes e(kls))\alpha_1(I_4 \otimes e(mls)^T)$$

$$\sum_{klsm} a_2(l,n)(I_4 \otimes e(kls))\alpha_2(I_4 \otimes e(kns)^T)$$

$$\sum_{klsp} a_3(s,p)(I_4 \otimes e(kls))\alpha_3(I_4 \otimes e(klp)^T) + m_0\beta \otimes I$$

where we may choose $e(mnp), m, n, p \geq 1$ as any orthonormal basis for $l^2(\mathbb{Z}_+)$, the Hilbert space of all one-sided square summable infinite sequences and define

$$c(t) = \sum_{mnp} c(mnp, t)e(mnp)$$

By orthonormal, we mean that

$$e(kls)^T e(mnp) = \delta_{km}\delta_{ln}\delta_{sp}$$

Thus, the free Dirac equation in the RDRA has been put in "standard" block matrix form:

$$i\frac{dc(t)}{dt} = H_0\boldsymbol{c}(t)$$

The general solution to which can be expressed as

$$\boldsymbol{c}(t) = \sum\nolimits_n d(n) \cdot \boldsymbol{c}_n \exp(-iE(n)t)$$

where $\boldsymbol{c}_n, n \geq 1$ forms an orthonormal basis for $l^2(\mathbb{Z}_+)$ and the $d(n)'$s are arbitrary complex numbers such that

$$\sum_n |d(n)|^2 = 1$$

$E(n)^{'}$s are the (energy) eigenvalues of infinite-dimensional Hermitian H_0:

$$\det(H_0 - E(n)\mathbf{I}) = 0$$

The average energy of the free Dirac field of electrons and positrons within the cavity is then

$$\langle \boldsymbol{c}(t), H_0\boldsymbol{c}(t)\rangle = \sum\nolimits_n E(n)d(n)^* d(n)$$

It is easy to see as in the case of the Dirac equation in free space that if $E(n)$ is an eigenvalue of H_0 then so is $-E(n)$ where the $E(n)'s$ may be taken as positive, Hence if $\boldsymbol{c}_{en}$ is an eigenvector of H_0 corresponding to the eigenvalue $-E(n)$ and $\boldsymbol{c}_{pn}$

is an eigenvector corresponding to the eigenvalue $-E(n)$, then the solution can be expressed as

$$c(t) = \sum_n \left[d_e(n)c_{en} \exp(iE(n)t) + d_p(n)^* c_{pn} \exp(iE(n)t)\right]$$

Therefore, it is plausible in the second quantized theory to look upon the $d_e(n)'s$ as annihilation operators of the electrons and the $d_p(n)^{*'}s$ as the creation operators of the positrons. The actual Dirac wave function $\psi(t,r)$ in the absence of electromagnetic interactions is then

$$\psi(t,r) = \sum\nolimits_{kmnp} \left[d_e(k)c_{ek}(mnp)u_{mnp}(r)\exp(-iE(k)t)\right.$$

$$\left. +d_p(k)^* c_{pk}(mnp)u_{mnp}(r)\exp(iE(k)t)\right]$$

A simple calculation then shows that the second quantized Hamiltonian of the free Dirac field of electrons and positrons within the cavity is given by

$$\begin{aligned} H_{D0} &= \int_B \psi(t,r)^*((\alpha, -i\nabla) + \beta_m)\psi(t,r)d^3r \\ &= \int_B \psi(t,r)^* i\partial_t \psi(t,r)d^3r \\ &= \sum_k E(k)\left(d_e(k)^* d_e(k) - d_p(k)d_p(k)^*\right) \end{aligned}$$

Now from the basic anticommutation relations for the Dirac field, we have

$$\{\psi(t,r), \psi(t,r')^*) = \delta^3(r-r')I\}$$

And this immediately implies the following anticommutation relations for the electron and positron creation and annihilation operators:

$$\{d_e(k)d_e(m)^*\} = \delta_{km}, \{d_p(k), d_p(m)^*\} = \delta_{km}$$

with all the other anticommutators vanishing. This completes our description of the free Dirac field of electrons and positrons within the RDRA. Using these anticommutation relations, we immediately get that the scalar potential is zero since there are no charges while the magnetic vector potential admits and expansion obtained from

$$\mathbf{E} = -\partial_t A$$

As

$$A(t,r) = \sum\nolimits_k [b(k)w_k(r) \exp(-i\omega(k)t) + b(k)^* w_k(r)^* \exp(i\omega(k)t))$$

where now $w_k(r)$ has three components that are calculated from the expansion of E_z and the relationship between the transverse and longitudinal components of the electric field within the cavity. We note that the third, i.e., z component of $w_k(r)$ where the index k is identified with the modal triplet (mnp) is proportional to

$$\sin\left(\frac{m\pi x}{a}\right)\sin\left(\frac{n\pi y}{b}\right)\cos\left(\frac{p\pi z}{d}\right)$$

in view of the boundary conditions on the electric field is the fact that each mode of the magnetic vector potential is proportional to the electric field $(-j\omega\mathbf{A}=\mathbf{E})\cdot b(k)=b(mnp)$ that is identified with a photon annihilation operator while $b(k)^*$ with a photon creation operator. They satisfy the canonical commutation relations

$$[b(k),b(m)^*]=\delta_{km}$$

Formally, we can compute both the free Dirac current density $\psi(t,r)^*\alpha^\mu\psi(t,r)$ of electrons and positrons with the cavity and the surface current density on the RDRA walls induced by the tangential components of the quantum magnetic $\boldsymbol{B}=\text{curl}\,\boldsymbol{A}$ and obtain the far-field radiation pattern generated by both of these cavity current components. Obviously, this far-field radiation pattern will have its first component being a quadratic form in the electron–positron creation and annihilation operators $d_e(k),d_p(k),d_e(k)^*,d_p(k)^*$ while the second component will be linear in the photon creation–annihilation operators $b(k),b(k)^*$ and therefore, in principle, we can compute all the statistical moments of the radiation field in a joint coherent state of the photons, electrons and positrons. However, this picture of the far-field quantum radiation pattern is incomplete because it does not take into account the cavity surface current density terms caused by perturbation in the Maxwell field caused by its interaction with the Dirac field. We shall now indicate an approximate first-order calculation between the Maxwell field and the Dirac field.

We denote the free Dirac field within the cavity derived before by $\psi^{(0)}(t,r)$ and the corresponding momentum space wave function $c(mnp\cdot t)$ by $c^{(0)}(mnp\cdot t)$. Likewise, we denote the free Maxwell field within the cavity by $\boldsymbol{A}^{(0)}$. Let $\delta\boldsymbol{A}$ denote the perturbation to the Maxwell field caused by the Dirac current and $\delta\psi,\delta c(mnp,t)$ the perturbation to the Dirac field caused by the Maxwell current. Then, clearly if $S(x-y)$ denotes the electron propagator and $D(x-y)$, the photon propagator, we have been using (9.1) and (9.2), approximately,

$$\delta A^\mu(t,r)=\mu_0 e\int D(t-t',r-r')\psi^{(0)*}(t',r')\alpha^\mu\psi^{(0)}(t',r')dt'd^3r'$$

$$\delta\psi(t,r)=e\int S(t-t',r-r')A_\mu^{(0)}(t',r')\alpha^\mu\psi^{(0)}(t',r')dt'd^3r'$$

where we substitute for $\psi^{(0)}$ and $A_\mu^{(0)}$ the expressions given in (9.3) and (9.4). Then,

$$\psi^{(0)*}\alpha^\mu\psi^{(0)}(t,r) = \sum_{kmnpk'm'n'p} [d_e(k)^*\bar{c}_{ek}(mnp)\exp(iE(k)t)$$

$$+ d_p(k)\bar{c}_{pk}(mnp)\exp(-iE(k)t))] \cdot \alpha^\mu$$

$$\cdot[d_e(k')c_{ek'}(m'n'p)\exp(-iE(k')t)$$

$$+d_p(k')^*c_{pk}(m'n'p)\exp(iE(k')t))u_{mnp}(r)u_{m'n'p'}(r)$$

$$= \sum [d_e(k)^*d_e(k')\bar{c}_{ek}(mnp)\alpha^\mu c_{ek'}(m'n'p')$$

$$\exp(iE(k) - E(k')t)u_{mnp}(r)u_{m'n'p'}(r)]$$

$$+\sum [d_e(k)^*d_p(k')\bar{c}_{ek}(mnp)\alpha^\mu c_{pk'}(m'n'p')$$

$$\exp(iE(k) + E(k')t)u_{mnp}(r)u_{m'n'p'}(r)]$$

$$\sum [d_p(k)d_e(k')\bar{c}_{pk}(mnp)\alpha^\mu c_{ek'}(m'n'p')$$

$$\exp(-iE(k) + E(k')t)u_{mnp}(r)u_{m'n'p'}(r)]$$

$$\sum [d_p(k)d_p(k)^*\bar{c}_{pk}(mnp)\alpha^\mu c_{ek'}(m'n'p')$$

$$\exp(-iE(k) - E(k')t)u_{mnp}(r)u_{m'n'p'}(r)]$$

We see that the frequencies of the Dirac current that generate the perturbation to the quantum electromagnetic field are $E(k) \pm E(k'), k, k' = 1, 2, \ldots$ or more precisely, these are divided by Planck's constant. Here $E(k)$ was obtained by solving the free Dirac eigenvalue equation inside the rectangular cavity with zero boundary conditions. The $E(k)'s$ were obtained as the eigen values of the Dirac Hamiltonian. From basic principles of special relativity, it is easy to see that these $E(k)'s$ are of the order

$$c\sqrt{m_0^2c^2 + P^2}$$

where

$$P^2 = \left(\frac{h}{2\pi}\right)^2 \left(\frac{m\pi}{a}\right)^2 + \left(\frac{n\pi}{b}\right)^2 + \left(\frac{p\pi}{d}\right)^2$$

with m, n, p being positive integers determined by the mode of oscillation of the field within the cavity. Now this current is of the general form

$$\psi^{(0)*}\alpha^{\mu}\psi^{(0)}(t,r)\sum_{k,k'}\left[d(k)^* d(k')f_{kk,}(t,r) + d(k)d(k')g_{kk,}(t,r) + d(k)^* d(k')^* \bar{g}_{k,}(t,r)\right]$$

where the $d(k)'s$ are the annihilation operators of the electrons and positrons and their adjoints $d(k)^*$ are the corresponding creation operators. The functions $f_{kk'}g_{kk'}$ are components that are easily seen to be expressible as superpositions of space–time sinusoids with the temporal frequencies being exp $(\pm i(E(k) \pm E(k'))u_{mnp}(r)u_{m'n'p'}(r)$ and these components are easily seen to be expressible as superpositions of space–time sinusoids with the temporal frequencies being $E(k) \pm E(k')$ or their negatives and the spatial frequencies, i.e., wave numbers being $m\pi/a, n\pi/b, p\pi/d, m'\pi/a, p'\pi/b, p'\pi/d$.

Let

$$\begin{aligned} J^{\mu}(k) &= e\int \psi^{(0)*}(x)\alpha^{\mu}\psi^{(0)}(x)\exp(-ik \cdot x)d^4x \\ &= \int \psi^{(0)*}(t,r)\alpha^{\mu}\psi^{(0)}(t,r)\exp\left(-i(k^0 t - K \cdot r)\right)dtd^3r \end{aligned}$$

where

$$k = (k^{\mu}) = (k^0, K)$$

denotes the space–time four-dimensional Fourier transform of the unperturbed Dirac four current density. Then, we can write down the space–time Fourier transform of the correction $\delta A^{\mu}(x), x = (t, r)$ to the electromagnetic four potential caused by this Dirac current as

$$\begin{aligned} \delta A^{\mu}(k) &= \int \delta A^{\mu}(x) \cdot \exp(-ik \cdot x)d^4x \\ &= \mu_0 J^{\mu}(k)/k^2, k^2 = k_{\mu}k^{\mu} = (k^0)^2 - |K|^2 \end{aligned}$$

in units where $c = 1$. It should be noted that by the convolution theorem for Fourier transforms, if $\psi^{(0)}(k)$ denotes the space–time Fourier transform of $\psi^{(0)}(x)$, then

$$J^{\mu}(k) = (2\pi)^{-4}\int \psi^{(0)*}(k' - k)\alpha^{\mu}\psi^{(0)}(k')d^4k'$$

and hence, the perturbation to the electromagnetic four potential in the space–time Fourier domain, i.e. in four momentum space of the photon, can be expressed as

$$\delta A^{\mu}(k) = \left(\frac{\mu_0}{(2\pi)^4 k^2}\right)\int \psi^{(0)*}(k'-k)\alpha^{\mu}\psi^{(0)}(k')d^4k'$$

The unperturbed electromagnetic fields is in the Coulomb gauge, i.e., $\operatorname{div} A^{(0)} = 0$ and also since there is no charge/current for the unperturbed field, the unperturbed electric scalar potential is a matter field that is identically zero, i.e., $A^{(0)0} = 0$. Hence we are guaranteed that the unperturbed electromagnetic potentials also satisfy the Lorentz gauge conditions, i.e., $\operatorname{div} A^{(0)} + \partial_t A^{(0)0} = 0$. This means that while computing the perturbations to the electromagnetic potentials caused by currents coming from the Dirac field, we can safely work in the Lorentz gauge. Likewise, the change in the Dirac field caused by interaction with the electromagnetic field within the cavity is given up to first-order perturbation theory by

$$\begin{aligned}
\delta\psi(x) = \delta\psi(x) &= e\int S(x-x')A_{\mu}^{(0)}(x')\alpha^{\mu}\psi^{(0)}(x')d^4x' \\
&= e\int S(x-x')A_{r}^{(0)}(x')\alpha^{r}\psi^{(0)}(x')d^4x' \\
&= -e\int S(x-x')\left(\alpha, A^{(0)}(x')\right)\psi^{(0)}(x')d^4x' \\
&= -e\int S\left(t-t', r-r'\right)\sum_k [b(k)\left(\alpha, \boldsymbol{w}_k\left(r'\right)\right)\exp(-i\omega(k)t') + b(k)^*\left(\alpha, \boldsymbol{w}_k\left(r'\right)^*\right)\exp\left(i\omega(k)t'\right)] \\
&\cdot\left[\sum_{kmnp}\left(d_e(k)c_{ek}(mnp)u_{mnp}\left(r'\right)\exp(-iEk)t'\right) + d_p(k)^* c_{pk}(mnp)u_{mnp}\left(r'\right)\exp\left(iE(k)t'\right)\right]dt'd^3r' \\
&-e\sum_{kk'mnp} b(k)d_e\left(k'\right)\int S\left(t-t', r-r'\right)\left(\alpha, \boldsymbol{w}_k\left(r'\right)\right)c_{ek'}(mnp)u_{mnp}\left(r'\right)\exp\left(-i\left(\omega(k)+E\left(k'\right)\right)t'\right)dt'd^3r' \\
&-e\sum_{kk'mnp} b(k)d_p\left(k'\right)^*\int S\left(t-t', r-r'\right)(\alpha, \boldsymbol{w}_k(r')^*)c_{pk'}(mnp)u_{mnp}\left(r'\right)\exp\left(-i\left(\omega(k)-E\left(k'\right)\right)t'\right)dt'd^3r' \\
&-e\sum_{kk'mnp} b(k)^* d_e(k)\int S\left(t-t', r-r'\right)(\alpha, \boldsymbol{w}_k(r'))c_{ek'}(mnp)u_{mnp}\left(r'\right)\exp\left(i\left(\omega(k)-E\left(k'\right)\right)t'\right)dt'd^3r' \\
&-e\sum_{kk'mnp} b(k)^* d_p(k')^*\int S\left(t-t', r-r'\right)\left(\alpha, \boldsymbol{w}_k(r')^*\right)c_{pk'}(mnp)u_{mnp}\left(r'\right)\exp\left(i\left(\omega(k)+E\left(k'\right)\right)t'\right)dt'd^3r'
\end{aligned}$$

From this expression, it is clear that the characteristic frequencies of the interaction term between the electromagnetic potentials and the Dirac field and hence the characteristic frequencies of the perturbation in the Dirac field caused by electromagnetic interaction are $\pm\omega(k) \pm E$. In terms of the compact notation introduced earlier, namely using the same symbol $d(k)$ for both electron and positron annihilation operators and likewise $d(k)^*$ for both electron and positron creation operators, we can write

$$\begin{aligned}
\delta\psi(x) = \int S\left(t-t', r-r'\right)&\left[\sum b(k)d\left(k'\right)h_1kk'\left(t', r'\right)\right. \\
&+ b(k)d\left(k'\right)^* h_2kk'\left(t', r'\right) + b\left(k\right)^* d\left(k'\right)h_3kk'\left(t', r'\right) \\
&\left.+ d\left(k'\right)^* d\left(k'\right)h_4kk'\left(t', r'\right)\right]dt'd^3r'
\end{aligned}$$

where the functions $h_m kk^{'}(t^{'}, r^{'})$ are built by superposing the functions $\exp(i \neq (\omega(k) \pm E(k^{'}))t)(\alpha, w_k(r)c_{k^{'}}(mnp)u_{mnp}(r))$ and the same expression with $w_k(r)$is replaced by its complex conjugate $w_k(r)^*$. Here, the symbol $c_{k^{'}}(mnp)$ stands for either $c_{ek^{'}}(mnp)\, or\, c_{pk^{'}}(mnp)$.

In particular, this expression shows that the perturbation to the Dirac field caused by electromagnetic interactions has frequencies $\pm\omega(k) \pm E(k^{'})$, namely linear combinations of the unperturbed electromagnetic characteristic frequencies and the unperturbed Dirac characteristic frequencies. This represents a new feature of our model.

Before proceeding further, observe that we can write in the dimensional momentum/space–time frequency domain:

$$\delta\psi(k) = S(k)\mathscr{F}\left(eA_{(\mu)}^{(0)}\alpha^{\mu}\psi^{(0)}\right)(k)$$

where

$$S(k) = \left(k^0 - (\alpha, \boldsymbol{K}) - \beta m_0 + i0\right)^{-1}$$

is the electron propagator in the four-momentum domain $k = (k^{\mu}) = (k^0, \boldsymbol{K})$and control of the quantum electromagnetic field and the Dirac field of electrons and positrons within the rectangular cavity by means of a classical electromagnetic field coming from a probe inserted into the cavity: let $A_{(\mu)}^{c}(x)$ denote the classical electromagnetic four potential from the laser and $J_{(\mu)}^{c}(x)$ be the classical current density coming from the probe insertion. The relevant equations are

$$A_{\mu} = -e\mu_0\psi^*\alpha_{\mu}\psi + \mu_0 J_{(\mu)}^{c}, ((\alpha, -i\nabla) + \beta m)\psi = [-e(\alpha, \boldsymbol{A}) - e(\alpha, \boldsymbol{A}^c)]\psi$$

The first-order perturbative solution to these equations with $x = (t, r)$ is

$$\begin{aligned} \psi(x) &= \psi^{(0)}(x) + \delta\psi(x), \\ A^r(x) &= A^{r(0)}(x) + \delta A^r(x), r = 1, 2, 3, \end{aligned}$$

where

$$\psi^{(0)}(x) = \sum_{kmnp} \left[d_e(k)c_{ek}(mnp)u_{mnp}(r)\exp\left(-iE(k)t\right)\right.$$

$$\left. + d_p(k)^* c_{pk}(mnp)u_{mnp}(r)\exp\left(iE(k)t\right)\right]$$

$$A^{r(0)}(x) = \sum_{k} \left[b(k)\omega_k^r(r)\exp\left(-iw(k)t\right) + b(k)^*\omega_k^{-r}(r)\exp\left(i\omega(k)t\right)\right]$$

$$
\begin{aligned}
\delta\psi(x) &= -e\int S_e(x-y)\left[\left(\alpha, \boldsymbol{A}^{(0)}(y)\right) + (\alpha, \boldsymbol{A}^c(y))\right]\psi^{(0)}(y)d^4y \\
&= -e\int S_e(x-y)\left[\alpha^r A_r^{(0)}(y) + \alpha^r A_r^c(y)\right]\psi^{(0)}(y)d^4y \\
&= \delta\psi_1(x) + \delta\psi_{ctr}(x), \\
\delta A_r(x) &= -e\mu_0\int D(x-y)(\psi^*\alpha_r\psi)(y)d^4y + \mu_0\int D(x-y)J_r^c(y)d^4y \\
&\approx -e\mu_0\int D(x-y)\left(\psi^{(0)*}\alpha_r\psi^{(0)}\right)(y)d^4y + \mu_0\int D(x-y)J_r^c(y)d^4y
\end{aligned}
$$

where the classically controllable part of the Dirac field as

$$
\psi_{ctr}(x) = -e\int S_e(x-y)\alpha^r A_r^{(0)}(y)\psi^{(0)}(y)d^4y
$$

And this component contains a classical field component A_r^c and a quantum field component $\psi^{(0)}$, while the part of the Dirac field perturbation that is not controllable is

$$
\delta\psi_1(x) = -e\int S_e(x-y)\alpha^r A_r^{(0)}(y)\psi^{(0)}(y)d^4y
$$

On the other hand, the controllable part of the electromagnetic field is purely classical:

$$
\delta A_{r,ctr}(x) = \mu_0\int D(x-y)J_r^c(y)d^4y
$$

If we go one step further in the perturbation series, then we get an additional term in the controllable part of the electromagnetic field so that the previous equation gets modified to

$$
\begin{aligned}
\delta A_{r,ctr}(x) = \mu_0&\int D(x-y)J_r^c(y)d^4y \\
-e\mu_0&\int D(x-y)\delta\psi_{ctr}(y)^*\alpha_r\left(\psi^{(0)} + \delta\psi_1\right)(y)d^4y \\
-e\mu_0&\int D(x-y)\left(\psi^{(0)*} + \delta\psi_1^*\right)(y)\delta\psi_{ctr}(y)d^4y
\end{aligned}
$$

Note that in this analysis, the perturbation parameter is the electron charge e and if we neglect $0(e^2)$ terms, then the previous expression for the controllable part of the electromagnetic field simplifies to

$$
\begin{aligned}
\delta A_{r,ctr}(x) = \mu_0&\int D(x-y)J_r^c(y)d^4y \\
-e\mu_0&\int D(x-y)\delta\psi_{ctr}(y)^*\alpha_r\psi^{(0)}(y)d^4y \\
e\mu_0&\int D(x-y)\psi^{(0)*}(y)\alpha_r\delta\psi_{ctr}(y)d^4y
\end{aligned}
$$

In the particular case of the RDRA considered here, we find that the controllable part of the Dirac field has the following expansion:

$$\delta\psi_{ctr}(x) = -e\int S_e(x-y)\alpha^r A_r^c(y)\psi^{(0)}(y)d^4y$$

$$= -e\int S_e\left(t-t^{'},r-r^{'}\right)\alpha^r A_r^c\left(t^{'},r^{'}\right)$$

$$\left[\sum_{kmnp}\left(d_e(k)c_{ek}(mnp)u_{mnp}\left(r^{'}\right)\exp\left(-iE(k)t^{'}\right)\right)\right.$$

$$\left.+\,d_p\left(k\right)^* c_{pk}(mnp)u_{mnp}\left(r^{'}\right)\exp\left(iE(k)t^{'}\right)\right]dt'd^3r'$$

Now define the following Fourier components of the control classical laser generated electromagnetic field w.r.t. the cavity boundary conditions and the energy spectrum of the free Dirac field in the cavity after

$$\int A_r^c\left(t^{'},r^{'}\right)u_{mnp}\left(r^{'}\right)\exp\left(-iK\cdot r^{'}\right)\exp\left(i\omega t^{'}\right)d^3r^{'}dt^{'} = C_{A,r}(\omega,K|m,n,p)$$

Then, we can express the previous controllable part of the Dirac field in the following form in the spatial–temporal Fourier domain:

$$\int\delta\psi_{ctr}(t,r)$$

$$\cdot\exp(i(\omega t-K\cdot r)dtd^3r$$

$$= \sum_{kmnp} -ed_e(k)S(\omega,K)c_{ek}(mnp)\alpha^r C_r(\omega$$

$$-\,E(k),K|mnp) - ed_e(k)^* S(\omega,K)c_{pk}(mnp)\alpha^r C_r(\omega+E(k),K|mnp)$$

$$= -eS(\omega,K)\sum_{kmnp} d_e(k)c_{ek}(mnp)\alpha^r C_r(\omega-E(k),K|mnp)$$

$$+\,d_p\left(k\right)^* c_{pk}(mnp)\alpha^r C_r(\omega+E(k),K|mnp)$$

The controllable part of the Dirac four current density is then given up to first-order perturbation terms by $(x=(t,r))$

$$\delta J^\mu(t,r) = -e\psi^{(0)*}(x)\alpha^\mu\delta\psi_{ctr}(x) - e\delta\psi_{ctr}(x)^*\alpha^\mu\psi^{(0)}(x)$$

And it is immediately clear from the previous expression that the far-field radiated electromagnetic potential generated by this controllable current field can be expressed in the following form:

$$\delta A^{\mu}_{R}(t,r) = \int D\left(t-t^{'},r-r^{'}\right)\delta J^{\mu}\left(t^{'},t^{'}\right)dt'd^{3}r'$$

$$\sum_{kmnp} d_e(k)^{*}d_e\left(k^{'}\right)\int C_r(\omega - E(k),K|mnp)\bar{C}_s\left(-\omega^{'} - E\left(k^{'}\right),K'|m'n'p'\right)$$

$$\times F^{\mu}rs\left(t,r|\omega,K,\omega^{'},K^{'},mnpk,m'n'p'k'\right)d\omega d^{3}Kd\omega' d^{3}K'$$

plus three other similar terms involving $d_e(k)^{*}d_p\left(k^{'}\right)^{*}, d_e\left(k^{'}\right), d_p(k)d_p\left(k^{'}\right)^{*}$. In compact notation, the expected value of this controllable far-field pattern can be expressed as a Hermitian quadratic form in the complex numbers $C_r(\omega,K|mnp), \omega\in\mathbb{R}, K\in\mathbb{R}^{3}, m,n,p\in\mathbb{Z}_{+}$. These complex numbers are controllable since they represent in some sense the spatiotemporal components of the Fourier components of the classical control electromagnetic field A^{c}_{μ}.

9.7 Conclusion

TDRAs are useful for high-speed communications. TDRAs are compactly efficient and they have ultralarge bandwidth. DRs are robust and reliable as they can operate at entire spectrum, i.e., microwave to optical regime. They have large design flexibility and customized aspect-ratio. At optical frequency, TDRAs can be used for retinal photoreceptors. LiDAR can be a beautiful application of terahertz antennas and sensors. Terahertz devices shall have compact size and less power consumption.

References

[1] R. D. Richtmyer, "Dielectric resonators," *J. Appl. Phys.*, vol. 10, pp. 391–398, 1939.

[2] S. A. Long, M. McAllister, and L. C. Shen, "The resonant cylindrical dielectric cavity antenna," *IEEE Trans. Antennas Propag.*, vol. 31, pp. 406–412, 1983.

[3] M. Gastine, L. Courtois, and J. J. Dormann, "Electromagnetic resonances of free dielectric spheres," *IEEE Trans. Microwave Theory Tech.*, vol. 15, no. 12, pp. 694–700, 1967.

[4] R. S. Yaduvanshi and H. Parthasarathy, *Rectangular Dielectric Resonator Antennas: Theory and Design*, Springer, New Delhi, 2016.

[5] J. Van Bladel, "On the resonances of a dielectric resonator of very high permittivity," *IEEE Trans. Microwave Theory Tech.*, vol. 23, no. 2, pp. 199–208, 1975.

[6] R. S. Yaduvanshi and V. Gaurav, *Nano Dielectric Resonator for 5G Applications*, CRC Press, Boca Raton, FL, 2020.

[7] K. M. Rajesh and B. Prakash, "Dielectric resonator antenna—A review and general design relation for resonant frequency and bandwidth," *Int. J. Microwave Millimetre Wave Comput. Aided Eng.*, vol. 4, no. 3, pp. 230–247, 1994.

[8] Y. M. Pan, K. W. Leung, and L. Guo, "Compact laterally radiating dielectric resonator antenna with small ground plane," *IEEE Trans. Antennas Propag.*, vol. 65, no. 8, pp. 4305–4310, 2017.
[9] Y. M. M. Antar, "Antennas for wireless communication: Recent advances using dielectric resonators," *IET Circuits Devices Syst.*, vol. 2, no. 1, pp. 133–138, 2008.
[10] A. Petosa and A. Ittipiboon, "Dielectric resonator antennas: A historical review and the current state of the art," *IEEE Antennas Propag. Mag.*, vol. 52, no. 5, pp. 91–116, 2010.
[11] K. W. Leung, E. H. Lim, and X. S. Fang, "Dielectric resonator antennas: From the basic to the aesthetic," *Proceedings of the IEEE*, vol. 100, no. 7, pp. 2181–2193, 2012.
[12] P. A. Martin and F. Rothen, *Many-Body Problems and Quantum Field Theory*, Springer, Berlin Heidelberg, 2002.
[13] P. Bharadwaj and L. Novotny, "Spectral dependence of single molecule fluorescence enhancement," *Opt. Express*, vol. 15, pp. 14266–14274, 2007.
[14] L. Novotny, "Optical antennas: A new technology that can enhance light-matter interactions," *Front. Eng.*, vol. 39, no. 4, pp. 100–120, 2012.
[15] S. Lepeshov and A. Gorodetsky, "All-dielectric optically tunable metasurface for terahertz phase and amplitude modulation," *J. Phys. Conf. Ser.*, vol. 1461, pp. 012203, 2020 IOP Publishing, doi:10.1088/1742-6596/1461/1/012203.
[16] M. Hayasi, *Quantum Information Theory*, Springer, Berlin Heidelberg, 2017.
[17] P. Bhardwaj, B. Deutsch, and L. Novotny, "Optical antennas," *Adv. Opt. Photonics*, vol. 1, pp. 438–483, 2009.
[18] M. Humayun and L. O. de Koo, *Retinal Prosthesis—A Clinical Guide to Successful Implementation*, Springer, 2018.
[19] Y. Yang, Y. Yamagami, X. Yu, *et al.*, "Terahertz topological photonics for on-chip communication," *Nature Photonics*, vol. 14, pp. 446–451, 2020.
[20] J. W. Choong, N. Nefedkin, and A. Krasnok, "Collectively driven optical nanoantennas," *Phys. Rev. A*, vol. 103, 043714, 2021.
[21] S. Weinberg, *The Quantum Theory of Fields*, vol. 1, Oxford University Press, New York, NY, 1995.

Chapter 10

Equivalent circuit analysis on terahertz and optical dielectric resonator antennas (DRAs)

Abstract

In this chapter, an equivalent *R–L–C* circuit for terahertz and optical dielectric resonator antenna (DRA) has been developed. The behavior of characteristics modes in response to resonance frequency, input impedance and quality factor has been formulated. Higher order resonant modes in terahertz DRA has been depicted with circuit concepts. The terms, such as quality factor, bandwidth, resonance frequency and selectivity in quantum circuits, have been comprehensively investigated in quantum circuits input impedance that is complex and nonlinear, the real part of which is a function of frequency. The dynamic impedance at terahertz frequency involves skin effect. The authors have developed *R*, *L*, *C* (resistor (*R*), inductor (*L*) and capacitor (*C*)) equivalent circuit for terahertz and optical DRAs. These investigations are useful for the theoretical modeling of optical DRAs that shall help designers during the design stage. Their comprehensive analysis and study of quality factor, bandwidth, resonant frequency and selectivity in quantum circuits have been illustrated. The mathematical formulations for antenna and absorber conditions have been worked out. Maximum resistance condition in *R*, *L*, *C* circuit-favored absorber designs and minimum resistance conditions have been found suitable for antenna functions. These investigations also depict the phenomenon of higher order resonant modes in terahertz DRA. These resonant modes can be modulated to get merged modes or controlled separated modes to provide wide bandwidth or narrow bandwidth in antenna designs.

10.1 Introduction

Optical antenna has a finite quantity of planks constant value. The quantum antennas' radiations can be correlated with black-body radiations or light emitted. An *RLC*-parallel circuit is equivalent for the classical dielectric resonator antenna (DRA), and the transfer function of the same is used to approximate a given function of frequency generated by a quantum antenna. The solution for error between quantum function and classical function can be found by minimizing the error function. Surface plasmon-mediated hot carrier generation is widely utilized for the manipulation of the electron–photon interactions in many types of optoelectronic devices. The spectrum of quantum antenna can be defined in terms of Planck's constant. The RLC circuit response of a classical circuit can be transformed to quantum circuit response by integrating over s-domain quantum spectrum omega–s-domain classical omega over a derivative of omega and differentiating with respect to R, L, C values. This will give us an average value, where mean and covariance are most important in a quantum antenna for radiated fields.

At quantum level, electromagnetic fields are realized by quantum wave operator fields, and it can be precisely defined as an ensemble of an infinite number of quantum harmonic oscillators. It can be analyzed as a superposition of constant plane wave with coefficients being creation and annihilation operators in a boson Fock space. At an optical spectrum, it is better to talk about quantum state of wave fields like coherent state of photons. It is a process of light–matter interaction in an optical antenna to introduce a new term known as SPP (surface plasmon polytrons). The light interacts with electrons, atoms and molecules of materials of a medium it travels through. At optical frequency, a material medium is treated as plasma having definite permittivity, permeability and conductivity for the propagation of electromagnetic fields in it.

The quantum antenna circuit, with input excitation, the potential drop across impedances Z_2 and Z_3 are the same since they are in parallel connection. Also, there is some potential drop in impedance Z_1. But currents in impedance Z_2 and Z_3 are different because Z_2 and Z_3 have different values:

$$Z_2 = R_2 + j\omega L_2;\ Z_3 = R_3 - \frac{j}{\omega C_3},$$
$$I_2 = \frac{V}{Z_2},\ I_3 = \frac{V}{Z_3}$$

where magnitudes of Z_2 and Z_3 impedances vary due to variable frequency term in Z_2 and Z_3. At a particular high frequency, the impedances Z_2 and Z_3 become equal; as a result, the current flowing through each of them also becomes equal. The currents in Z_2 and Z_3 remain equal for a very small duration of time as it happens only at particular high frequency. This phenomenon is due to the Dirac-delta function. This high frequency at which currents in Z_2 and Z_3 are the same is called first tuning frequency. After this frequency, the currents I_2 and I_3 in impedances Z_2 and Z_3, respectively, are the same in magnitude but the opposite in phase. So, the resultant of I_2 and I_3 is zero and the net current only flows through impedance Z_1, given by $Z_1 = R_1 - j/\omega C_1$ (Z_1 branch is frequency dependent). At high frequency, the magnitude of impedance Z_1 is very low resulting in very high current and power. The circuit impedance is dependent only upon Z_1, which is then called the dynamic impedance of the circuit. The current flowing through Z_1 is dependent upon potential drop across Z_1 and given by $i_1 = C(dV_1/dt)$ since dt is very small time, so the current is very high resulting in a high selectivity of the circuit. This current i_1 is known as surface current or displacement current of the circuit.

In the *RLC* circuit of quantum antenna, the input impedance of the circuit is frequency dependent and also of very small-time interval due to high-frequency analysis. This is due to the phenomena of the Dirac-delta function. In this time duration, current is maximum and depends upon impedance Z_1 that is frequency dependent. Since the impedance Z_1 is very small at high frequency, so the corresponding current through Z_1 is also very high. As a result, power is very high and signal strength is very high. Both signal capturing and delivering are very high. This is the main advantage of a quantum antenna circuit. During the initial tuning of this circuit, the current is the same but the opposite in phase, so their resultant is zero. In this condition, the inductor current behaves like a surface current that is proportional to area. The surface current in inductor is present due to skin effect. In the quantum antenna, the input impedance and the dynamic impedance of the circuit are frequency dependent. As the frequency increases, the dynamic impedance decreases, which further reduces the power loss. Since the real part of input impedance of a quantum antenna circuit consumes the real power, imaginary term of input impedance does not consume any power at all. And the other elements like L and C transfer power from one cycle to other cycle among each other. These circuits are very effective for 5G technologies because these technologies need very less power loss and high selectivity at high frequency. As L and C store more power, the quality factor of the circuit is very high: $Q = (f_r/\text{bandwidth})$. The quality factor is very high, which implies the bandwidth of the system is very low. But in a quantum antenna circuit, the resonance frequency is very high due to which the bandwidth of the system is not very small. The design analysis of an antenna needs a strong mathematical background. An *RLC* circuit can be realized by a simple solution for any antenna [1]. In this chapter, HFSS NDRA modeling has been developed with equivalent circuit design approach [2–8]. An approximate *RLC* circuit model has been introduced [9]. This narrates the main mode of propagating as an equivalent transmission line by making use of an aperture-coupled slot. Lumped impedances have been derived to accurately describe the functions from source to end points. The reactance's account for the reactive power due to feed and the termination and resistances exactly show the radiated waves in space. This technique can be used to compute an input impedance of NDRA with aperture-coupled loading [10]. Very less research on modeling an NDRA equivalent circuit has been done so far; however, most of the literature is available on a patch antenna equivalent circuit [8,11–13]. The equivalent circuit representation gives correct results for both the load and the internal impedance of NDRA. The modeling of NDRA impedances and radiation fields has been presented with higher order and fundamental modes. Circuit bandwidth, resonance and other radiation field parameters have been represented using an equivalent circuit model. Higher order modes can be accurately modeled with this approach. In this approach, circuit models of resonant modes can be used to predict radiation patterns and other field behavior of NDRA [14]. In this chapter, a physics-based circuit for resonant modes has been developed with simple and absolutely accurate form. This equivalent circuit model can extrapolate bandwidth, resonant frequency, impedance, quality factor and radiation pattern of NDRA. This is beneficial in designing and analyzing of different NDRA structures. The study of resonance can simplify designing of NDRAs. This method of analysis has linked the NDRA circuit models to its radiated fields for the first time in NDRA research [15].

10.2 Quantum DRA-equivalent circuit mathematical analysis for mixed circuits

10.2.1 Impedance (Z_{in})

Equivalent circuits of optical antenna are given in Figure 10.1.

In the previous circuit, the resistance and reactance values are as follows:

$$X_1 = -\frac{1}{\omega C_1},\ X_2 = \omega L_2 \text{ and } X_3 = -\frac{1}{\omega C_3}$$

Input impedance of this circuit

$$Z_{in} = Z_1 + Z_2 \| Z_3 = Z_1 + \frac{Z_2 Z_3}{Z_2 + Z_3}$$

where $Z_1 = R_1 + jX_1$, $Z_2 = R_2 + jX_2$ and $Z_3 = R_3 + jX_3$

$$Z_{in} = R_1 + jX_1 + \frac{(R_3 + jX_3)(R_2 + jX_2)}{R_2 + jX_2 + R_3 + jX_3} \tag{10.1}$$

$$Z_{in} = R_1 + jX_1 + \frac{R_2R_3 - X_2X_3 + j(X_2R_3 + R_2X_3)}{R_2 + R_3 + j(X_3 + X_2)}$$

$$Z_{in} = R_1 + jX_1 + \frac{\{R_2R_3 - X_2X_3 + j(X_2R_3 + R_2X_3)\}\{R_2 + R_3 - j(X_3 + X_2)\}}{(R_2 + R_3)^2 + (X_3 + X_2)^2}$$

$$Z_{in} = R_1 + jX_1 + \frac{(R_2R_3 - X_2X_3)(R_2 + R_3) + (X_2R_3 + R_2X_3)(X_3 + X_2) + j\{(R_2 + R_3)(X_2R_3 + R_2X_3) - (X_3 + X_2)(R_2R_3 - X_2X_3)\}}{(R_2 + R_3)^2 + (X_3 + X_2)^2}$$

$$Z_{in}(\text{real part}) = \left[R_1 + \frac{(R_2R_3 - X_2X_3)(R_2 + R_3) + (X_2R_3 + R_2X_3)(X_3 + X_2)}{(R_2 + R_3)^2 + (X_3 + X_2)^2}\right] \tag{10.2}$$

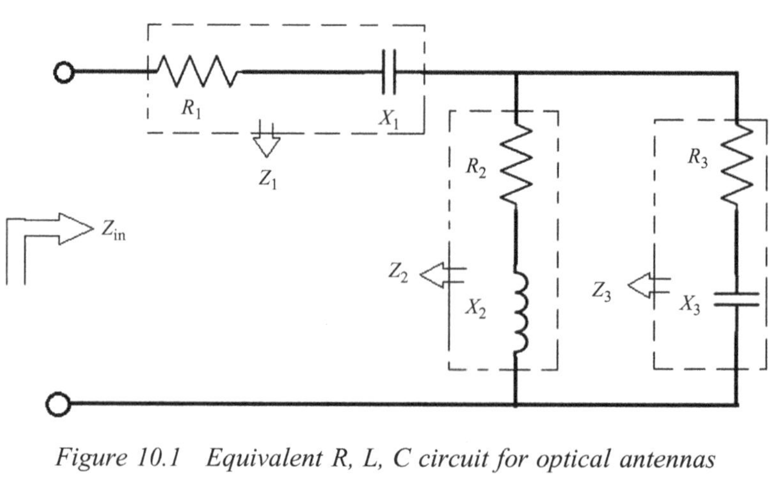

Figure 10.1 Equivalent R, L, C circuit for optical antennas

$$Z_{in}(\text{imaginary part}) = \left[X_1 + \frac{(R_2+R_3)(X_2R_3+R_2X_3)-(X_3+X_2)(R_2R_3-X_2X_3)}{(R_2+R_3)^2+(X_3+X_2)^2}\right] \tag{10.3}$$

Resonance frequency imaginary part of input impedance is equal to zero:

$Z_{in} = Z_{in}(\text{real part}) + Z_{in}(\text{imaginary part})$

At resonance imaginary part of $Z_{in} = 0$

$$X_1 + \frac{(R_2+R_3)(X_2R_3+R_2X_3)-(X_3+X_2)(R_2R_3-X_2X_3)}{(R_2+R_3)^2+(X_3+X_2)^2} = 0 \tag{10.4}$$

$$R_2R_3X_2+R_2^2X_3+R_3^2X_2+R_2R_3X_3-(R_2R_3X_2-X_2^2X_3+R_2R_3X_3-X_3^2X_2)+X_1\left\{(R_2+R_3)^2+(X_3+X_2)^2\right\}=0$$

$$R_2^2X_3+R_3^2X_2+X_2^2X_3+X_3^2X_2+X_1(R_2^2+R_3^2+2R_2R_3+X_2^2+X_3^2+2X_2X_3)=0$$

$$R_2^2(X_1+X_3)+R_3^2(X_1+X_2)+X_2^2(X_1+X_3)+X_3^2(X_1+X_2)+2X_1(R_2R_3+X_2X_3)=0$$

$$(R_2^2+X_2^2)(X_1+X_3)+(R_3^2+X_3^2)(X_1+X_2)+2X_1(R_2R_3+X_2X_3)=0$$

Substituting the value of X_1, X_2 and X_3 in the previous equation,

$$\left(\omega L_2 - \frac{1}{\omega C_1}\right)\left(R_3^2 + \frac{1}{\omega^2 C_3^2}\right) + \left(-\frac{1}{\omega C_1} - \frac{1}{\omega C_3}\right)(R_2^2 + \omega^2 L_2^2) - \frac{2}{\omega C_1}\left(R_2R_3 - \frac{L_2}{C_3}\right) = 0$$

$$\left(\frac{\omega^2 L_2 C_1 - 1}{\omega C_1}\right)\left(\frac{R_3^2\omega^2 C_3^2 + 1}{\omega^2 C_3^2}\right) + \frac{1}{\omega}\left(-\frac{1}{C_1} - \frac{1}{C_3}\right)(R_2^2 + \omega^2 L_2^2) - \frac{2}{\omega C_1}\left(R_2R_3 - \frac{L_2}{C_3}\right) = 0$$

Multiplying ω^3 both sides in previous equation,

$$\frac{(\omega^2 L_2 C_1 - 1)(R_3^2\omega^2 C_3^2 + 1)}{C_1 C_3^2} + \omega^2\left(-\frac{1}{C_1} - \frac{1}{C_3}\right)(R_2^2 + \omega^2 L_2^2) - \frac{2\omega^2}{C_1}\left(R_2R_3 - \frac{L_2}{C_3}\right) = 0$$

$$\frac{(\omega^4 L_2 C_1 R_3^2 C_3^2 - 1 + \omega^2(L_2C_1 - R_3^2C_3^2))}{C_1C_3^2} + \omega^4 L_2^2\left(-\frac{1}{C_1} - \frac{1}{C_3}\right) + \omega^2 R_2^2\left(-\frac{1}{C_1} - \frac{1}{C_3}\right) - \frac{2\omega^2}{C_1}\left(R_2R_3 - \frac{L_2}{C_3}\right) = 0$$

$$\omega^4\left[L_2R_3^2 - L_2^2\left(\frac{1}{C_1} + \frac{1}{C_3}\right)\right] + \omega^2\left[\frac{L_2}{C_3^2} - \frac{R_3^2}{C_1} - R_2^2\left(\frac{1}{C_1} + \frac{1}{C_3}\right) - \frac{2}{C_1}\left(R_2R_3 - \frac{L_2}{C_3}\right)\right] - \frac{1}{C_1C_3^2} = 0 \tag{10.5}$$

$$\omega^4[A] + \omega^2[B] - C = 0 \tag{10.6}$$

Comparing the previous two equations, we get A, B and C as

$$A = L_2R_3^2 - L_2^2\left(\frac{1}{C_1} + \frac{1}{C_3}\right), B = \frac{L_2}{C_3^2} - \frac{R_3^2}{C_1} - R_2^2\left(\frac{1}{C_1} + \frac{1}{C_3}\right) - \frac{2}{C_1}\left(R_2R_3 - \frac{L_2}{C_3}\right) \text{ and}$$

$$C = \frac{1}{C_1C_3^2}$$

Let $\omega^2 = p$

$$p^2[A] + p[B] - C = 0$$

$$P = \frac{-B \pm \sqrt{B^2 + 4AC}}{2A}; \text{ where } \omega^2 = p$$

$$\omega=\sqrt{\frac{-B\pm\sqrt{B^2+4AC}}{2A}}$$
$$=\sqrt{\frac{-\left(\frac{L_2}{C_3^2}-\frac{R_3^2}{C_1}-R_2^2\left(\frac{1}{C_1}+\frac{1}{C_3}\right)-\frac{2}{C_1}\left(R_2R_3-\frac{L_2}{C_3}\right)\right)\pm\sqrt{\left(\frac{L_2}{C_3^2}-\frac{R_3^2}{C_1}-R_2^2\left(\frac{1}{C_1}+\frac{1}{C_3}\right)-\frac{2}{C_1}\left(R_2R_3-\frac{L_2}{C_3}\right)\right)^2+4\left\{L_2R_3^2-L_2^2\left(\frac{1}{C_1}+\frac{1}{C_3}\right)\right\}\frac{1}{C_1C_3^2}}}{2\left\{L_2R_3^2-L_2^2\left(\frac{1}{C_1}+\frac{1}{C_3}\right)\right\}}} \tag{10.7}$$

Thus, resonant frequency of this circuit is depending on R, L and C, but input impedance of this circuit is a function of R, L, C and frequency.

Hence, impedance is frequency dependent:

$$Z_{\text{in}}=f(R,L,C,\omega) \tag{10.8}$$

$$Z_{\text{in}}=Z_{\text{inreal}}(\omega)+Z_{\text{inimaginary}}(\omega) \tag{10.9}$$

Hence, terahertz DRA input impedance of the circuit,

$$Z_{\text{in}}=Z_{\text{inreal}}+Z_{\text{inimaginary}}(\omega) \tag{10.10}$$

$$Z_{\text{inreal}}\neq f(\omega) \tag{10.11}$$

where Z_{inreal} means only real part of input impedance that is called only resistive part of input impedance.

On the other hand, terahertz DRA resistive part of input impedance is a function of frequency:

$$Z_{\text{inreal}}=\text{resistance}=f(\omega) \tag{10.12}$$

10.2.2 The frequency-dependent resistance is also called dynamic resistance of the circuit

$$Z_d=R_1+\frac{(R_2R_3-X_2X_3)(R_2+R_3)+(X_2R_3+R_2X_3)(X_3+X_2)}{(R_2+R_3)^2+(X_3+X_2)^2} \tag{10.13}$$

$X_1=\frac{1}{-\omega C_1}$, $X_2=\omega C_2$ and $X_3=\frac{1}{-\omega C_3}$ in the previous equation

$$Z_d=R_1+\frac{\left(R_2R_3-\frac{L_2}{C_3}\right)(R_2+R_3)+\left(\omega L_2R_3-\frac{R_2}{\omega C_3}\right)\left(\omega L_2-\frac{1}{\omega C_3}\right)}{(R_2+R_3)^2+\left(\omega L_2-\frac{1}{\omega C_3}\right)^2}$$

$$Z_d=R_1+\frac{\left(R_2R_3-\frac{L_2}{C_3}\right)(R_2+R_3)+\left(\frac{\omega^2L_2R_3C_3-R_2}{\omega C_3}\right)\left(\frac{\omega^2L_2C_3-1}{\omega C_3}\right)}{(R_2+R_3)^2+\left(\frac{\omega^2L_2C_3-1}{\omega C_3}\right)^2}$$

$$Z_d = R_1 + \frac{\frac{\left(R_2R_3-\frac{L_2}{C_3}\right)(R_2+R_3)\omega^2C_3^2+\left(\omega^2L_2R_3C_3-R_2\right)\left(\omega^2L_2C_3-1\right)}{\omega^2C_3^2}}{\frac{\omega^2C_3^2(R_2+R_3)^2+(\omega^2L_2C_3-1)^2}{\omega^2C_3^2}}$$

$$Z_d = \frac{R_1\left\{\omega^2C_3^2(R_2+R_3)^2+(\omega^2L_2C_3-1)^2\right\}+\left(R_2R_3-\frac{L_2}{C_3}\right)(R_2+R_3)\omega^2C_3^2+(\omega^2L_2R_3C_3-R_2)(\omega^2L_2C_3-1)}{\omega^2C_3^2(R_2+R_3)^2+(\omega^2L_2C_3-1)^2}$$

$$Z_d = f(\omega) \;\; or \;\; R_{\text{in}} = f(\omega)$$

10.2.3 Two resonant modes, i.e. fundamental and higher order

$X_1 = -\frac{1}{\omega C_1}$, $X_2 = \omega L_2$ and $X_3 = -\frac{1}{\omega C_3}$

$$Y_l = Y_2 + Y_3$$

$$Y_l = \frac{1}{R_2 + jX_2} + \frac{1}{R_3 + jX_3}$$

$$Y_l = \frac{1}{R_2 + j\omega L_2} + \frac{1}{R_3 - \frac{j}{\omega C_3}}$$

$$Y_l = \frac{R_2 - j\omega L_2}{R_2 + j\omega L_2} + \frac{R_3 + \frac{j}{\omega C_3}}{R_3 - \frac{j}{\omega C_3}}$$

$$Y_l = \frac{R_2}{R_2^2 + \omega^2 L_2^2} + \frac{R_3}{R_3^2 + \frac{1}{\omega^2 C_3^2}} + j\left[\frac{\frac{1}{\omega C_3}}{R_3^2 + \frac{1}{\omega^2 C_3^2}} - \frac{\omega L_2}{R_2^2 + \omega^2 L_2^2}\right] \tag{10.14}$$

$$Y_l = G_l + jB_l$$

For the first or fundamental resonant mode imaginary part of $(Y_l) = 0$; it means

$$B_l = 0$$

$$\frac{\frac{1}{\omega C_3}}{R_3^2 + \frac{1}{\omega^2 C_3^2}} - \frac{\omega L_2}{R_2^2 + \omega^2 L_2^2} = 0$$

$$\frac{\frac{1}{\omega C_3}}{R_3^2 + \frac{1}{\omega^2 C_3^2}} = \frac{\omega L_2}{R_2^2 + \omega^2 L_2^2}$$

$$\frac{\omega C_3}{R_3^2\omega^2 C_3^2 + 1} = \frac{\omega L_2}{R_2^2 + \omega^2 L_2^2}$$

$$C_3\left(R_2^2+\omega^2L_2^2\right)=L_2\left(R_3^2\omega^2C_3^2+1\right)$$

$$\omega^2\left(L_2^2C_3-L_2R_3^2C_3^2\right)=L_2+R_2^2C_3$$

$$\omega_1=\sqrt{\frac{L_2+R_2^2C_3}{L_2^2C_3-L_2R_3^2C_3^2}} \tag{10.15}$$

10.2.4 Second resonant mode

$$Y_T=Y_1+Y_l$$

$$Y_T=Y_1+(G_l+jB_l)$$

$$Y_T=\frac{1}{R_1-\frac{j}{\omega C_1}}+(G_l+jB_l)$$

$$Y_T=\frac{R_1+\frac{j}{\omega C_1}}{R_1^2+\frac{1}{\omega^2C_1^2}}+(G_l+jB_l)$$

$$Y_T=\left(\frac{R_1\omega^2C_1^2}{R_1^2\omega^2C_1^2+1}+G_l\right)+j\left(B_l+\frac{\frac{1}{\omega C_1}}{R_1^2+\frac{1}{\omega^2C_1^2}}\right)$$

For the second resonant mode or over all resonant mode imaginary part of $Y_T=0$

$$B_l+\frac{\omega C_1}{R_1^2\omega^2C_1^2+1}=0$$

$$\frac{\omega C_3}{R_3^2\omega^2C_3^2+1}-\frac{\omega L_2}{R_2^2+\omega^2L_2^2}+\frac{\omega C_1}{R_1^2\omega^2C_1^2+1}=0$$

$$\frac{C_3}{R_3^2\omega^2C_3^2+1}-\frac{L_2}{R_2^2+\omega^2L_2^2}+\frac{C_1}{R_1^2\omega^2C_1^2+1}=0$$

On solving the previous equation the value of resonant frequency for higher order mode (ω_r) (Figure 10.2):

$$\omega_r=\sqrt{\frac{-\left(\frac{L_2}{C_3^2}-\frac{R_3^2}{C_1}-R_2^2\left(\frac{1}{C_1}+\frac{1}{C_3}\right)-\frac{2}{C_1}\left(R_2R_3-\frac{L_2}{C_3}\right)\right)\pm\sqrt{\left(\frac{L_2}{C_3^2}-\frac{R_3^2}{C_1}-R_2^2\left(\frac{1}{C_1}+\frac{1}{C_3}\right)-\frac{2}{C_1}\left(R_2R_3-\frac{L_2}{C_3}\right)\right)^2+4\left\{L_2R_3^2-L_2^2\left(\frac{1}{C_1}+\frac{1}{C_3}\right)\right\}\frac{1}{C_1C_3^2}}}{2\left\{L_2R_3^2-L_2^2\left(\frac{1}{C_1}+\frac{1}{C_3}\right)\right\}}} \tag{10.16}$$

$$\text{Quality factor }(Q)=2\pi\times\frac{\text{maximum energy stored per cycle}}{\text{power dissipatetd per cycle}}$$

For the series case $Z=R+jX\cdots=R+j(X_L-X_C)\ \cdots$

$$Q=\frac{|X_L|}{R}=\frac{|X_C|}{R}$$

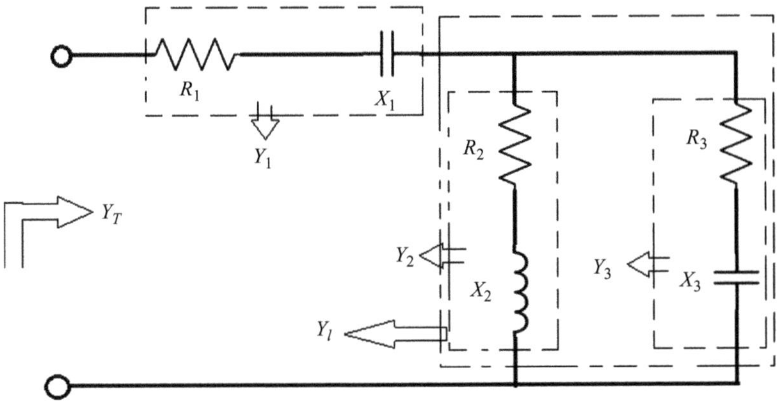

Figure 10.2 Higher order modes analysis

For the parallel case

$$Y = G + jB = G + j(B_c - B_L)$$

$Q = \frac{|B_L|}{G} = \frac{|B_C|}{G} \cdots$. According to this,

$$Z_{\text{in}} = R_{\text{in}} + jX_{\text{in}} = Z_{\text{in}}(\text{real}) + Z_{\text{in}}(\text{imaginary})$$

$$\text{Quality factor } (Q) = \left| \frac{X_1 + \frac{(R_2+R_3)(X_2R_3+R_2X_3)-(X_3+X_2)(R_2R_3-X_2X_3)}{(R_2+R_3)^2+(X_3+X_2)^2}}{R_1 + \frac{(R_2R_3-X_2X_3)(R_2+R_3)+(X_2R_3+R_2X_3)(X_3+X_2)}{(R_2+R_3)^2+(X_3+X_2)^2}} \right| \tag{10.17}$$

$$Q = \frac{(R_2+R_3)(X_2R_3+R_2X_3)-(X_3+X_2)(R_2R_3-X_2X_3)+X_1\left\{(R_2+R_3)^2+(X_3+X_2)^2\right\}}{(R_2R_3-X_2X_3)(R_2+R_3)+(X_2R_3+R_2X_3)(X_3+X_2)+R_1\left\{(R_2+R_3)^2+(X_3+X_2)^2\right\}}$$

$$Q = \frac{R_2^2X_3+R_3^2X_2+X_2^2X_3+X_3^2X_2+X_1\{R_2^2+R_3^2+2R_2R_3+X_2^2+X_3^2+2X_2X_3\}}{R_2^2R_3+R_3^2R_2-R_2X_2X_3-X_2R_3X_3-(X_2^2R_3+X_3^2R_2+X_2X_3R_2+X_2X_3R_3)+R_1\{R_2^2+R_3^2+2R_2R_3+X_2^2+X_3^2+2X_2X_3\}}$$

$$Q = \frac{R_2^2(X_3+X_1)+R_3^2(X_2+X_1)+X_2^2(X_3+X_1)+X_3^2(X_2+X_1)+2X_1\{R_2R_3+X_2X_3\}}{R_2^2(R_3+R_1)+R_3^2(R_2+R_1)+X_2^2(R_3+R_1)+X_3^2(R_2+R_1)+2R_1(R_2R_3+X_2X_3)-2X_2X_3R_2-2X_2X_3R_3}$$

$$Q = \frac{(R_2^2+X_2^2)(X_3+X_1)+(R_3^2+X_3^2)(X_2+X_1)+2X_1(R_2R_3+X_2X_3)}{(R_3+R_1)(R_2^2+X_2^2)+(R_2+R_1)(R_3^2+X_3^2)+2R_2(R_1R_3-X_2X_3)+2X_2(X_3R_1-X_3R_3)}$$

$$Q = \frac{\left(-\frac{1}{\omega C_1}-\frac{1}{\omega C_3}\right)(R_2^2+\omega^2L_2^2)+\left(\omega L_2-\frac{1}{\omega C_1}\right)\left(R_3^2+\frac{1}{\omega^2C_3^2}\right)-\frac{2}{\omega C_1}\left(R_2R_3-\frac{L_2}{C_3}\right)}{(R_3+R_1)(R_2^2+\omega^2L_2^2)+(R_2+R_1)\left(R_3^2+\frac{1}{\omega^2C_3^2}\right)+2R_2\left(R_1R_3-\frac{L_2}{C_3}\right)+2\omega L_2\left(\frac{-R_1}{\omega C_1}+\frac{R_3}{\omega C_3}\right)}$$

$$Q = \frac{-\frac{1}{\omega}\left(\frac{1}{C_1}+\frac{1}{C_3}\right)(R_2^2+\omega^2L_2^2)+\left(\frac{\omega^2L_2C_1-1}{\omega C_1}\right)\left(\frac{\omega^2C_3^2R_3^2+1}{\omega^2C_3^2}\right)-\frac{2}{\omega C_1}\left(R_2R_3-\frac{L_2}{C_3}\right)}{(R_3+R_1)(R_2^2+\omega^2L_2^2)+(R_2+R_1)\left(\frac{\omega^2C_3^2R_3^2+1}{\omega^2C_3^2}\right)+2R_2\left(R_1R_3-\frac{L_2}{C_3}\right)+2L_2\left(\frac{-R_1}{C_1}+\frac{R_3}{C_3}\right)}$$

Multiplying ω^3 numerator and denominator,

$$Q = \frac{-\omega^2\left(\frac{1}{C_1}+\frac{1}{C_3}\right)(R_2^2+\omega^2L_2^2)+\frac{1}{C_1C_3^2}(\omega^2L_2C_1-1)(\omega^2C_3^2R_3^2+1)-\frac{2\omega^2}{C_1}\left(R_2R_3-\frac{L_2}{C_3}\right)}{\omega^3(R_3+R_1)(R_2^2+\omega^2L_2^2)+\omega\left(\frac{R_2+R_1}{C_3^2}\right)(\omega^2C_3^2R_3^2+1)+2\omega^3R_2\left(R_1R_3-\frac{L_2}{C_3}\right)+2\omega^3L_2\left(\frac{-R_1}{C_1}+\frac{R_3}{C_3}\right)}$$

$$Q = \frac{\omega^4\left[L_2R_3^2-L_2\left(\frac{1}{C_1}+\frac{1}{C_3}\right)\right]+\omega^2\left[\frac{L_2}{C_3^2}-R_2^2\left(\frac{1}{C_1}+\frac{1}{C_3}\right)-\frac{R_3^2}{C_1}-\frac{2}{C_1}\left(R_2R_3-\frac{L_2}{C_3}\right)\right]-\frac{1}{C_1C_3^2}}{\omega^5\left[L_2^2(R_3+R_1)\right]+\omega^3\left[R_2^2(R_3+R_1)+R_3^2(R_2+R_1)+2R_2\left(R_1R_3-\frac{L_2}{C_3}\right)+2L_2\left(\frac{-R_1}{C_1}+\frac{R_3}{C_3}\right)\right]+\omega\left[\frac{R_2+R_1}{C_3^2}\right]} \tag{10.18}$$

10.3 Higher order resonant modes

Terahertz antenna circuit behaves like a high-pass filter because the circuit output zero for low frequency and output is finite due to high frequency. Only one-stage two-mode exists in antenna circuit, but the higher stage odd and even harmonics of ω increase (Figure 10.3).

For the first stage of a high-pass filter, we calculate first mode and second mode resonator values. Consider that the previous circuit input and output are represented by X_i and X_o:

Using the Laplace domain $Z_1 = \frac{R_1sC_1+1}{sC_1}$, $Z_2 = R_2 + sL_2$ and $Z_3 = \frac{R_3sC_3+1}{sC_3}$

$$\frac{X_o(s)}{X_i(s)} = \frac{Z_p}{Z_1 + Z_p} = \frac{Z_2Z_3}{Z_1(Z_2 + Z_3) + Z_2Z_3} \quad \text{where} \quad Z_p = \frac{Z_2Z_3}{(Z_2 + Z_3)}$$

$$\frac{X_o(s)}{X_i(s)} = \frac{1}{1 + \frac{Z_1}{Z_p}} = \frac{1}{1 + \frac{\frac{R_1sC_1+1}{sC_1}}{\frac{(R_2+sL_2)\left(\frac{R_3sC_3+1}{sC_3}\right)}{(R_2+R_3+sL_2)sC_3+1}}} = \frac{1}{1 + \frac{(R_1sC_1+1)\{(R_2+R_3+sL_2)sC_3+1\}}{sC_1(R_2+sL_2)(R_3sC_3+1)}}$$

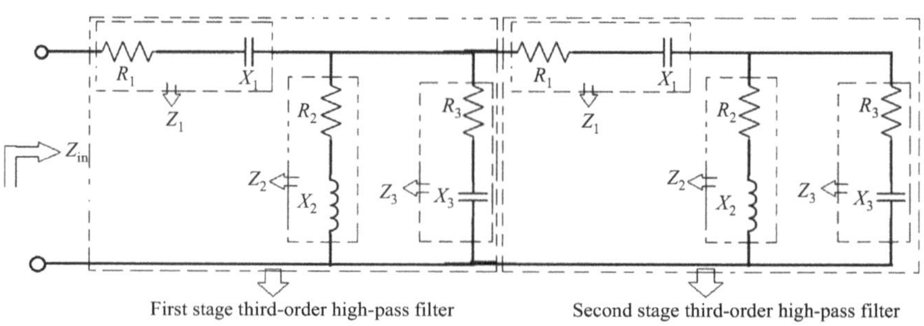

Figure 10.3 Higher order resonant modes, i.e., sixth-order high-pass filter (multistage high-pass filter)

$$\frac{X_o(s)}{X_i(s)} = \frac{1}{1 + \frac{s^3R_1C_1L_2C_3 + s^2\{L_2C_3 + R_1C_1C_3(R_2+R_3)\} + R_1C_1s + 1}{s^3R_3C_1L_2C_3 + s^2(L_2C_3 + R_2R_3C_1C_3) + R_2C_1s}}$$

The value of $s = j\omega$, transfer function

$$\frac{X_o(j\omega)}{X_i(j\omega)} = \frac{1}{1 + \frac{1-\omega^2\{L_2C_3 + R_1C_1C_3(R_2+R_3)\} + j(R_1C_1\omega - \omega^3R_1C_1L_2C_3)}{-\omega^2(L_2C_3 + R_2R_3C_1C_3) + j(\omega R_2C_1 - \omega^3R_3C_1L_2C_3)}}$$

In the previous transfer function for very low frequency (ideally $= 0$), gain is zero.

And for very high frequency (ideally $= \infty$), gain is a finite value, it means that for low frequency it does not pass the signal but for the high frequency, it passes the signal. Hence it behaves as a high-pass filter.

For the calculation of different modes of operation to simplify the transfer function,

$$\frac{X_o(j\omega)}{X_i(j\omega)} = \frac{1}{1 + \frac{a+jb}{c+jc}} = \frac{1}{1 + \frac{a+c+j(b-d)}{c^2+d^2}}$$

where

$$a = 1 - \omega^2\{L_2C_3 + R_1C_1C_3(R_2 + R_3)\},$$

$$b = R_1C_1\omega - \omega^3R_1C_1L_2C_3$$

$$c = -\omega^2(L_2C_3 + R_2R_3C_1C_3),\ d = \left(\omega R_2C_1 - \omega^3R_3C_1L_2C_3\right)$$

$$\left|\frac{X_o(j\omega)}{X_i(j\omega)}\right| = \frac{1}{\sqrt{\left(1 + \frac{a+c}{c^2+d^2}\right)^2 + \left(\frac{b-d)}{c^2+d^2}\right)^2}}$$

For a high value of gain $b - d = 0$ and $a + c = 0$.

There two frequencies are arising:

$$\left(R_1C_1\omega - \omega^3R_1C_1L_2C_3\right) - \left(\omega R_2C_1 - \omega^3R_3C_1L_2C_3\right) = 0$$

And the second condition is $1 - \omega^2\{L_2C_3 + R_1C_1C_3(R_2 + R_3)\}$

$$- \omega^2(L_2C_3 + R_2R_3C_1C_3)$$

$$= 0$$

$$\omega_1 = \sqrt{\frac{R_1C_1 - R_2C_1}{R_1C_1L_2C_3 - R_3C_1L_2C_3}}$$

$$\omega_2 = \sqrt{\frac{1}{L_2C_3 + R_2R_3C_1C_3 + \{L_2C_3 + R_1C_1C_3(R_2 + R_3)\}}}$$

10.4 Bandwidth (BW) of terahertz DRA

$$(\omega_2 - \omega_1) = \sqrt{\frac{1}{L_2C_3 + R_2R_3C_1C_3 + \{L_2C_3 + R_1C_1C_3(R_2 + R_3)\}}} - \sqrt{\frac{R_1C_1 - R_2C_1}{R_1C_1L_2C_3 - R_3C_1L_2C_3}}$$

The input impedance depends only on real part of Z_{in} and the real part of Z_{in} is dependent on frequency of the signal.

$$Q = \frac{\omega_r}{(\omega_2 - \omega_1)} = \frac{\sqrt{\dfrac{-\left(\frac{L_2}{C_3^2} - \frac{R_3^2}{C_1} - R_2^2\left(\frac{1}{C_1} + \frac{1}{C_3}\right) - \frac{2}{C_1}\left(R_2R_3 - \frac{L_2}{C_3}\right)\right) \pm \sqrt{\left(\frac{L_2}{C_3^2} - \frac{R_3^2}{C_1} - R_2^2\left(\frac{1}{C_1} + \frac{1}{C_3}\right) - \frac{2}{C_1}\left(R_2R_3 - \frac{L_2}{C_3}\right)\right)^2 + 4\left\{L_2R_3^2 - L_2^2\left(\frac{1}{C_1} + \frac{1}{C_3}\right)\right\}\frac{1}{C_1C_3^2}}}{2\left\{L_2R_3^2 - L_2^2\left(\frac{1}{C_1} + \frac{1}{C_3}\right)\right\}}}}{\sqrt{\frac{1}{L_2C_3 + R_2R_3C_1C_3 + \{L_2C_3 + R_1C_1C_3(R_2 + R_3)\}}} - \sqrt{\frac{R_1C_1 - R_2C_1}{R_1C_1L_2C_3 - R_3C_1L_2C_3}}} \tag{10.19}$$

10.5 Simulated results based on MATLAB®

Simulated results based on MATLAB are given in Figures 10.4–10.8.

10.6 Design development and evaluation of NDRA

Nano rectangular DRA design and implementation: the design step of nano-DRA is given in Figure 10.9.

10.6.1 Resonant frequency of TRDRA formulations

Resonant frequencies of TRDRA formulations are shown in Figures 10.10–10.14.

$$f_{mn\ell} = \frac{ck_{mn\ell}}{2\pi\sqrt{\mu_r\epsilon_r}} = \frac{c}{2\pi\sqrt{\mu_r\epsilon_r}}\sqrt{\left(\frac{m\pi}{a}\right)^2 + \left(\frac{n\pi}{b}\right)^2 + \left(\frac{\ell\pi}{d}\right)^2}.$$

10.7 Synthesis of NDRA radiation theory

1. $[q_a, p_b\ ] = \frac{ih}{2\pi}\delta(a, b)$
   where q is the position vector, p is the time derivative moment vector.

   $$\beta = \frac{1}{kT},$$ where T is the temperature.

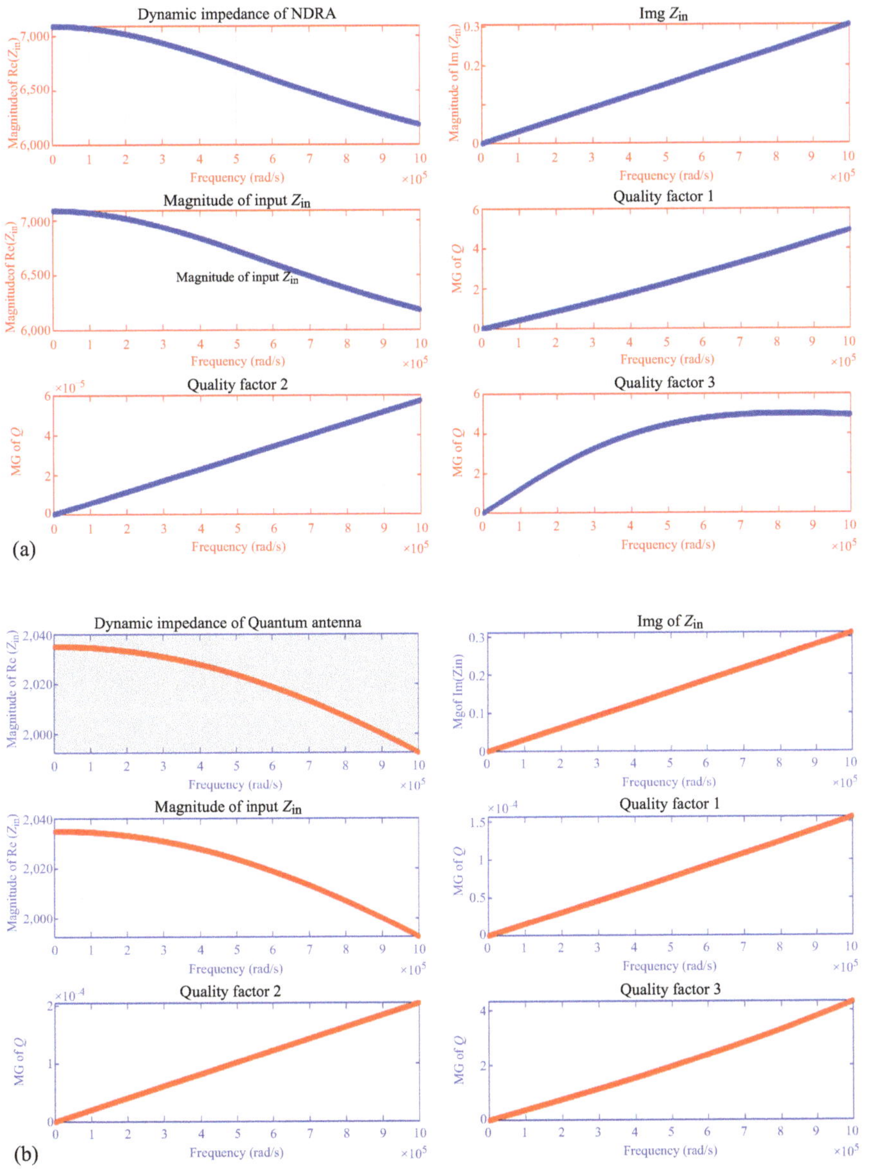

Figure 10.4 Magnitude plot of all factors of a quantum antenna for the values of: (a) $R_1 = 3\ 530\ \Omega$, $R_2 = 3\ 560\ \Omega$, $R_3 = 3\ 590\ \Omega$; (b) $R_1 = 1\ 000\ \Omega$, $R_2 = 1\ 035\ \Omega$, $R_3 = 1\ 065\ \Omega$; (c) $L_2 = 1\ 210e{-}6H$, $C_1 = 25e{-}15F$, $C_3 = 143e{-}15F$, $R_1 = 3\ 530\ \Omega$, $R_2 = 3\ 560\ \Omega$, $R_3 = 3\ 590\ \Omega$; (d) $L_2 = 1\ 210e{-}6H$, $C_1 = 25e{-}15F$, $C_3 = 143e{-}15F$ with $R_1 = 1\ 000\ \Omega$, $R_2 = 1\ 035\ \Omega$, $R_3 = 1\ 065\ \Omega$

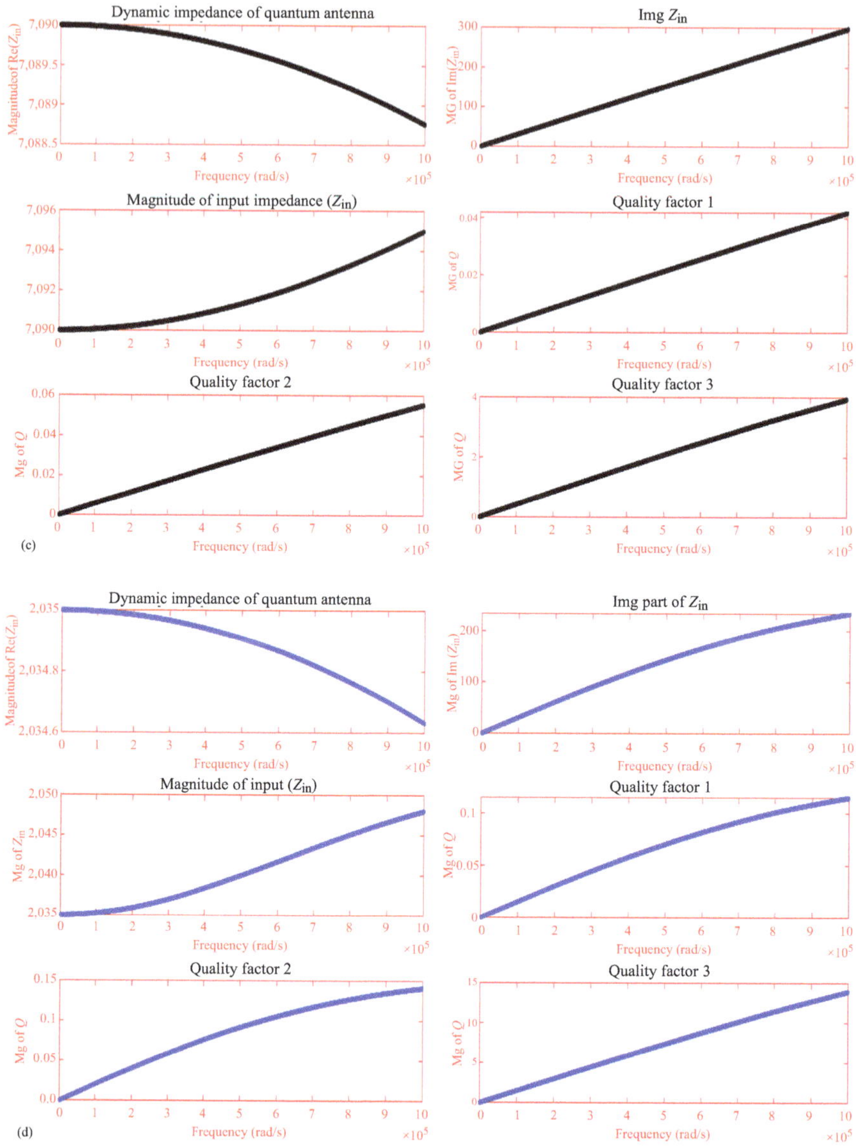

Figure 10.4 (Continued)

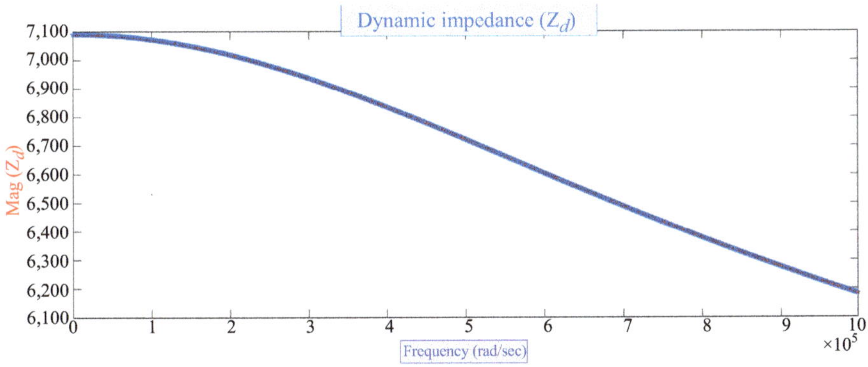

Figure 10.5 Dynamic impedance

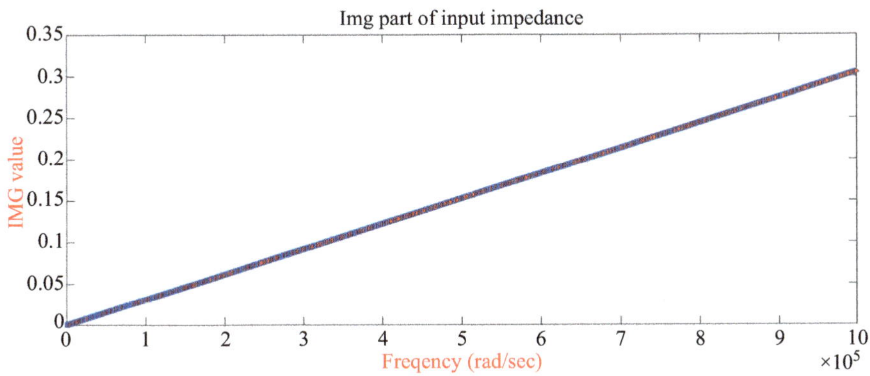

Figure 10.6 Dynamic impedance (Img)

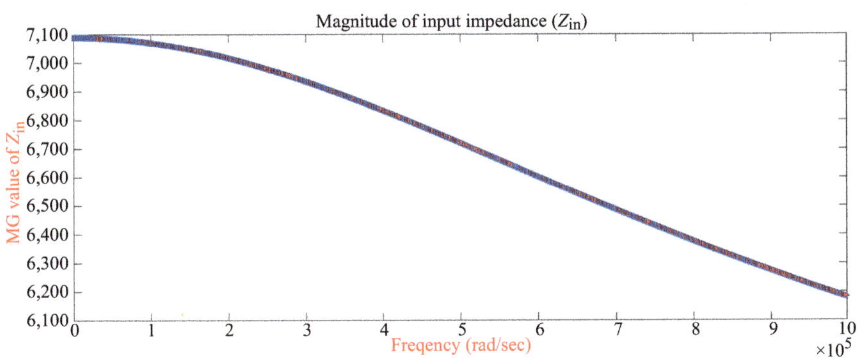

Figure 10.7 Dynamic impedance vs frequency

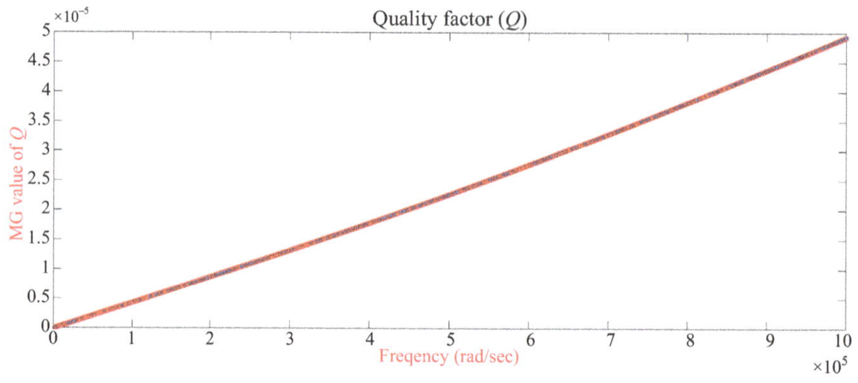

Figure 10.8 Quality factor vs frequency

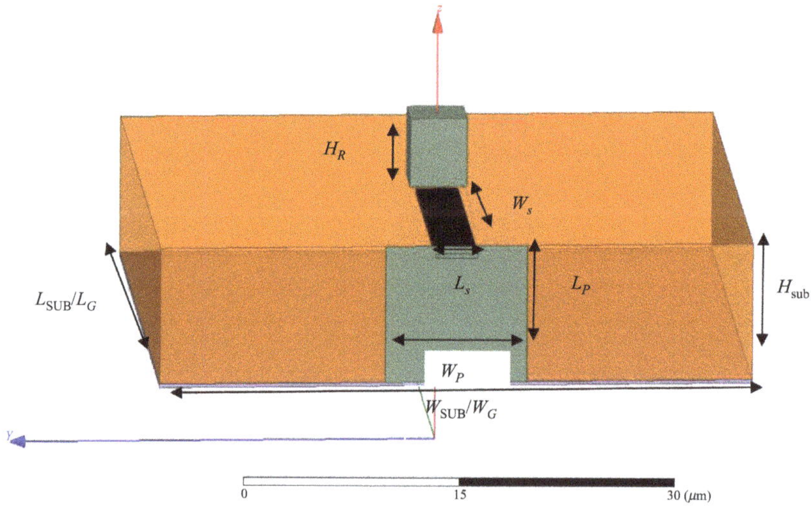

Figure 10.9 Nano-RDRA (rectangular DRA)

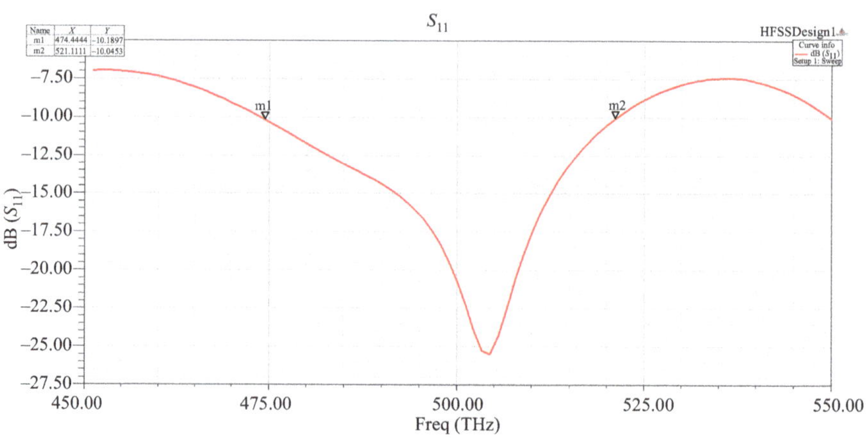

Figure 10.10 S_{11}

2. $\rho = \frac{\exp(-\beta H)}{T_r(\exp(-\beta H))}$, mean-field energy in Gibbs state $= \rho$ is given by $T_r(\rho H)$ and the $T_r(\rho p_\epsilon)$ classically we would get

$$\frac{\int \exp(-\beta H) p_a d^N q d^N p}{\int \exp(-\beta H) d_q^N d_p^N} = 0$$

$$a_k = \frac{a_k + i p_k}{\sqrt{2}} \text{annihilation and creation operators}$$

$$a_k^* = \frac{a_k - i \quad p_k}{\sqrt{2}}, \; k = 1, 2, 3 \ldots$$

$$\left[a_k, \; a_m^*\right] = \frac{h}{2\pi} \delta(k, m), \text{ Heisenberg commutation relation}$$

$$\text{Electromagneticfield} = H = \frac{1}{2} \sum_{k=1}^{N} a_k^* a_k + \frac{Nh}{4\pi}; \text{field energy}$$

Quantam states or eigen states of H

$$H|\overline{n}\rangle = \left|\overline{n_1, n_2, n_3, \ldots, \; n_N}\right\rangle, n_1, n_2, n_3, \ldots, n_N = 0, 1, 2, 3, \ldots$$

$$H|\overline{n}\rangle = \frac{h}{2\pi} \left(\frac{N}{2} + n_1, + n_2 + \ldots + n_N \right)$$

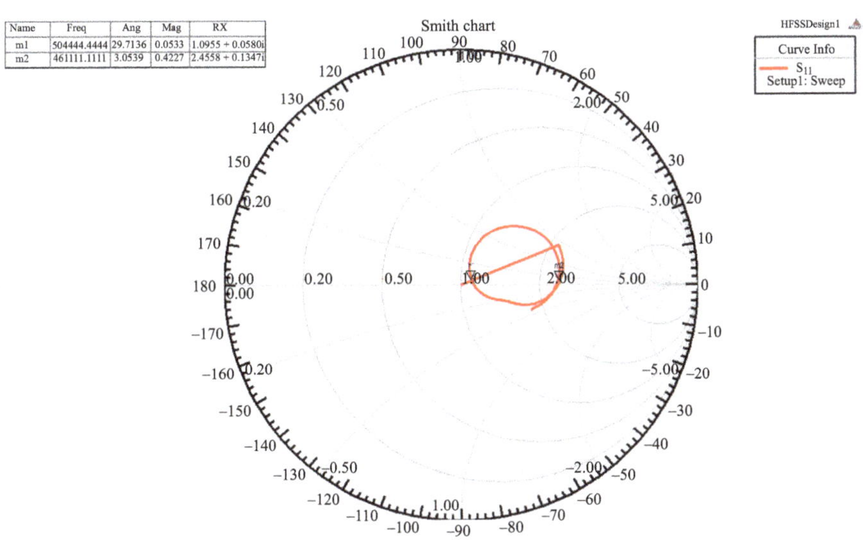

Figure 10.11 Smith chart showing impedance

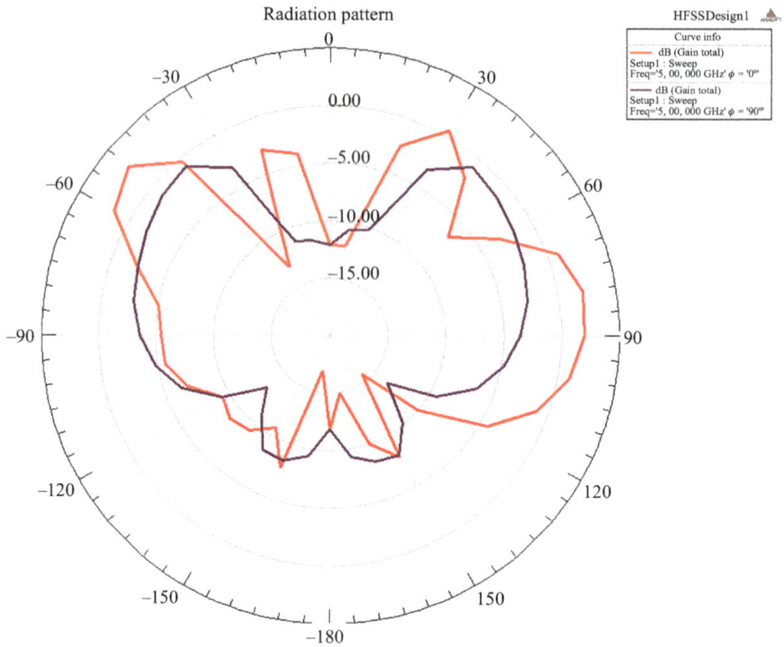

Figure 10.12 Radiation pattern

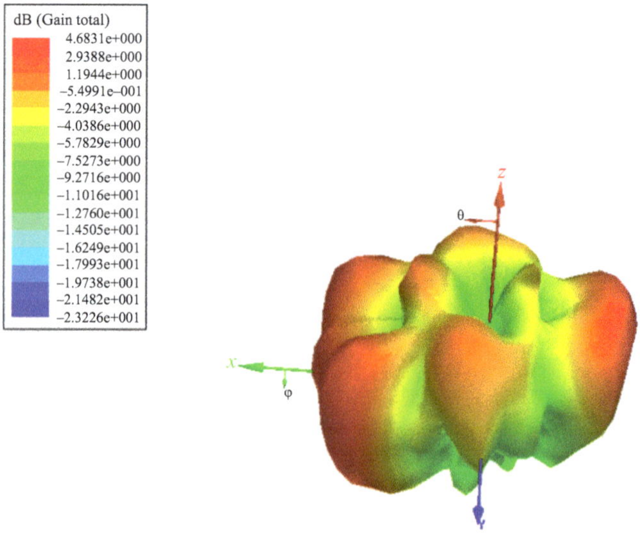

Figure 10.13 Three-dimensional radiation pattern

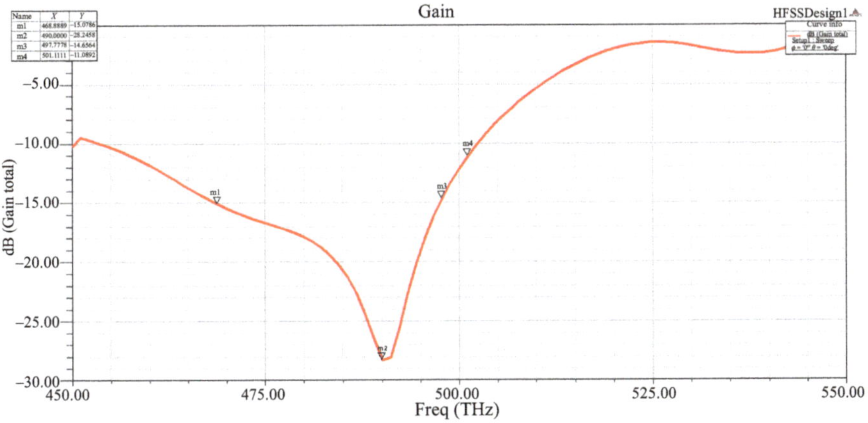

Figure 10.14 Gain vs frequency

Thus, the quantum a_k in the Gth state is

$$\exp\frac{(-\beta Nh)}{4\pi}n|a_k\cdot\exp\left(\frac{(-\beta h)}{2\pi}\sum_m a_m^* a_m\right)$$

$$\langle n\rangle = \exp\frac{(-\beta Nh)}{4\pi}\sum\nolimits_n \exp\left(\frac{(-\beta h)}{2\pi}\sum_m n_m\right)\delta(n, m - e_k) = 0$$

$$e_k = [0, 0, \ldots, \ 0, 1, \ldots, 0]$$

Here

$$T_r\left(p_k^2\exp(-\beta h)\right) = -\frac{\frac{1}{2}T_r\left[\left(p_k^2 + p_k^{*2} - a_k a_k^* - a_k^* a_k\right)e^{-\beta h}\right]}{z_q(\beta)}$$

$z_q(\beta) =$ Quantam partition function

$$= \frac{1}{2z_q(\beta)}T_r\left[\left(a_k^* a_k + \frac{h}{4\pi}\right)\exp(-\beta h)\right]$$

$$= \frac{1}{2z_q(\beta)}\exp\left(\frac{(-\beta Nh)}{4\pi}\right)\sum\nolimits_n\left(n_k + \frac{1}{2}\right)\frac{h}{2\pi}\exp\left(-\frac{\beta h}{4\pi}\sum_m n_m\right)$$

$$\int \exp\left(-\beta\sum_k\frac{(a_k^2 + p_k^2)}{2}\right)a_k^2 d_q^N d_p^N / z_c(\beta)$$

where $z_c(\beta)$ is a quantum part.

10.8 Drude's model

$$\varepsilon_{\text{Ag}} = \varepsilon_0 \left[\varepsilon_a + \frac{f_p^2}{f(f + i\gamma)} \right]$$

γ−collision frequency
$\varepsilon_a = 5$ (offset of real frequency)
f_p−plasmon frequency
$\frac{\gamma}{\pi} = 4.35$ THz
ε_0−absolute permitvity
k_0—free-space wave number
ε_r−dielectric constant
c—velocity of light.

In order of Planck's constant in quantum antennas, radiation energy comes in quanta. Thus, discrete particle distributions of electrons, positrons and photons do matter, and this discreetness is also responsible for producing quantum fluctuation and the radiation field pattern waves are

$$\varepsilon_{\text{Ag}} = \varepsilon_0 \left[\varepsilon_a + \frac{f_p^2}{f(f + i\gamma)} \right] = 129.17 + j3.28$$

$$\varepsilon_0 = 8.85 \times 10^{-12} \text{F/m}$$

$$\varepsilon_a = 5$$

$\varepsilon_r = 1.4 \times 10^{16}$ rad/s (high) due to collision photo gas particle

$\varepsilon_r = 1.4 \times 10^{16}$ rad/s (low) (collective motion of electron)

Larentz disperstion model

Drude–Larentz model for dispersion in dielectrics,

$$\varepsilon_p = \sqrt{\frac{ne^2}{\varepsilon_0 m}}$$

Optical antennas, sub-wavelength devices that directly convert an electric mode of energy ω, where is the reduced Planck constant and $\omega = 2\pi f$ is the particular frequency, which are the modes [16–40].

10.9 MATLAB program

```
r1=3530;
r2=3560;
r3=3590;
```

```
l2=1210e-9;
c1=25e-12;
c3=143e-12;
w=1:1:1000;
p1=r1*(w.^2*c3^2.*(r2+r3)^2+(w.^2*l2*c3-1).^2)
p2=((r2*r3-l2/c3).*(r2+r3)*w.^2*c3^2+(w.^2*l2*r3*c3-r2).*(w.^2*l2*c3-1))
p3=(w.^2*c3^2*(r2+r3)^2+(w.^2*l2*c3-1).^2)
r=(p1+p2)./p3
% plot(w,r)
x1=-1./w*c1;
x2=w*l2;
x3=-1./w*c3;
q1=((r2+r3).*(r3*x2+r2*x3))-((x2+x3).*(r2*r3-x2.*x3));
q2=(r2+r3).^2+(x2+x3).^2;
q=x1+(q1./q2)
% plot(w,q)
z=abs(r+1i*q)
% plot(w,z)
g=q./r
% plot(w,g)
k1=((r.^2+x2.^2).*(x1+x3))+((r3.^2+x3.^2).*(x1+x2))+(2*x1).*(r2*r3+x2.
*x3);
k2=((r1+r3)*(r2.^2+x2.^2))+((r1+r2)*(r3.^2+x3.^2))+2*r2*(r1*r3-x2.*x3)+
(2*x2).*(r1*x3-x3*r3);
k=k1./k2
% plot(w,k)
h1=w.^4*(l2*r3^2-l2*(1/c1+1/c3))+w.^2*(l2/(c3^2)-r2^2*(1/c1+1/c3)-r3^2/
(c1)-2/c1*(r2*r3-l2/c3)-1/c1*c3^2);
h2=w.^5*((l2^2*(r1+r3)))+w.^3*(r2^2*(r1+r3)+r3^2*(r1+r2)+2*r2*(r1*r3-l2/
c3)+2*l2*(-r1/c1+r3/c3))+w*(r2+r1/c3^2)
h=abs(h1./h2)
plot(w,h)
```

10.10 Conclusion

Reflection coefficient, radiation pattern and impedance of NDRA have been evaluated using *R*, *L*, *C* circuit as an equivalent circuit of NDRA. The measured and simulated results of NDRA have also been obtained. The work done on equivalent circuit configuration has been carried out for the first time in this manner and theoretical model thus developed will help designers in the realization of NDRA circuits.

References

[1] M. Alam and Y. Massoud, "RLC ladder model for scattering in single metallic nanoparticles," *IEEE Trans. Nanotechnol.*, vol. 5, no. 5, pp. 491–498, 2006.

[2] R. K. Mongia and P. Bhartia, "Dielectric resonator antennas—Are view and general design relations for resonant frequency and bandwidth," *Int. J. Microwave Millimeter Wave Comput. Aided Eng.*, vol. 4, no. 3, pp. 230–247, 1994.
[3] M. Alam and Y. Massoud, "An accurate closed-form analytical model of single nanoshells for cancer treatment," Proc. 2005 48th Midwest Symp. Circuits Syst., pp. 1928–1931, 2005.
[4] J. L. West and N. J. Halas, "Applications of nanotechnology to biotechnology: Commentary," *Curr. Opin. Biotechnol.*, vol. 11, pp. 215–217, 2000.
[5] C. Oubre and P. Nordlander, "Optical properties of metallodielectric nanostructures calculated using the finite difference time domain method," *J. Phys. Chem. B*, vol. 108, pp. 17740–17747, 2004.
[6] P. J. Flatau, "Fast solvers for one dimensional light scattering in the discrete dipole approximation," *Opt. Express*, vol. 12, pp. 3149–3155, 2004.
[7] M. Alam and Y. Massoud, "A closed-form analytical model for single nanoshells," *IEEE Trans. Nanotechnol.*, vol. 5, no. 3, pp. 265–272, 2006.
[8] J. L.-W. Li, Z.-C. Li, H.-Y. She, S. Zouhdi, J. R. Mosig, and O. J. Martin, "A new closed-form solution to light scattering by spherical nanoshells," *IEEE Trans. Nanotechnol.*, vol. 8, no. 5, pp. 617–626, 2009.
[9] L. A. Ambrosio and H. E. Hernández-Figueroa, "RLC circuit model for the scattering of light by small negative refractive index spheres," *IEEE Trans. Nanotechnol.*, vol. 11, no. 6, pp. 1217–1222, 2012.
[10] D. C. Tzarouchis, P. Ylä-Oijala, and A. Sihvola, "Resonant scattering characteristics of homogeneous dielectric sphere," *IEEE Trans. Antennas Propag.*, vol. 65, no. 6, pp. 3184–3191, 2017.
[11] M. Farhat, C. Rockstuhl, and H. Bağcı, "A 3D tunable and multi-frequency grapheme plasmonic cloak," *Opt. Express*, vol. 21, pp. 12592–12603, 2013.
[12] S. Ghadarghadr and H. Mosallaei, "Coupled dielectric nanoparticles manipulating metamaterials optical characteristics," *IEEE Trans. Nanotechnol.*, vol. 8, no. 5, pp. 582–594, 2009.
[13] C. F. Bohren and D. R. Huffman, *Absorption and Scattering by a Sphere*, Wiley, New York, NY, 1983.
[14] A. Garcıa-Etxarri, R. Gómez-Medina, L. S. Froufe-Pérez, *et al.*, "Strong magnetic response of submicron silicon particles in the infrared," *Opt. Express*, vol. 19, pp. 4815–4826, 2011.
[15] Y. Guo, T. Zhang, W.-Y. Yin, and X.-H. Wang, "Improved hybrid FDTD method for studying tunable graphene frequency-selective surfaces (GFSS) for THz-wave applications," *IEEE Trans. THz Sci. Technol.*, vol. 5, no. 3, pp. 358–367, 2015.
[16] T. Bian, X. Gao, S. Yu, L. Jiang, J. Lu, and P. Leung, "Scattering of light from graphene-coated nanoparticles of negative refractive index," *Optik*, vol. 136, pp. 215–221, 2017.
[17] A. Vakil, *Transformation Optics Using Graphene: One-Atom-Thick Optical Devices Based on Graphene*, University of Pennsylvania, Philadelphia, PA, 2012.

[18] Y. S. Cao, L. J. Jiang, and A. E. Ruehli, "An equivalent circuit model for graphene-based terahertz antenna using the PEEC method," *IEEE Trans. Antennas Propag.*, vol. 64, no. 4, pp. 1385–1393, 2016.

[19] T. Christensen, A.-P. Jauho, M. Wubs, and N. A. Mortensen, "Localized plasmons in graphene-coated nanospheres," *Phys. Rev. B*, vol. 91, Art. no. 125414, 2015.

[20] M. Abramowitz and I. A. Stegun, *Handbook of Mathematical Functions: With Formulas, Graphs, and Mathematical Tables*, vol. 55, Courier Corporation, North Chelmsford, MA, 1964.

[21] L. Novotny and B. Hecht, *Principles of Nano-Optics*, Cambridge University Press, Cambridge, 2006.

[22] H. W. Kroto, J. R. Heath, S. C. O'Brien, R. F. Curl, and R. E. Smalley, "C60: Buckminsterfullerene," *Nature*, vol. 318, pp. 162–163, 1985.

[23] H. Nishihara, T. Simura, S. Kobayashi, *et al.*, "Oxidation-resistant and elastic mesoporous carbon with single-layer graphene walls," *Adv. Funct. Mater.*, vol. 26, pp. 6418–6427, 2016.

[24] H. Yang, Z. Hou, N. Zhou, B. He, J. Cao, and Y. Kuang, "Graphene encapsulated SnO_2 hollow spheres as high-performance anode materials for lithium ion batteries," *Ceram. Int.*, vol. 40, pp. 13903–13910, 2014.

[25] J.-S. Lee, S.-I. Kim, J.-C. Yoon, and J.-H. Jang, "Chemical vapor deposition of mesoporous graphene nanoballs for supercapacitor," *ACS Nano*, vol. 7, pp. 6047–6055, 2013.

[26] M. Naserpour, C. J. Zapata-Rodríguez, S. M. Vuković, H. Pashaeiadl, and M. R. Belić, "Tunable invisibility cloaking by using isolated graphene-coated nanowires and dimers," *Sci. Rep.*, vol. 7, 2017, Art. no. 12186.

[27] A Santos and V Garjo, *An exact Solution of the Boltzmann Equation for a Binary Mixture*, Taylor and Francis, 2006. (doi.org/10.1080/00411459208203789).

[28] L. Novotny and N. van Hulst, "Antennas for light," *Nat. Photonics*, vol. 5, no. 2, pp. 83–90, 2011.

[29] P. Biagioni, J.-S. Huang, and B. Hecht, "Nano antennas for visible and infrared radiation," *Rep. Prog. Phys.*, vol. 75, p. 024402, 2012.

[30] L. Zou, W. Withayachumnankul, C. Shah, A. Mitchell, M. Bhaskaran, S. Sriram, and C. Fumeaux, "Dielectric resonator nano antennas at visible frequencies," *Opt. Express*, vol. 21, no. 1, pp. 1344–1352, 2013.

[31] G. N. Malheiros-Silveira, G. S. Wiederhecker, and H. E. Hernández-Figueroa, "Dielectric resonator antenna for applications in nano photonics," *Opt. Express*, vol. 21, no. 1, pp. 1234–1239, 2013.

[32] M. W. Knight, H. Sobhani, P. Norlander, and N. J. Halas, "Photodetection with active optical antennas," *Science*, vol. 332, pp. 702–704, 2011.

[33] P. Mühlschlegel, H.-Eisler, O. J. F. Martin, B. Hecht, and D. W. Pohl, "Resonant optical antennas," *Science*, vol. 308, no. 5728, pp. 1607–1609, 2005.

[34] Y. Zhao, N. Engheta, and A. Alù, "Effects of shape and loading of optical nano antennas on their sensitivity and radiation properties," *J. Opt. Soc. Am. B*, vol. 28, no. 5, pp. 1266–1274, 2011.

[35] E. Ozbay, "Plasmonics: merging photonics and electronics at nanoscale dimensions," *Science*, vol. 311, p. 189, 2006.

[36] R. S. Yaduvanshi and H. Parthasarathy, "Coupled solution of Boltzmann transport equation, Maxwell's and Navier Stokes equations," *IJACSA*, vol. 3, 2010.

[37] L. Zou, W. Withayachumnankul, C. Shah, *et al.*, "Efficiency and scalability of dielectric resonator antennas at optical frequencies," *IEEE Photon. J.*, vol. 6, pp. 1–10, 2014.

[38] R. S. Yaduvanshi and H. Parthasarathy, *Rectangular DRA Theory and Design*, Springer, 2016.

[39] R. S. Yaduvanshi and V. Gaurav, *Nano Dielectric Resonator for 5G Applications*, CRC Press, 2020.

[40] A. Bonakdar and H. Mohseni, "Impact of optical antennas on active optoelectronic devices," *Nanoscale*, vol. 6, pp. 10961–10974, 2014.

Chapter 11

Optical DRA for retinal applications—next generation DRAs

Abstract

Human eye retinas have millions of cones and rods cells, also known as photoreceptors, to receive photons from the nature. There are arrays of photoreceptors on retina periphery and central part of retina. These cells perform functions of photonic antennas, for example they receive light particles and convert it into electrical impulses. In this chapter, cones and cylindrical rods have been studied and analyzed as photonic wavelength antennas. These terahertz dielectric resonator antennas (DRAs) have been developed using HFSS and CST simulators. Their simulation modeling and theoretical modeling have been presented in this chapter. The work carried out in this chapter is much useful for the development of artificial retinal antennas for possible use in retinal prosthesis using terahertz DRAs and photodiodes as bio-photoreceptors. These DRAs at terahertz frequencies can also be used in applications of scanning, imaging and wideband ultrahigh-speed communication systems.

11.1 Introduction

Light is reflected by natural objects when falls on any object, later it reaches eye cornea, retina, optical nerves for image formation into brain; thus reflected light particles are being captured by retina in the vision spectrum. Light enters through the eye lens and is focused on to retina through the white color ball known as a vitreous humor, which is a gel-like material and consists of 98% water. An eye retina has 1.26 million rods and approximately 5 million cones known as photoreceptors. Rods in retina have 496-nm wavelengths, which are responsible for black and white vision. Cone are color vision cells and they are in three wavelengths and corresponds to R, G, B basic color wavelengths. Cones have three different wavelengths of 420, 530 and 560 nm, respectively. These photoreceptors can absorb light in the frequency spectrum from 430 to 750 THz, also known as a vision spectrum [1,2].

These photoreceptors have capability to transduce light (photons) into electrical impulses (neurons). Photoreceptor cells have plasmonic membranes and vision pigment molecules called rhodopsin. These photons react with rhodopsin, phenomenon of electro-chemistry takes place and electrical current of very small magnitude is generated in the form of neurons. Neurons can communicate in chemical form as well as in electrical form. There are networks of millions of optic nerves that carry these neuron signals from retinal surface to brain. Brain performs processing at a very fast rate to develop images corresponding to visuals or scene captured by a human eye. Each photon-receptor cells constitute one pixel to develop pixel formats in the brain, and this will enable to develop an array of pixels for transforming it to a two-dimensional image. These photons are collected by both the retinas simultaneously and synchronously processed by the brain. These captured scenes or images are of temporal and spatial in nature. The phenomenon of real-time processing takes place at a rapid rate.

Rods cells are responsible for creating black and white vision, whereas cones cells have capability to create color vision. Thus, rod cells create photopic vision and cone cell create scotopic vision. Rods provide peripheral vision and cones provide central vision. The diameter of rods cell is 1–3 μm and the length of rod is 6–50 μm in retina. Rods are in cylindrical shape and cones are in conical shape. The refraction index of eye tissue is 1.336, in a vision spectrum. The specific absorption rate of tissue is measured as 2 W/kg on 10 g. Permittivity of rod tissues is 69 F/m and conductivity is 1.53 S/m. Technical specifications of retinal membranes for rods and cones are given in Tables 11.1 and 11.2.

Photoreceptors form the first part of vision process, for sending signals progressively to different cells before reaching the brain. Bipolar cells and ganglion cells form the important part of vision process. Bipolar cells are induced graded potential based on photoreceptor vision signal and ganglion cells process it as gated signal and transverse it to a bundle of optic nerves, which get terminated on a lateral geniculate nucleus (LGN). From an LGN the visual signal is sent to the primary visual cortex of brain to generate two-dimensional image that is also called visual scene. Ganglion cells play a crucial part in vision process to detect object movements contrast and intensity part of the seen image due to visual pigment present as then.

Table 11.1 Rod cell parameters

Count and shape	**1.26 millions, cylindrical**
Vision spectral	430–750 THz
Diameter	1–3 μm
Length	5–60 μm
Photocurrent	1 pA
Refraction index	1.336
Wavelength	496 nm
Vision	Peripheral vision
Specific absorption rate (SAR)	2 W/kg
Vitreous humor (ϵ_r)	67 F/m
Photopigment	Rhodopsin
Density	1.60×10^3/mm^2

Table 11.2 Retina parameters

Number of layers	10
Number of incident photons	500 photons/μm/s
Gated potential	Alternate signal
Grade potential	Bipolar cell 0/P (−10 to −70 mV)
Incident power	7 mW
Photoreceptors in center	Cones
Photoreceptors in periphery	Rods

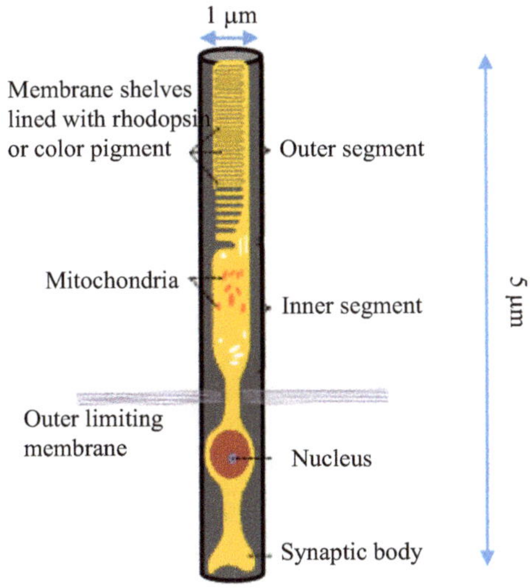

Figure 11.1 Quantum antenna as a 5-μm length cylindrical rod

There are two types of retinal diseases: one is age-related macular degeneration and the other is retinitis pigments. These are caused by a degeneration of photopigment in old age. Worldwide, millions of people have been affected by these two abovementioned diseases resulting into visual impairment or blindness. This chapter proposes a technique of vision restoration by suggesting a quantum antenna as an artificial rod cell to be used as eye prosthesis (Figure 11.1). A quantum antenna is a unique technique with an integration of photodiode to provide required graded potential to bipolar cells, and the rest of the vision process will remain unaltered. The resolution obtained by using a quantum antenna will be higher than the previous work available. In our proposed work, capture of the image scene requirement of external camera has been eradicated. As the per literature survey Argus I, Argus II and Alpha eye prosthetics have been found very bulky and with low resolution. Cylindrical, spherical and conical antennas in

MIMO or array pattern can be suitable candidates for designing artificial photoreceptors. These are given in Figures 11.2–11.4.

Working principle of optical antennas:

Light in (source LASER) light–matter interaction: the electrons **and positrons interaction with photons or the quantum light–matter field in a fermionic bosonic coherent state (scattering, distortions, SPP (surface plasmon polytron) and Hamiltonian as the amplitudes for scattering)**
Define the **fermionic and bosonic annihilation and creation operators.**
Describe the **fermionic and bosonic as Dirac field** (momentum-spin, photonic wave field).
Compute the Dirac current (displacement type, tangential magnetic field, combined effect of electrons positrons and photons).

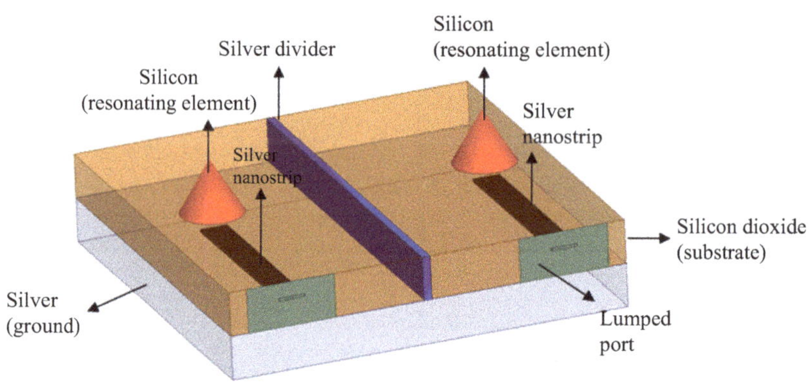

Figure 11.2 MIMO conical DRA at photonic wavelength

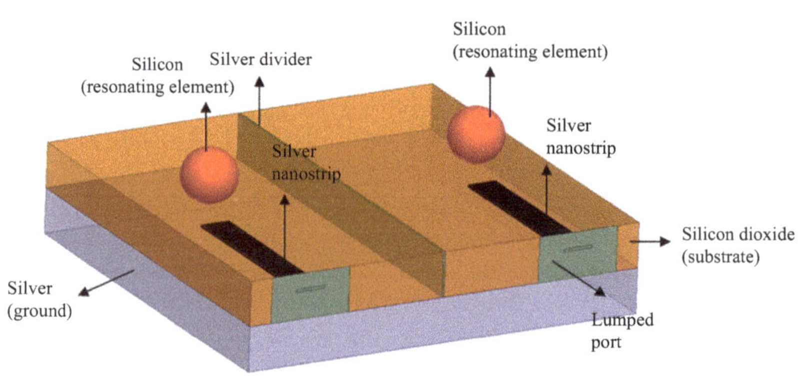

Figure 11.3 MIMO spherical DRA at photonic wavelength

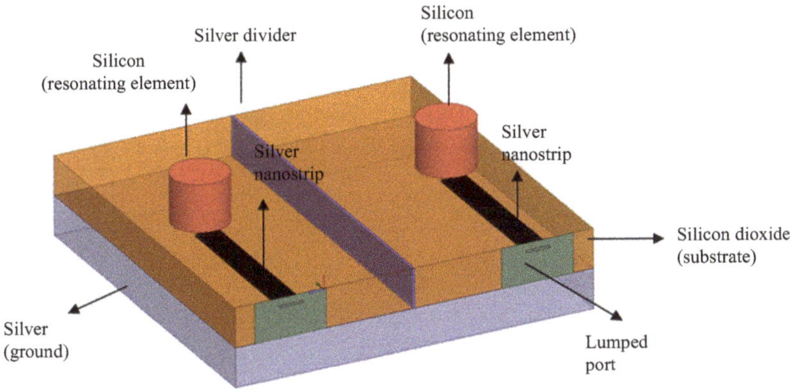

Figure 11.4 MIMO CDRA at photonic wavelength

Determine eigen solution to the Dirac wave equation using 4×4 Dirac-α-matrices satisfying the canonical anticommutation relation for fermions and canonical commutation relation (CCR) for bosons:

Express the **Dirac current density** (electromagnetic (e.m.) wave field pattern).
Find **far-field magnetic vector potential** (retardation potentials).
Determine **far-field radiation pattern** (quantum wave operator field).
Berezin weight function parameterized by $\underline{\theta}$ (relates to Grassmann variables).
Minimize the **quantum fluctuation correlation** in the pattern for possible matching the **average antenna pattern to the desired pattern** (boson Fock space).
The mean and mean-square fluctuations of quantum e.m. fields (antenna and free photons).
In the super directive (emitted vector potential, photons with different momenta and helicities, e.m. four potential fields), **radiated field is a state of jointly coherent for the bosons (i.e. photons) and for the fermions (positrons–electrons).**

11.2 Optical antenna arrays basic requirements

1. Study of vision process.
2. Use of biocompatible electronic devices for prosthesis.
3. To develop an electronic device for a possible replacement of dead photoreceptors in retinal layer.
4. Compatibility of electronic device output with bipolar cell graded potential.
5. Major part of image processing involves cornea, lens, vitreous humor, bipolar cells, ganglion cells, rods, cones (photoreceptors) with epithelium as absorber, optic nerves and brain.
6. Reflected light from natural objects falls on cornea travels through vitreous humor reaches to photoreceptors gets converted into electrical signal.

7. These photoreceptors are formed by 126 million rods and cones in retina.
8. Cones and rods have several layers of plasma membranes, and it has photosensitive vision pigment molecules. It has capability to absorb light by making use of rhodopsin and convert into electrical signal.
9. Rods are responsible for black and white vision that is also known as peripheral vision.
10. Rods are located at periphery of the retina.
11. Cones are responsible for colored vision and located at the center of retina.
12. Photoreceptors absorb, amplify incident light, transduce it to neural response for further communication to brain through network of optic nerves.
13. Image is developed from the scenes in the brain, i.e. photons get converted into pixels.
14. Photoreceptors are at initial stage in the vision process. In the second stage these signals are progressively processed by bipolar cells as graded potential. At the next stage, ganglion cells participate to convert the graded potential signals into gated potentials signals. Now these signals can be termed neural signals.
15. Each ganglion cell gated potential signal travels to brain through optic nerves.
16. Neurons have capability to communicate through chemical potential as well as through electrical potential.
17. Temporal and spatial phenomena on real-time basis take place in ten layers of retina before developing an image in the brain.
18. The resolution of eye cell depends on the number of photons received by quantum antennas in prosthesis.
19. Array of quantum antennas can be used as retinal implants.
20. Vision in human retinal is circularly polarized.
21. Optical antenna: Light–matter interaction on the nanoscale, occurrence of plasmon resonances, volume currents, kinetic inductance, effective wavelength scaling, near-field intensity enhancement and Lorentzian resonance phenomenon takes place.
22. The e.m. field is only one possible physical observable operator of the quantum state.
23. The desired higher order modes can be excited by surface plasmon resonance (SPR), which is a sub-wavelength phenomenon and allows coexistence between different modes.
24. Incident field, scattering field, total field and directivity are calculated. The permittivity of the metal can be described using Drude's model.
25. SPPs or resonance frequency, the quantum efficiency and the enhancement of field are major objectives of an optical antenna.
26. We can incorporate some unknown real parameters into the Berezin linear combination of **coherent states and estimate these parameters by minimizing the distance between the average value of the e.m. field generated by the fermions and the desired e.m. field pattern.** If need be, we may modify this cost function to be minimized by **constraining the higher order quantum statistical moments of the generated quantum e.m. field** to be

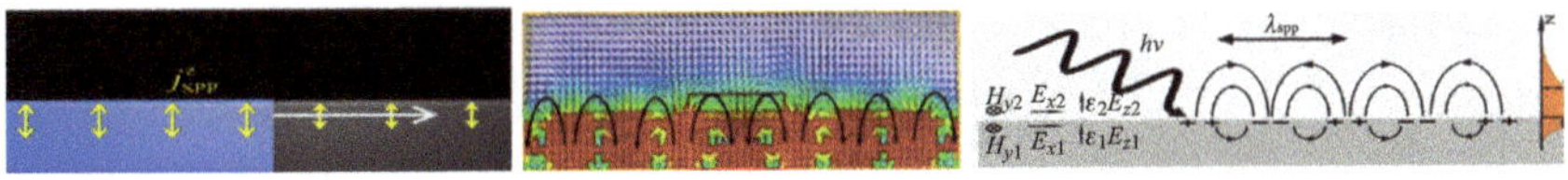

Figure 11.5 Wave vector of SPP ($k_{SPP} = 1{:}1k_0$)

specified. An example of an application of this circle of ideas is to use a quantum antenna to generate a set of desired spatial patterns at a given set of frequencies.

27. Let γ be a Grassmannian variable that will be used to specify the coherent state of this fermion just as a complex number z is used to specify a coherent state of a single boson.

    γ Anticommutes with itself, γ^* and with a, a^* just as in bosonic situation, the complex number z that specifies the coherent state commutes with itself, with $\bar{z}$ and with the boson creation and annihilation operators.
28. The current density field generated by this field as according to Dirac's theory, a quadratic function of these operators, and hence the e.m. field generated by this current density according to the retarded potential formula, is also a quadratic function of these operators.
29. Figure 11.5 shows a rectangular silicon terahertz dielectric resonator antenna (DRA) with LASER and silver FEED (side and top view).

Due to interaction between the surface charge density and the e.m. field results in the momentum of the SP mode in terahertz DRA cavity. Wave vector (SPP) being greater than that of a free-space photon of the same frequency is the free-space wave vector ($k_0 = \omega/c$). SPPs are generated on a metal–dielectric interface and a feedback mechanism allows SPP to resonate. The interaction of surface charge on the metal and dielectric at interface generates a strong capacitive region.

The previous phenomenon is expressed as follows using flowchart mechanism:

The terahertz antenna far field radiations
Light in Gaussian pulse for phase matching.

⇓

The electrons **and positrons or the quantum matter field in a fermionic coherent state.**

⇓

Define the **fermionic annihilation operator.**

⇓

Describe the **fermionic Dirac field.**

⇓

Compute the Dirac current.

⇓

Determine eigen solution to the Dirac wave equation using 4×4 Dirac-α-matrices satisfying the anticommutation relation.

⇓

Express the **Dirac current density.**

⇓

Find **far-field magnetic vector potential.**

⇓

Determine **far-field radiation pattern.**

⇓

Berezin weight function parameterized by $\underline{\theta}$.

⇓

Minimize the **quantum fluctuation correlation** in the pattern for possible matching the **average antenna pattern to the desired pattern.**

⇓

e.m. radiated fields (super directive; Figure 11.6)

The enhancement effect is attributed to the resonance of SPPs. Near field is due to a resonant excitation of SPPs. The electric dipole source, SPP, is first launched by an emitter in the nanogap and then scattered, e.m. radiation, emission rates and the far-field radiation pattern; SPP imposes fundamental dominant mode and bounded. The electrooptical antennas working principal is based on electron plasmon coupling. Localized fields near optical antenna structures have spatial dimensions that approach the length scale of atomic/molecular quantum wave functions. The localization of optical antenna gives rise to photon momenta that are of the order of the momenta of electrons in matter and can therefore give rise to traditionally momentum-forbidden transitions use of optical sensing. It can absorb all of the incident radiation if each molecule is coupled to an optical antenna. The penetration of radiation into metals takes place. Owing to the finite electron

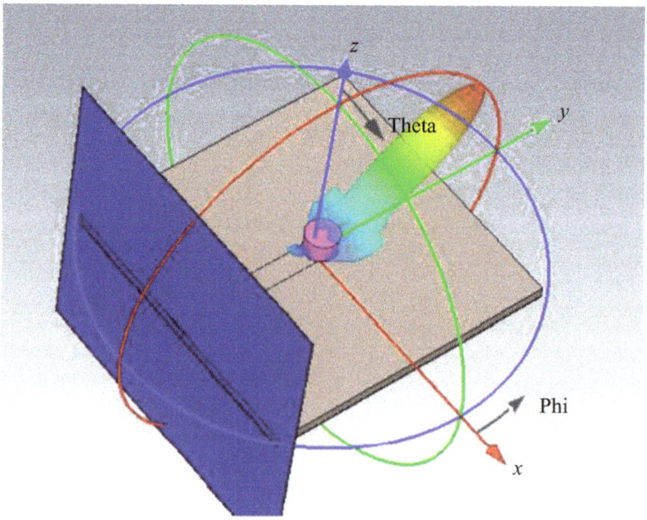

Figure 11.6 Super-directive radiations

density, there is delay between the driving field and the electronic response, resulting in a skin depth that is typically larger than the diameter of the antenna elements. As a consequence, electrons in metals do not respond to the wavelength λ of the incident radiation but to an effective wavelength λ_{eff}, which is determined by a simple linear scaling rule:

$$\lambda_{\text{eff}} = n_1 + n_2\left(\frac{\lambda}{\lambda_p}\right)$$

where n_1 and n_2 are geometric constants and λ_p is the plasma wavelength; according to this wavelength scaling rule, an optical half-wave antenna is not $\lambda/2$ in length but a shorter length of $\lambda_{\text{eff}}/2$. The difference between λ and λ_{eff} depends on geometric factors but is typically in the range of 2–5 for most of the metals used as optical antennas. Reducing the dimensions even further reaches the molecular scale, where the design of optical antennas could draw inspiration from biology. For example, in light-harvesting proteins, chlorophyll molecules arrange.

Nonlinear optical antennas hold promise for generating and controlling this nonlinear response on a sub-wavelength scale. For example, it has been demonstrated that controlling the gap between the nanoparticles in a gold nanoparticle dimer allows the intensity of the nonlinear frequency conversion to be tuned over **four orders** of magnitude.

11.3 Optical antenna design

To control and manipulation of optical fields at the nanometer scale, SPP mechanism is used. This is generated through proximity-coupled feed. The optical antennas are similar to their microwave antennas. Optical antennas have small size with resonant properties of nanostructures. Optical antennas surpass the diffraction limit in optical imaging. SPRs make optical antennas particularly efficient at selected frequencies. These are useful in biological sensing and detection applications.

They have a feature of photo detection, light emission, sensing, heat transfer and spectroscopy.

A terahertz DRA here is a cavity resonator consisting of an ensemble of electrons, positrons and photons (Figure 11.7). The electron, positron field is the Dirac second quantized wave function $\psi\left(t, \underline{r}\right)$, i.e. a fermionic operator field, while the second quantized photon field is the Maxwell potential field. The first-order change in the current density $\delta J_\mu(x)$ caused by the laser source and the interaction between the Dirac field with the quantum photon field is

$$\delta J_\mu(x) = -e\psi_{(x)}^{(o)^+}\alpha_\mu\delta\psi(x) - e\delta\psi(x)^+\alpha_\mu\psi_{(x)}^{(o)}$$

The notion of a fermionic coherent state and its application to the computation of the quantum statistical moments of the quantum e.m. field is generated by electrons and positrons in terahertz DRA.

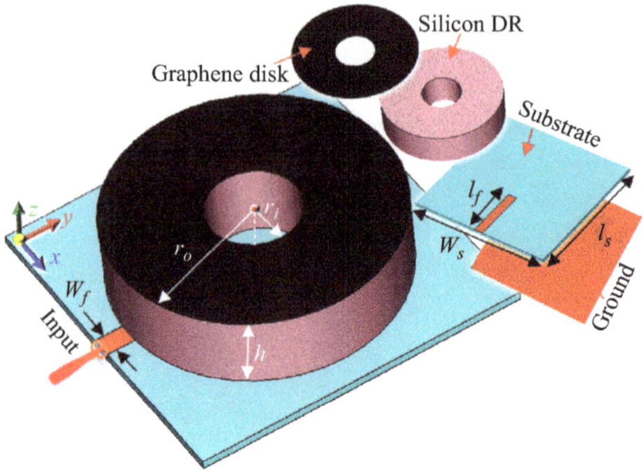

Figure 11.7 Terahertz DRA design using graphene disk and silicon annular DRA

11.4 Entanglement

It is said to be in entanglement when states of photons are in coherent or synchronous spin. Particles remain connected so that actions performed on one instantaneously affect the state of the other, even when they are separated by large distances. Convert entanglement of photons into the entanglement of the atoms that absorb those photons. Let two ground-state atoms separated by a small distance excite those atoms to a very high-energy state, they interact over longer ranges and can thus shift each other's energy levels. The two atoms are now entangled if both are in superposition due to laser pulse, i.e. Rydberg level and other in ground state. Hence, an interaction in two atoms that leads to indeterminate but correlated final states. How to identify SPP in optical antenna response is reflectance (reflected power) measurement on wavelength, which provide two dips: larger dip is antenna resonance (lower frequency) and smaller dip as SPR (at higher frequency). An optical antenna can provide focused optical energy when absorption quality factor is equal to radiation quality factor. This is achieved by an SiO_2 surface on silver metal by substrate thickness-controlled radiations through constructive interference. Coupled mode theory is involved for field enhancement.

11.5 Modeling of optical antennas

Laser output of coherent states is given inputs to optical antenna. Due to laser inputs, SPP phenomenon takes place and generates SPR. Multimode phenomenon as well as coupling of resonant modes takes place. Due to the interaction of light with matter scattering, absorption, i.e. creation and annihilation, happens. SPR is now coupled to DRA for possible radiations. This process can be described in more detail as given next.

Creation and annihilation (add particle and remove particles from state) antenna resonance is dependent upon dimensions, material and loading, effective mode volume, effective antenna aperture to match laser input resulting into enhanced directivity, minimal loss (silver) at optimum quality factor, i.e. $Q_{\text{rad}} = Q_{\text{abs}}$ (SiO_2 thickness); also less thickness more radiation and more thickness, more absorption and less radiation.

It is possible to shape the second and higher order correlation functions of the field emitted by a quantum antenna in the far-field zone by designing its initial state. The designing states with two initial excitations lead to highly directional emission of photon pairs for the same antenna or even produce the effect of no radiating sources by suppressing the field in the far-field zone. The quantum antenna can produce multi-photon momentum entangled states. The desired spatial patterns of the correlation functions, also patterned higher order correlation functions of the emitted field, are introduced by a quantum antenna. The quantum antennas can be realized in a number of different ways, such as chains of semiconductor quantum dots, cold atoms in optical lattices.

Laser-generated coherent photon: the second-order correlation function of the two-particle entangled state can produce photon pairs that are strongly correlated in momentum. We describe the procedure for designing the antenna state with the required correlation functions.

11.6 Light–matter interaction

Localization of the emitted field inside a finite volume is something that one would really expect. In twin-photon generation, the coefficients provide a desired spatial pattern of the second-order correlation function. Effective excitation and photon extraction are particularly important, weak signals emitted by a single atom or molecule, and large efficiencies can be achieved by either planar dielectric antennas beaming with nanostructures that beam emission into a narrow angular distribution, accurate positioning of the quantum emitter, planar optical antenna that beams light emitted by a single molecule, high numerical aperture. The initial state corresponds to excited pairs of the dipoles located symmetrically on the opposite sides of the antenna. The wave function of the emitted field state is computed. The dipole moments are orthogonal to the antenna plane. Quantum interferences are essential for shaping the correlation functions (spatial correlation functions). Capture the mechanism of spontaneous emission and the effects stemming from it. Field correlation effects can still be successfully captured. The best known example is super radiance. The onset of cooperative effects and phase correlations leading to the formation of the super radiant field pulse can be quite accurately described. The correlations self-establish at the initial stage of cooperative emission. Simultaneously changing the shape and the initial state of the antenna, one can get a high directivity of the correlation function, which is a simple way to obtain directional correlations of emitted photons that are localized in a narrow region. In dipole orientation to produce polarization, spatiotemporal current (J) is produced or displacement current, inductive impedance. Power is mostly transferred to the metal by absorption or excitation of surface plasmon polaritons, or the emission is suppressed by destructive interference with the image dipole. Excited electrons jump to higher state and come to lower state, and these are Rabi oscillations or surface plasmonic resonance. Matching the spatial profile of the emitter with that of a focused laser beam, the antenna allows for lower pump powers, which in turn implies that a better signal-to-noise ratio is achieved.

Results can be explained in terms of the image dipole induced in the reflector by the source. When the two dipoles radiate with an appropriate phase difference, which depends on the optical constants, they constructively interfere in the forward direction, which results in a beaming effect (beaming occurs in the range of $\lambda/6n \leq d1 \leq \lambda/4n$). In momentum space structure or gradient field operator or field enhancement, constructive interference gives rise to the beaming effect. The simultaneous second-order correlation functions in the far-field zone for excitation

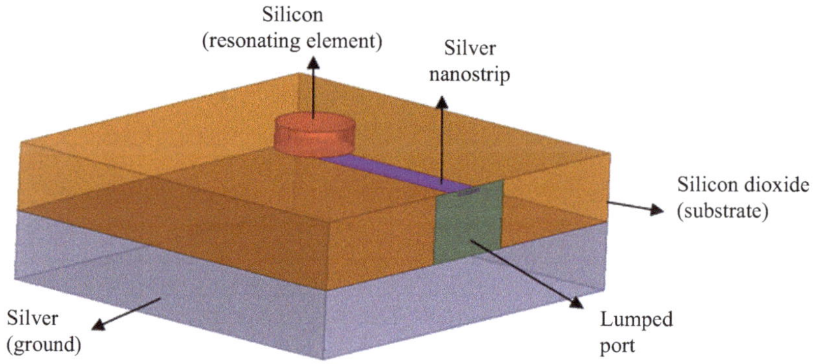

Figure 11.8 Terahertz cylindrical DRA design and analysis

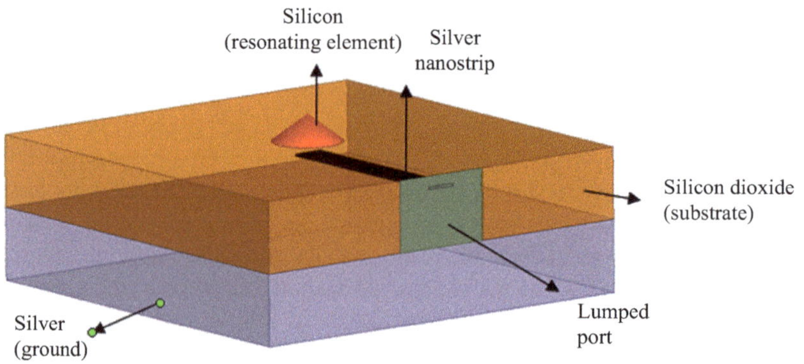

Figure 11.9 Terahertz conical DRA design and analysis

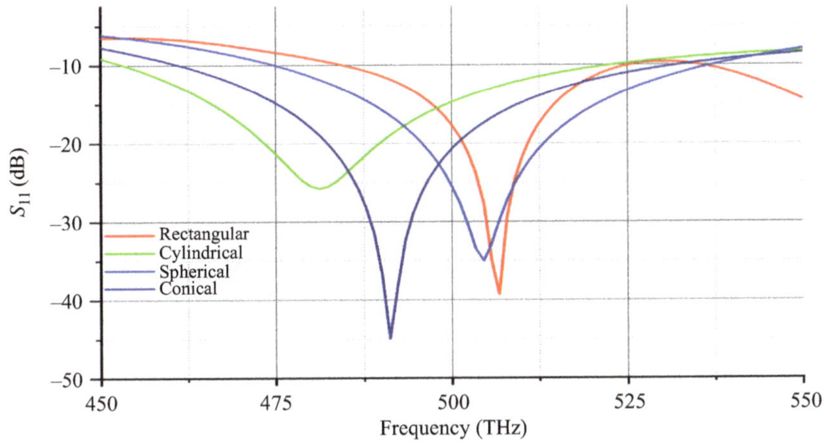

Figure 11.10 Reflection coefficient (S_{11}) simulation results on different geometries of terahertz DRAs

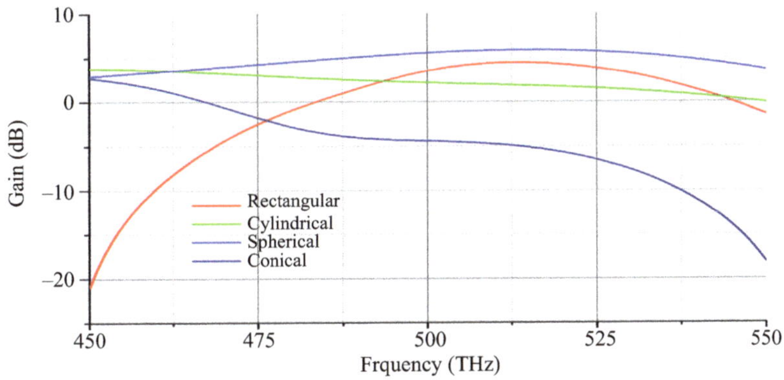

Figure 11.11 Gain simulation results

of two states. The beaming of light from a single molecule in a semi-cone is only 20° wide, the emission pattern resembles a weakly focused Gaussian laser beam.

11.7 Theory of coupled resonant modes

The coupling of cavity modes based on structure takes place. The field enhancement is another important parameter. The emitted power is inversely proportional to the excited state lifetime. The optical antenna can strongly direct the radiation in a particular direction. An equivalent circuit of a quantum antenna can be developed on the basis of coupled Boltzmann transport equation (BTE) solution. Nanoantennas are operating in the terahertz, infrared and visible spectral ranges.

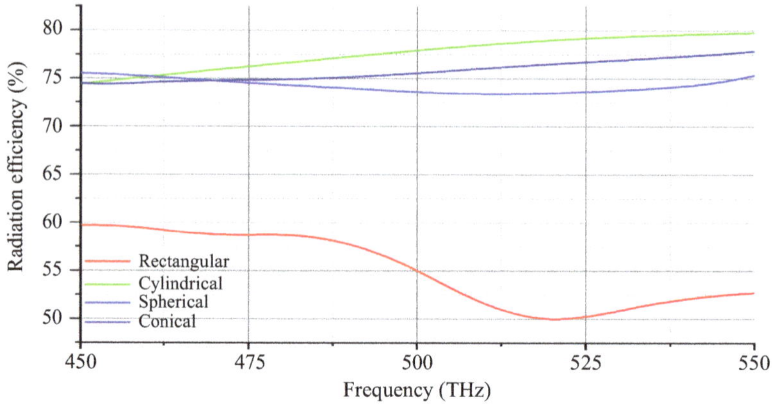

Figure 11.12 Radiation efficiency simulation results

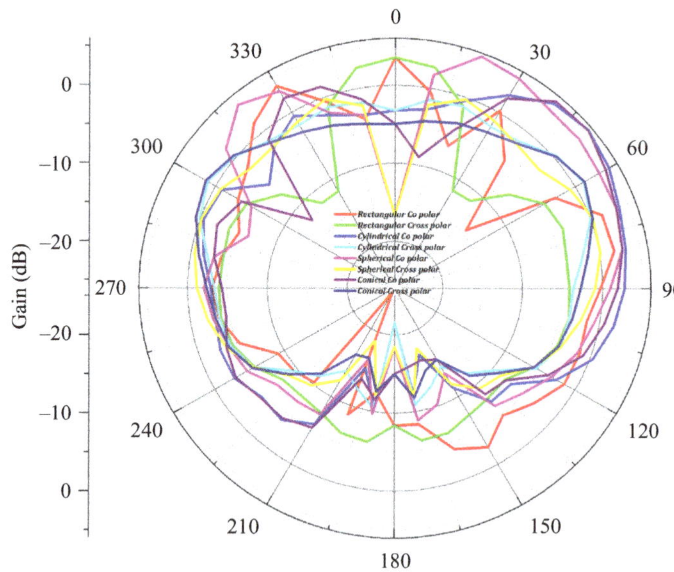

Figure 11.13 Radiation pattern co- and cross-radiation pattern simulation results

It is well known that an entangled state of emitters can lead to an entanglement of the emitted photons (i.e. the state of the field can be mapped into the emitter's state and vice versa). This effect was suggested as a basis for a quantum memory device capable of storing entangled states of light. The entanglement of emitters in antennas can lead to intensity distributions. So-called timed Dick states bear information about the location of emitters and provide a special quantum mechanism that introduces a nonreciprocity of the antenna.

11.8 Designs of terahertz DRAs simulation results for various shapes

Designs of terahertz DRAs simulation results for various shapes are described in Figures 11.8–11.13.

11.9 Conclusion and applications

Terahertz DRAs are most efficient antenna. They offer low loss at higher frequencies. They can provide super directivity and ultralarge bandwidth for wireless communications. Optical antennas are useful in sensing, scanning and they are super directive. They are also used in quantum information processing, spectroscopy, optics and secure communication. Higher order modes generation has many advantages such as mode merging for wideband, highly directive or super directive for higher gain and directivity with less beam width. Super-directive antennas are much useful for communication and radars. TDRA can be the best candidate for photoreceptors in applications to retinal prosthesis.

References

[1] M. Humayun and L. Olmos de Koo, *Retinal Prosthesis—A Clinical Guide to Successful Implementation*, Springer, Cham, 2018.

[2] Y. Yang, Y. Yamagami, X. Yu, *et al.*, "Terahertz topological photonics for on-chip communication," *Nat. Photonics*, vol. 14, pp. 446–451, 2020.

Chapter 12
Conclusion and futuristic vision

Abstract

The dielectric resonator antennas (DRAs) have capabilities to operate at low frequencies, microwave frequencies, terahertz frequencies as well as optical frequencies. These DRAs possess low loss, design flexibilities and efficient radiation mechanism even at high temperatures. DRAs have been recently been integrated with an Apple i-phone 12 mobile phone antenna at millimetric-wave frequencies for 5G applications, i.e., 28 GHz. These DRAs can operate higher order modes in addition to fundamental modes. They can provide reconfigurable antenna characteristics and wide bandwidths. DRAs have mode control mechanism. They can provide physical insights during design stage itself. They are versatile due to availability of many dielectric constants materials. They can use DR values ranging from 10 to 1,600. These DRAs can have two different aspect ratios to extend design flexibilities. Resonant modes merging is another important feature available to designer.

DRA absorbers are also designed using the same concept of DR materials. Absorbers are used as artificial photosynthesis and energy-harvesting solar curtains can be developed using DR absorbers. These DRs can have materials such as TMM, PDMS, TiO_2, graphene, graphite, sapphire, polymers and polymers composites. TMM materials are mostly suitable for millimetric-wave frequencies and microwave frequencies. Graphite, graphene and polymers composites are being used at terahertz frequencies. TiO_2, sapphire and PDMS are used at optical frequencies. They are in use for developing photonic devices. Sapphire antennas can be used for esthetic design and used for wearable sensors such as rings. DRAs can also be used as LiDAR in autonomous cars and self-driving vehicles. DRAs have been used for artificial leaves for photosynthesis and rectennas for energy-harvesting application. They are being used as absorbers. Solar curtain is another application for energy harvesting. These solar curtains shall become boon to green buildings, smart homes and smart cities. They can provide clean energy for domestic use.

Designing retinal DRAs is another era of Hi-Fi. These can become futuristic DRAs/antennas arrays as a substitute of rods and cones in human eye retina as implants. They can very well be used for sight restoration or eye prosthesis. The other field of applications of DRAs can be agriculture and environmental. Namely, moisture sensor, rain sensor, temperature sensor, carbon monoxide gas sensor can be developed using DRA. Scanning and imaging is another field where terahertz DRA can be used for service to the society.

12.1 Introduction

In this book, design simulations and implementations have been included in microwave, millimetric wave, terahertz and optical spectrum using different DR materials and different shapes. Their theoretical background using mathematical derivations has also been comprehensively developed and nicely explained. These derivations will provide physical insight to design engineer, application engineer, industry personal, new reader, researcher, professional, academicians and UG/PG students. They have been simulated on computer simulation technology (CST) to provide graphical insights and few have been converted into prototype models to have feel of physical device and instrumentation testing. Figure 12.1 is the implementation of MIMO dielectric resonator antenna (DRA) to be used in X, K and Ku band radars and satellites. MIMO DRAs can be developed for diverse applications such as high gain and high isolation for diversity requirements. Many such MIMO DRAs can be developed for 5G and beyond applications by choosing a right kind of aspect-ratio and dimension of DRAs to achieve desired gain and isolation along with other

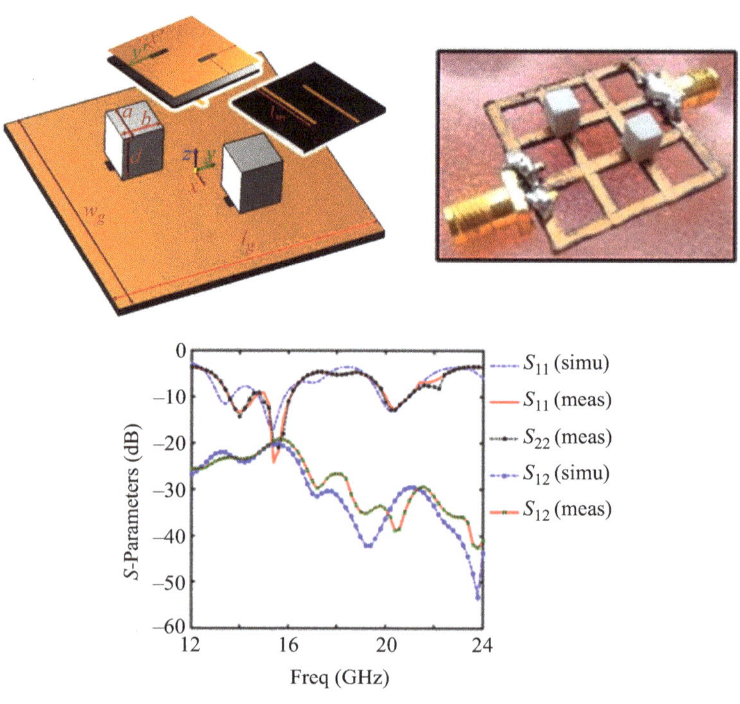

Figure 12.1 MIMO DRA prototype developed for 12–24 GHz, gain 5.7 dBi for radar and satellite applications

radiation parameters. Here, terahertz DRAs have been designed to operate at 511, 541 and 10 THz in different shapes such as rectangular DRA, spherical DRA, conical DRA and cylindrical DRA. They have been broadly supported with mathematical formulation on radiation mechanism and effective control on beam orientation. MIMO DRAs are very useful in high data rate communications.

12.2 Patient-centric healthcare system outline

Patient-centric healthcare system is innovative approach developed for the wellbeing people living in society. In this system, various health parameters are being proposed to be monitored at regular interval of time or randomly. These sensors can be integrated as wearable just like a wrist watch. These measured parameters through a sensor are being wireless transmitted to a server for analytics. Wi-Fi connectivity is maintained between wearables and server. The server has microcontroller-based Arduino Uno system for acquiring biomarker data such as pulse, temperature or glucose level, etc. Human health is continuously monitored by using these sensors thus it develops biomedical history of person. The proposed model is developed for blood glucose monitoring. Thumb based tissue model DRAs have been proposed for glucose level monitoring. Due to change in glucose concentrations levels, shift in radiated resonant frequency is observed. The change in concentrations of glucose level is calibrated with frequency shifts. For data acquisition, Arduino-based Internet of Things (IoT) system is also integrated with machine learning algorithm for analytic purpose. Using thumb DRA with Arduino health monitoring system, alerts on biomarker deviations shall be recorded and reported to patients as well as doctors. Medical history using data acquisition system can be used for better treatment and illness diagnosis. With this system a patient is well educated about any illness beforehand as preliminary information received as biomarker alerts.

12.3 Thumb DRA sensors integrated with patient-centric healthcare system

Use of an oximeter for checking oxygen saturation is common nowadays. Pulse rate is measured by electronic sensors. Body temperature is monitored using an IR sensor. Breathe analysis can be monitored using sensors. A moisture sensor is used for detecting rain, etc. Blood pressure using cuffless system can be developed. Blood glucose is an important biomarker for monitoring health of the human being. The idea to develop patient-centric healthcare system was inspired by student-centric education system. Now students can select their choice of subjects, projects and teacher and college. Internet has created revolution in education system. Libraries have been integrated for facilitating and sharing of resources. Media and open network became discussion platforms like Quora, YouTube and may be other platforms. This could happen because of IT developments. This has embedded collaborations approaches. In this model, a patient and a doctor both are sharing database of healthcare system. Doctors and patients shall share information from the history and records of biomarkers along with preinformation to a doctor and a patient both. The doctor will be able to understand the reason and cause of illness. It will help to localize and better diagnose any kind of illness and execute treatment fast. A patient can also take precautions in one's life-style based on history biomarkers. Machine learning algorithm integration shall help to determine normal and deviated parameters. Machine learning model was trained on data acquisition parameters to get predictions about health or illness in future. Sensed output is DC signal and converted into DC values to be integrated further with Arduino Uno. A block diagram of Arduino connected sensors is given in Figures 12.2–12.4. The sensors attached as wearables transmit using Wi-Fi frequency are received by Arduino Wi-Fi modules. A thumb DRA is developed using permittivity analysis based on shift in frequency due to change in frequency. The permittivity of glucose shall change when the concentration of glucose level changes. This change in permittivity will cause shift in frequency. Hence based on calibration and signal processing, glucose level can be monitored and displayed.

12.4 Thumb DRA design and implementations

The DRAs have been designed based on Cole–Cole model of a bio-characteristic of tissue dielectric properties, and DRA properties have been used in this CST model. Figure 12.5 shows a simulation diagram of a thumb DRA and it monitors for shift in frequency due to change in dielectric constant. Figures 12.2–12.4 have been developed for showing total patient-centric healthcare system using an Arduino microcontroller and a sensor connected with it using Wi-Fi links for periodic data acquisition, and machine learning algorithm is for data analysis and predictions to predetect any possible disease or illness in the patient. This data can be used by a concerned doctor for possible medication and timely treatment. DRAs with and without thumb have been simulated and results have been given in Figures 12.6–12.11 with a proper label marked below of each figure.

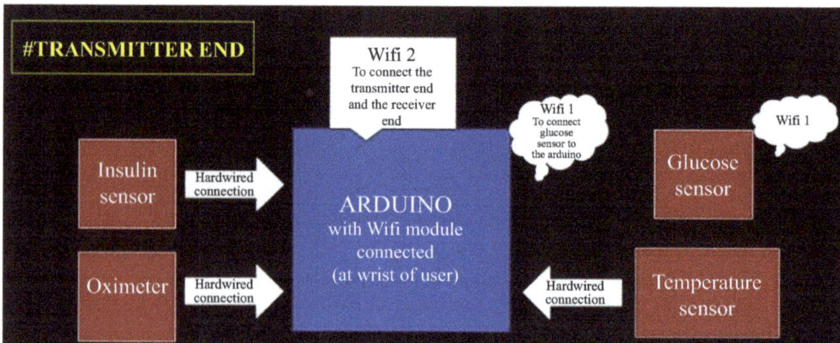

Figure 12.2 Sensors developed are connected to Arduino microcontroller with wireless system

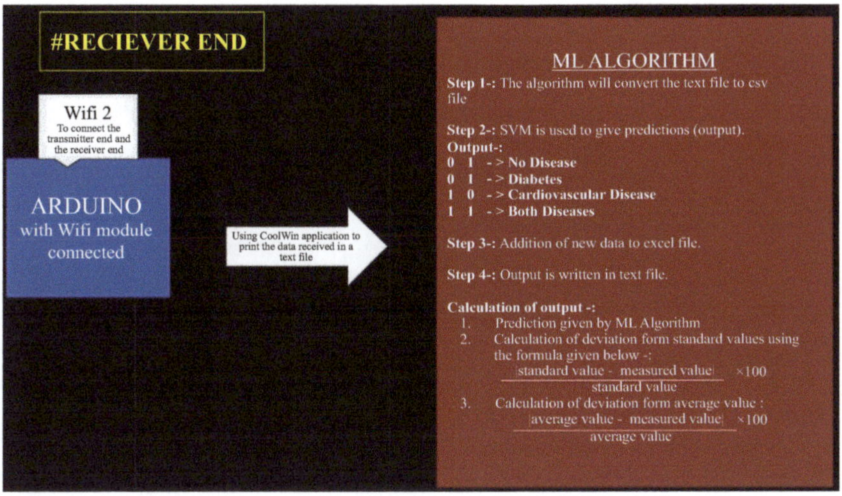

Figure 12.3 Server connected to Arduino microcontroller with machine learning environment for feedback to patient

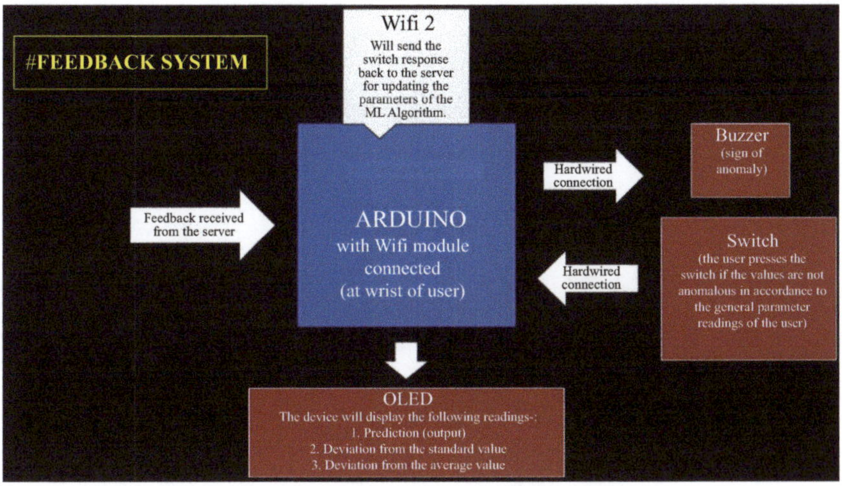

Figure 12.4 Patient information system for health alerts

Tissue details are given in Figure 12.11 and an equivalent circuit of thumb tissue is given in Figure 12.10. The effects of glucose concentrations to provide shift in frequency have been given by S_{11} parameters and radiation parameters are given in Figures 12.6 and 12.9. The dimensions and material details have been

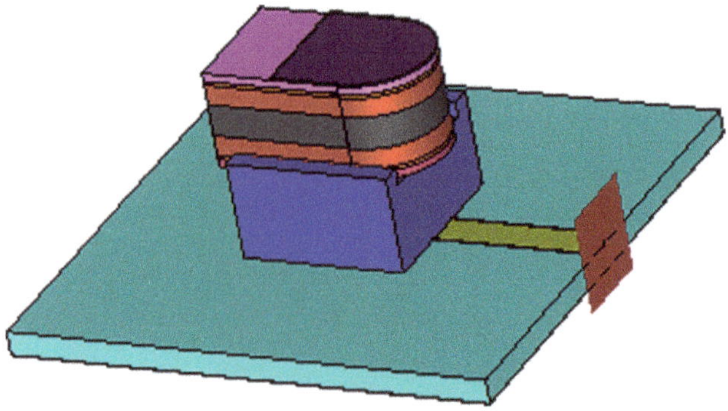

Figure 12.5 *Design of thumb shape tissue DRA for blood glucose monitoring (TMM DR, tissue electrical properties have been used in design based on Cole–Cole model)*

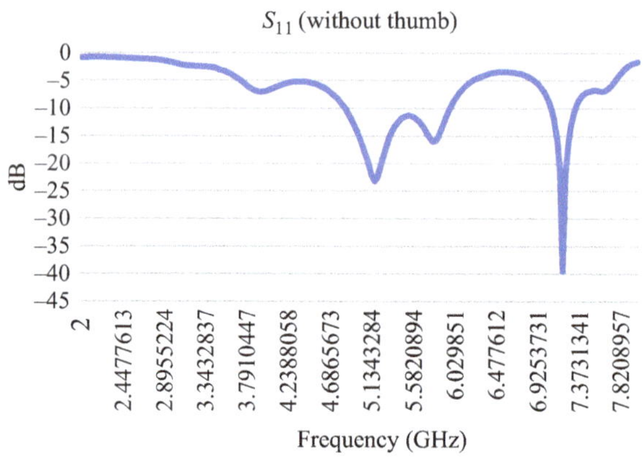

Figure 12.6 *DRA S_{11} at frequency 7.37 GHz*

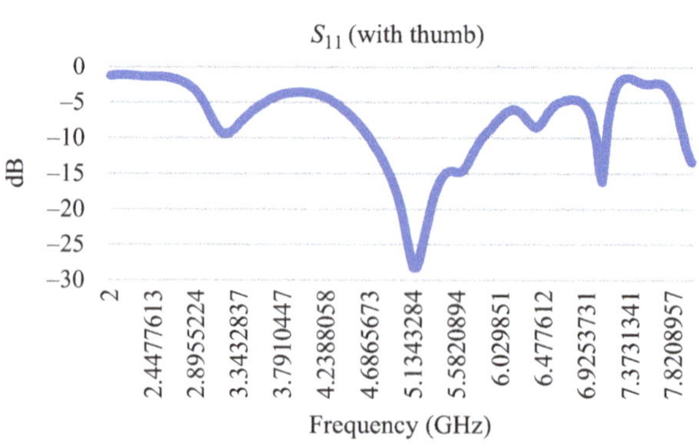

Figure 12.7 *S_{11} of thumb DRA sensor at frequency 5.13 GHz*

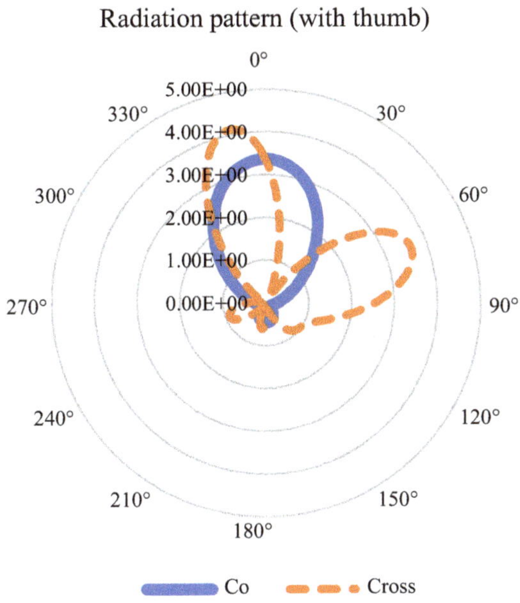

Figure 12.8 Radiation pattern of thumb DRA sensor

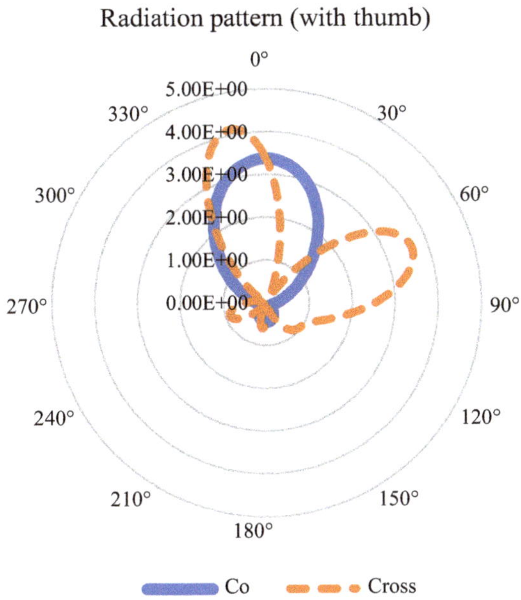

Figure 12.9 Thumb DRA sensor 3D radiation pattern

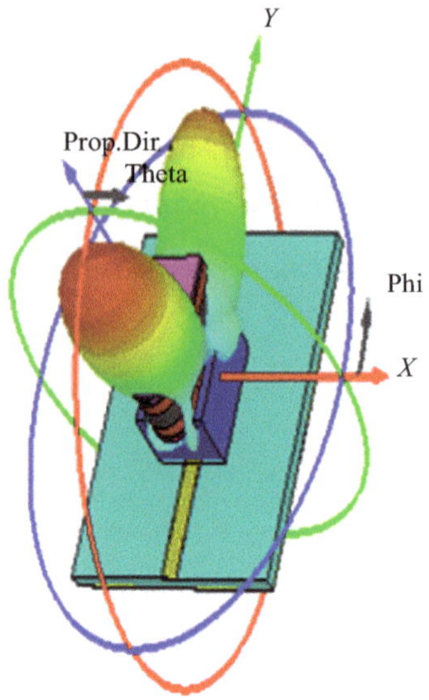

Figure 12.10 Equivalent circuit of thumb DRA tissues (showing electrical properties of skin, fat and muscle)

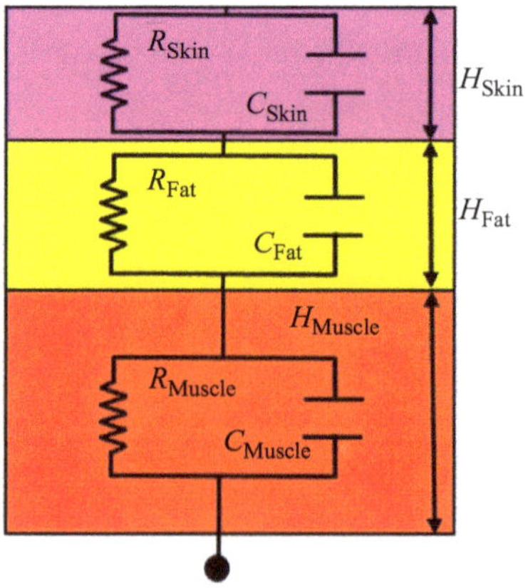

Figure 12.11 Actual view of thumb tissue skin, fat and muscles details

provided in Tables 12.1 and 12.2. Hence, other sensors can be developed to monitor real environments and real health biomarkers using DRAs. Hence, shift in frequency is calibrated with glucose concentration levels of human being in the proposed healthcare system.

Table 12.1 DRA without thumb material details

Component name	Material	Length (*x*-axis) (mm)	Breath (*y*-axis) (mm)	Height (*z*-axis) (mm)
Ground	Copper	60	60	0.035
Substrate	Rogers RT5880	60	60	3.04
Feed	Copper	4.65	40	0.7
RDRA	Alumina	20	20	12
Groove	Alumina	15	20	2

Table 12.2 Thumb (tissue) DRA sensor material details

Component name	Permittivity	Length (*x*-axis) (mm)	Breath (*y*-axis) (mm)	Height (*z*-axis) (mm)
Skin	30	15	24	1
Fat	5.5	15	24	0.5
Blood	Cole–Cole	15	24	2.5
Bone	20	15	24	4
Nail	3	15	15.5	0.3

12.5 Conclusion

This thumb DRA can wirelessly communicate Arduino environment based healthcare system for monitoring and acquisition of biomedical data of any person. In this manner biomarker deviations can be real-time monitored and communicated to any remote location. Thus, proposed healthcare system can be used for remote health monitoring and medication.

Future plans are to develop a similar thumb DRA for sensing other noninvasive biomarkers sensors such as blood pressure, breathe analyzer, breathe pattern, heart rate, oxygen saturation, thyroid and other environmental gas-monitoring sensors.

Appendix A
Case studies

In this annexure, DRAs prototyped have been developed for operating at different frequencies of applications, i.e. Wi-Fi, ISM band, WiMAX, communication satellites, and millimeter-wave frequencies. These DRAs have been developed using Roger substrates, Sapphire and TMM DR materials and SMA connectors for feeding RF inputs through feed. These prototyped DRA products have been fabricated, tested and then validated in echo-free environments. Anechoic chamber testing results are matched with simulated results (Cases A.1–A.15).

Case A.1 DRA implementation at 3.5 GHz for WiMAX applications

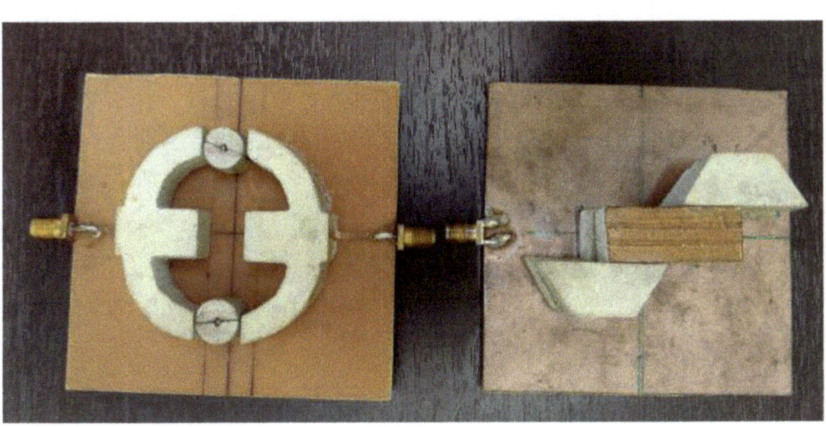

Case A.2 MIMO DRA at 4 and 3.5 GHz for WiMAX and satellite communication applications

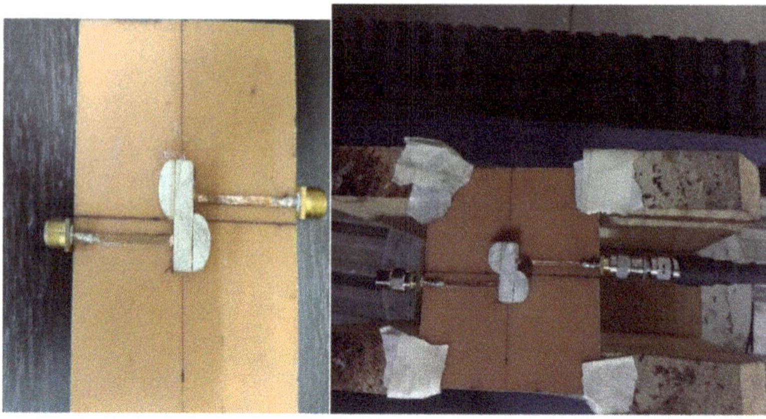

Case A.3 DRAs at 3.5 GHz for WiMAX applications

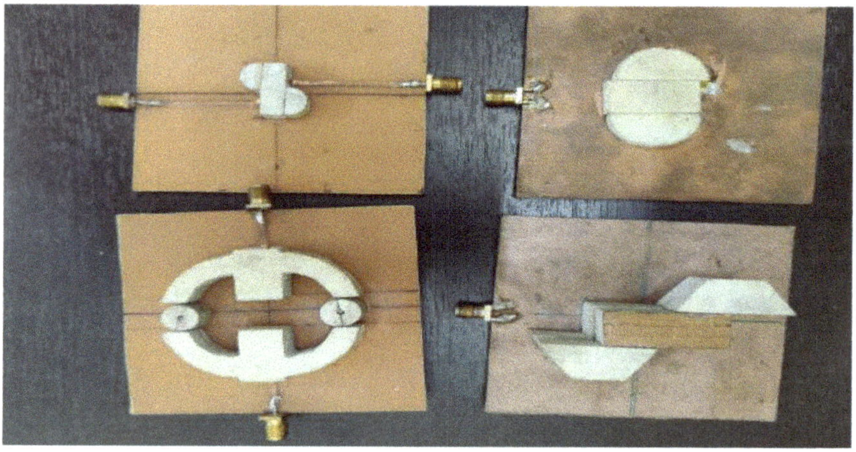

Case A.4 DRAs at 3.5 and 4 GHz mobile communication applications

Case A.5 DRA MIMO at 4 GHz for applications satellite communication applications

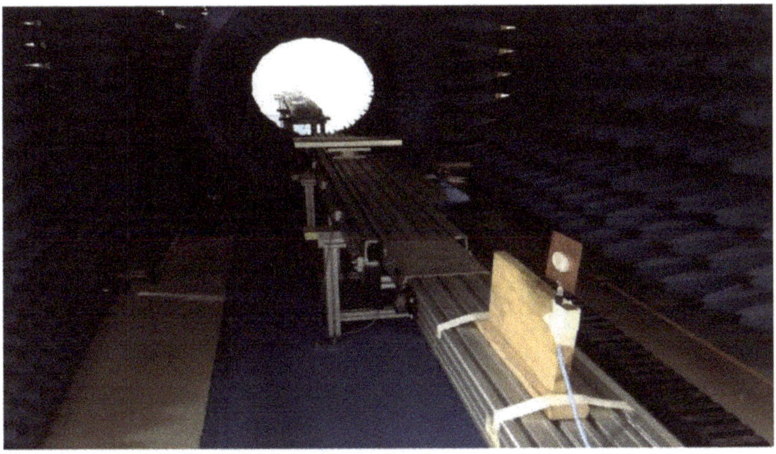

Case A.6 DRAs under an anechoic chamber at 5.8 GHz for Wi-Fi applications

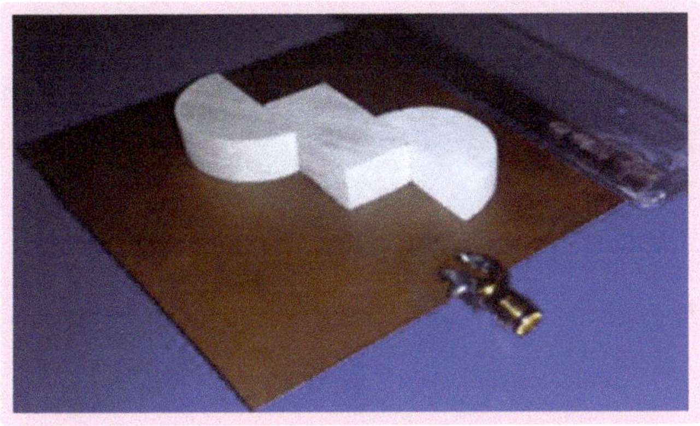

Case A.7 DRAs at 3.5 GHz for WiMAX applications

Case A.8 DRAs at 12 GHz for satellite communication applications

Case A.9 DRAs at 6.5 GHz for satellite communication applications

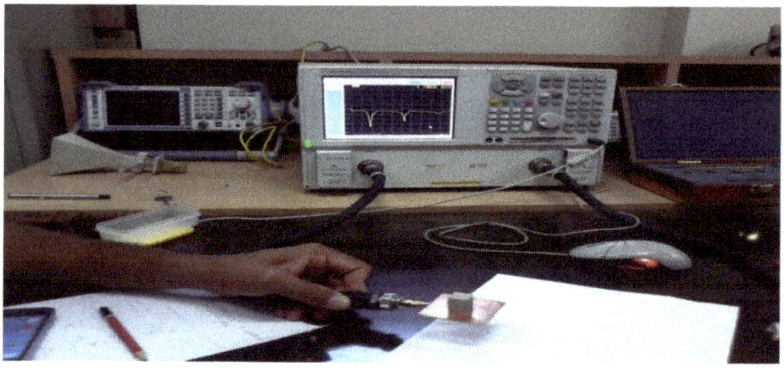

Case A.10 DRAs under measurement with VNA at 4.6-GHz satellite communication applications

Case A.11 Rectangular DRAs at 22.5- and 2.4-GHz radar and Wi-Fi applications

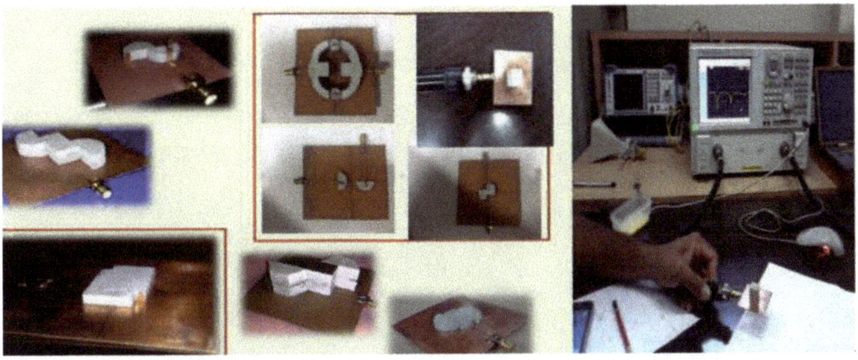

Case A.12 *DRAs designed from 2.4 to 22.5 GHz frequencies for mobiles, satellites and radar applications*

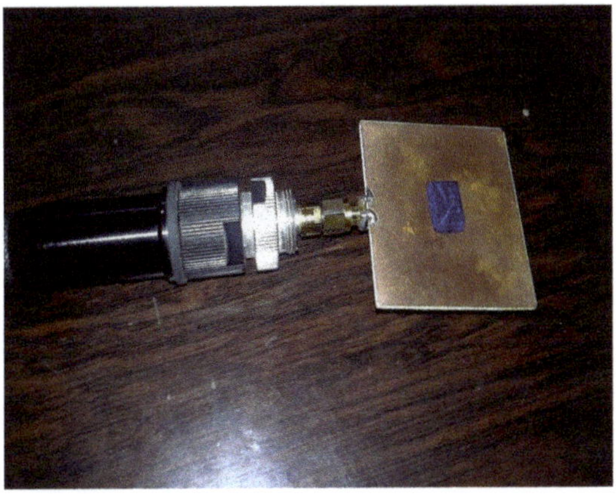

Case A.13 *Sapphire DRAs at 28 GHz for millimeter-wave high data rate transmission*

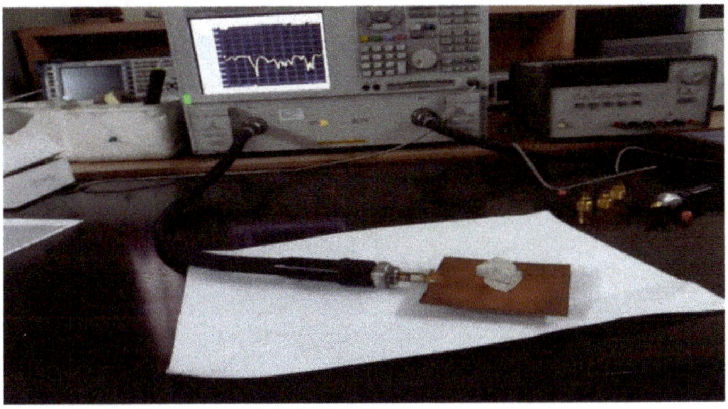

Case A.14 *Vehicle DRA products prototyped at 3.6 GHz*

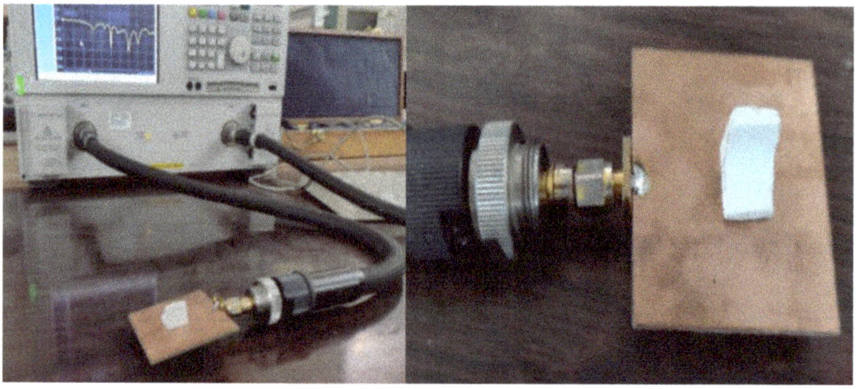

Case A.15 DRAs at 6-GHz frequency for sub-6 GHz, 5G communications

Appendix B

Terahertz absorbers

Absorbers have many applications for energy harvesting and metamaterials devices designs. Many optical sensors are optical absorbers that are used in our daily life. Retinal cones and rods are excellent examples of photonic device absorbers. Optical absorbers are new era technology electronic devices. These devices can generate renewable energy. Artificial photo synthesis is a new field of research. Electric leaves are new types of electronics for use energy resources. They can be embedded into high rise glass buildings for generating electrical energy. Graphene materials have many such applications of absorbers to be used in day-to-day life. These graphene curtains can thus generate electrical energy just like solar cells. The quality factor can be expressed as $Q = (\omega W_e)/(P_{\text{rad}})$. Hence, DRA forming standing waves due to boundary conditions can act as absorbers. They have high-Q factor and high-energy storage; if accelerating charges are created in DRA absorbers, they behave like DRAs due to energy leakage into air. Thus the quality factor of absorbers is reduced to act as antennas. Hence, the Q factor of absorbers is high and that of DRAs is low. The eigen state at resonance is absorber cavity, and excited state of cavity is an antenna due to accelerating charge careers and simultaneously decreasing the quality factor as well. The feed used at THz DRA is Gaussian beam and feed is wave port for absorber. For making absorbers, imaginary part must be high. The second derivate of image impedance w.r.t. frequency must be negative.

B.1 Absorber characteristics

High-Q factor and plane wave incidence are two important features of absorbers. Polarization sensitivity and incidence angle are two important properties of absorbers. Cavity, multilayer architecture, plane wave input excitations, absorptance, resonance are few important features. Standing wave formation, SPP help to develop absorbers. They are used for sensing, security, reducing radar cross section, energy harvesting, solar fuel, solar cells, solar house curtain, graphene curtains, graphene batteries and other applications. The complex imaginary part of impedance is made high to increase loss in all absorbers. Absorbers are frequency dependent and resonance is based on cavity dimensions. The complex permittivity and permeability play an important role to develop absorbers. Addition of magnetic materials will significantly increase complex permeability of the absorber thus enhances capability of absorber. Reflection loss is measured in absorbers. The metallic ground is used for obtaining the reflection and then the electromagnetic wave is absorbed in the materials and frequency-selective resonant cells. A graphite/graphene disk stack is placed on the low-permittivity dielectric resonant cavity for obtaining the multiple generated resonance peaks at the nearby frequencies. The response of absorbers has been tuned by varying the chemical potential of graphene for generating higher order modes and playing with resonant modes (Figures B.1–B.3).

This is directly related to dielectric loss value, sample thickness and impedance matching. Impedance matching is a concept used to maximize energy absorption by the load and minimize refection. This occurs when the input impedance at the air and composite interface is equivalent to that of free space; $Z_{\text{in}} = Z_0$, where it is clear that the input impedance Z_{in} is dependent on the composite electric and magnetic properties, ε_r and μ_r, respectively, the sample thickness, d, and the frequency, ω. The attenuation of the electromagnetic energy inside the sample is controlled by the attenuation factor e^{-ad}, which works on attenuating (reducing) the wave amplitude with respect to traveling distance, d. The attenuation constant, α, is also dependent on the electric and magnetic properties of the material. The dielectric loss is the reason of absorption. The large height of a substrate or composites can absorb more power. Air impedance and absorber impedance must be matched for loss or absorption.

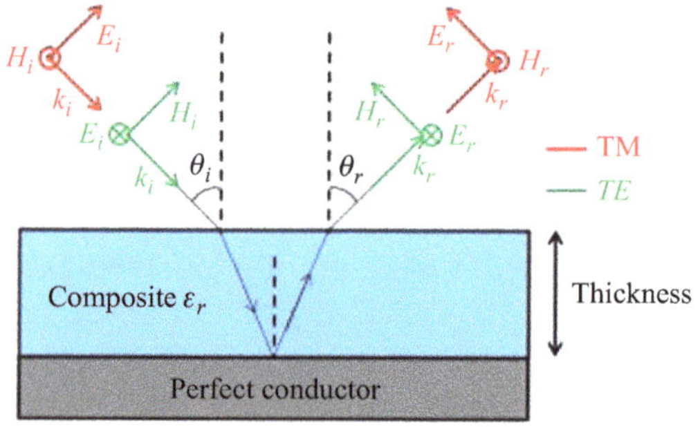

Figure B.1 Reflection loss in absorber

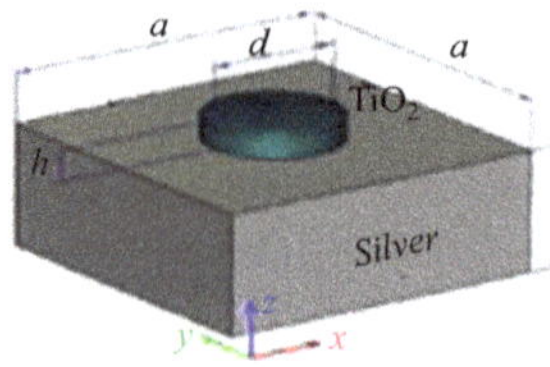

Figure B.2 TiO_2 DRAs as optical absorbers

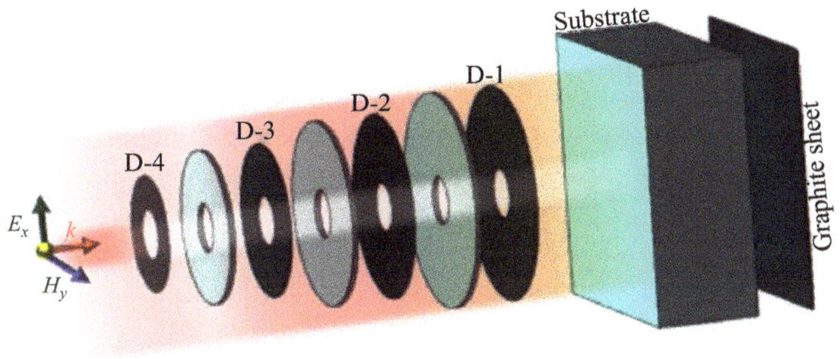

Figure B.3 Terahertz absorbers (9.5 THz) using multilayers of graphite and graphene

B.2 Absorbers mathematical analysis

The relationship is given as follows: loss tangent = tan(∂) = $\varepsilon''/\varepsilon'$; where ε'' is the imaginary part of permittivity and ε' is the real part. Generally capacitive part is imaginary and loss tangent must be high for the absorber in dielectrics. The loss tangent is also dependent on the conductivity of the material if metals are used. High conductivity shall lead to high-loss tangent, hence a good absorber. High-quality factor is desired for good absorbers.

$$\frac{Z_{\text{in}}}{Z_0} = \sqrt{\frac{\mu_r}{\varepsilon_r}} \tan\text{h}\left[j\left(\frac{\omega d}{c}\right)\sqrt{\mu_r \varepsilon_r}\right], \; \alpha = \omega\left\{\frac{\mu\varepsilon}{2}\left[1 + \left(\frac{\sigma}{\omega\varepsilon}\right)^2\right]^{1/2} - 1\right\}^{1/2};$$

$$RL(\text{dB}) = 20 \log\left|\frac{Z_{\text{in}} - Z_0}{Z_{\text{in}} - Z_0}\right|.$$

Absorption loss depends upon loss tangent, and loss tangent depends upon complex permittivity, frequency, conductivity and dielectric constant; tan(δ) = permittivity second derivative upon permittivity first derivative = σ upon ω into permittivity. The absorber design can be with perfect match and dielectric constant layer can be placed from low to high and from high to low for perfect absorption. The absorber design can be built with a ground plane as metals and layers of dielectric materials in stacking. Zero reflection is absorber requirements.

B.3 Optical absorbers applications

RF energy harvesting, imaging, scanning, rectenna, solar curtains, solar cells, attenuators, microwave absorbers, sensing, isolation, detection, etc. are applications. Graphene is a highly conductive material for electrical and thermal properties: light weight, transparent, flexible materials made of carbon single atoms and two-dimensional materials. This is the graphene material that we are using as the tip of a pencil to write on a piece of paper. Graphene is used to design absorbers at THz frequencies. The absorber has broad applications of sensing, imaging, isolation, energy harvesting, etc. High-dielectric constant materials have high tangent loss, and negative imaginary parts of permittivity offer huge dielectric loss and positive imaginary parts of permittivity offer electrical energy to storage. Magnetic properties enhance absorber properties of materials used in design. The cavity gives plane wave inputs to all absorbers against antenna input that is Gaussian beam at THz. The resonance in cavity occurs at plane wave inputs, standing waves are formed and electrical energy is absorbed or stored into cavity. Boundary conditions are maintained for no leakage of power. Loss tangent increases in the case of an absorber and it is minimized in the case of antennas. The electrical energy stored in an absorber may be converted into heat due to the law of conservation of energy.

At an optical window or a vision spectrum, human retina has cones and rods of multilayers that accept photons and convert these photons into neuron impulses, which are electrical in nature due to absorption properties of light. Hence multilayered absorbers are generally designed. Absorption high- and zero reflection coefficients are two important measurement parameters for optical absorbers.

Appendix C
Antenna measured values in anechoic chamber

Figures C.1 is hardware of 5.8 GHz antenna used in mobile communication.

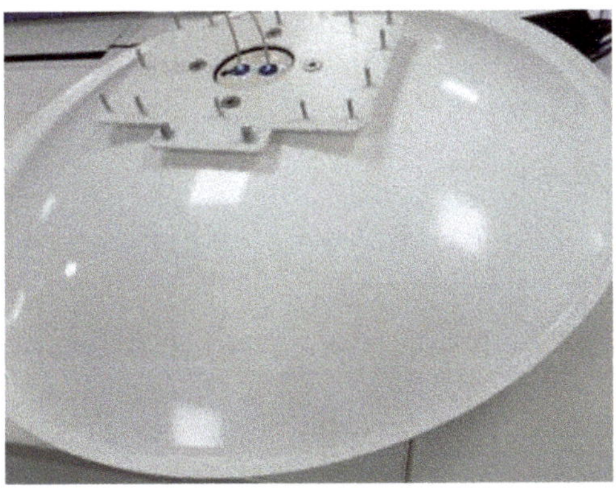

Figure C.1 0.46 m, 25 dBi, antenna used for mobile towers

Radiation plot of antenna at 5.8 GHz (Figures C.2–C.4).

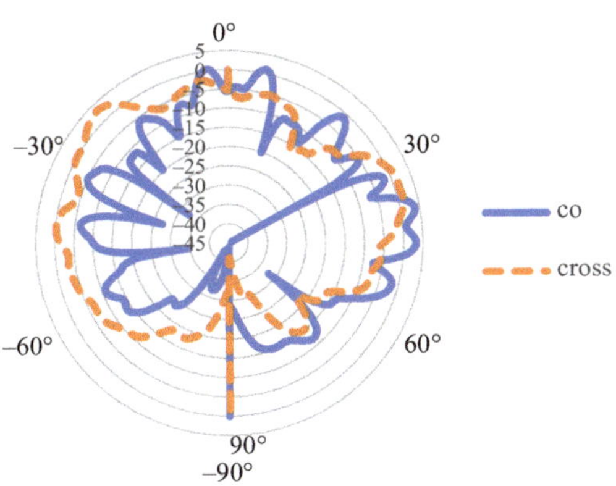

Figure C.2 Radiation pattern polar plot at 5.8 GHz

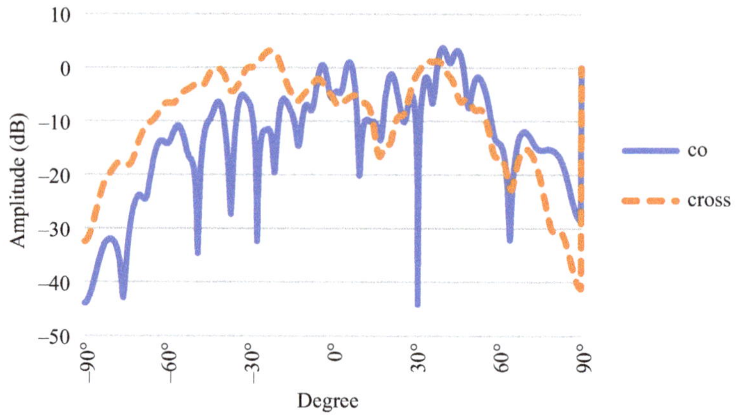

Figure C.3 Radiation pattern 2D plot at 5.8 GHz

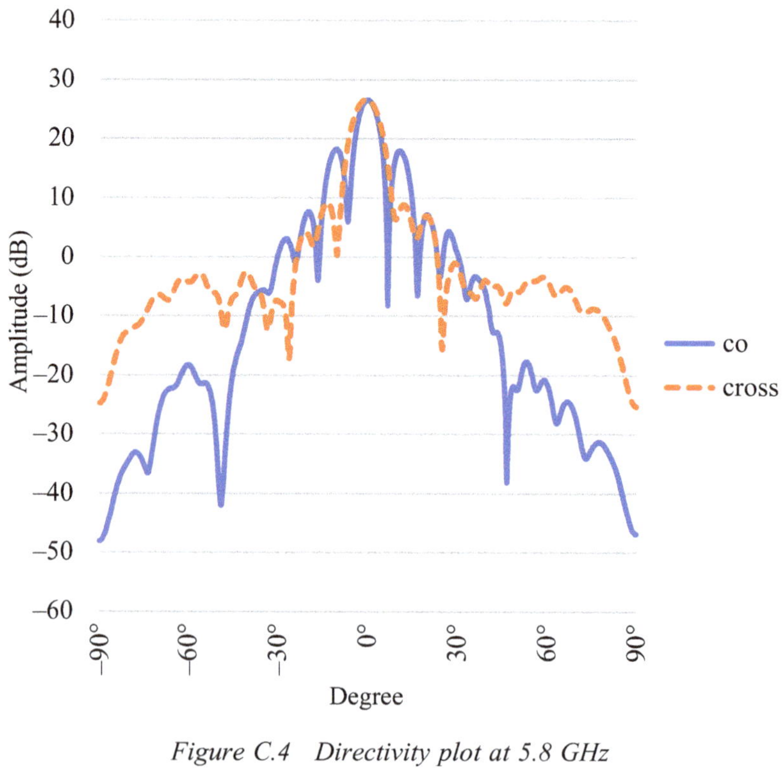

Figure C.4 Directivity plot at 5.8 GHz

Vertical port reading for directivity: Frequency 5.8 GHz		
Elevation (°)	**Amplitude (dB)**	**Phase (°)**
−90.0000	−45.2769	−153.6596
−89.9100	−45.2706	−153.6586
−89.8199	−45.2515	−153.6555
−89.7299	−45.2199	−153.6504
−89.6398	−45.1758	−153.6434
−89.5498	−45.1193	−153.6344
−89.4597	−45.0505	−153.6234
−89.3697	−44.9698	−153.6107
−89.2796	−44.8774	−153.5960
−89.1896	−44.7734	−153.5797
−89.0995	−44.6583	−153.5616
−89.0095	−44.5323	−153.5419
−88.9195	−44.3958	−153.5206
−88.8294	−44.2491	−153.4979
−88.7394	−44.0927	−153.4738
−88.6493	−43.9269	−153.4484
−88.5593	−43.7521	−153.4219
−88.4692	−43.5687	−153.3943
−88.3792	−43.3772	−153.3657
−88.2891	−43.1781	−153.3362
−88.1991	−42.9716	−153.3060
−88.1091	−42.7582	−153.2751
−88.0190	−42.5385	−153.2437
−87.9290	−42.3127	−153.2119
−87.8389	−42.0813	−153.1798
−87.7489	−41.8447	−153.1476
−87.6588	−41.6033	−153.1153
−87.5688	−41.3575	−153.0830
−87.4787	−41.1077	−153.0510
−87.3887	−40.8542	−153.0192
−87.2987	−40.5974	−152.9879
−87.2086	−40.3377	−152.9570
−87.1186	−40.0753	−152.9269
−87.0285	−39.8106	−152.8974
−86.9385	−39.5439	−152.8689
−86.8484	−39.2756	−152.8414
−86.7584	−39.0058	−152.8149
−86.6683	−38.7349	−152.7897
−86.5783	−38.4632	−152.7657
−86.4882	−38.1908	−152.7432
−86.3982	−37.9181	−152.7222
−86.3082	−37.6453	−152.7028
−86.2181	−37.3726	−152.6851
−86.1281	−37.1002	−152.6693
−86.0380	−36.8283	−152.6554
−85.9480	−36.5571	−152.6435
−85.8579	−36.2869	−152.6337

(Continues)

(*Continued*)

Vertical port reading for directivity: Frequency 5.8 GHz		
Elevation (°)	**Amplitude (dB)**	**Phase (°)**
−85.7679	−36.0177	−152.6262
−85.6778	−35.7498	−152.6210
−85.5878	−35.4833	−152.6182
−85.4977	−35.2184	−152.6180
−85.4077	−34.9553	−152.6204
−85.3177	−34.6941	−152.6255
−85.2276	−34.4349	−152.6335
−85.1376	−34.1779	−152.6445
−85.0475	−33.9232	−152.6584
−84.9575	−33.6709	−152.6756
−84.8674	−33.4212	−152.6961
−84.7774	−33.1742	−152.7200
−84.6873	−32.9301	−152.7474
−84.5973	−32.6888	−152.7785
−84.5073	−32.4506	−152.8134
−84.4172	−32.2156	−152.8522
−84.3272	−31.9838	−152.8952
−84.2371	−31.7554	−152.9424
−84.1471	−31.5305	−152.9940
−84.0570	−31.3092	−153.0502
−83.9670	−31.0916	−153.1113
−83.8769	−30.8778	−153.1773
−83.7869	−30.6680	−153.2485
−83.6968	−30.4622	−153.3251
−83.6068	−30.2605	−153.4073
−83.5168	−30.0631	−153.4955
−83.4267	−29.8701	−153.5899
−83.3367	−29.6816	−153.6908
−83.2466	−29.4978	−153.7984
−83.1566	−29.3186	−153.9132
−83.0665	−29.1444	−154.0354
−82.9765	−28.9752	−154.1656
−82.8864	−28.8111	−154.3041
−82.7964	−28.6524	−154.4514
−82.7064	−28.4991	−154.6080
−82.6163	−28.3514	−154.7745
−82.5263	−28.2096	−154.9514
−82.4362	−28.0736	−155.1393
−82.3462	−27.9439	−155.3391
−82.2561	−27.8205	−155.5515
−82.1661	−27.7036	−155.7773
−82.0760	−27.5936	−156.0175
−81.9860	−27.4906	−156.2731
−81.8960	−27.3948	−156.5453
−81.8059	−27.3066	−156.8353
−81.7159	−27.2259	−157.1445

(Continues)

(*Continued*)

Vertical port reading for directivity: Frequency 5.8 GHz		
Elevation (°)	**Amplitude (dB)**	**Phase (°)**
−81.6258	−27.1495	−157.4743
−81.5358	−27.0757	−157.8248
−81.4457	−27.0047	−158.1956
−81.3557	−26.9362	−158.5866
−81.2656	−26.8704	−158.9976
−81.1756	−26.8071	−159.4285
−81.0855	−26.7463	−159.8790
−80.9955	−26.6879	−160.3489
−80.9054	−26.6319	−160.8382
−80.8154	−26.5783	−161.3465
−80.7254	−26.5269	−161.8738
−80.6353	−26.4777	−162.4198
−80.5453	−26.4307	−162.9843
−80.4552	−26.3858	−163.5672
−80.3652	−26.3429	−164.1681
−80.2751	−26.3020	−164.7869
−80.1851	−26.2631	−165.4233
−80.0950	−26.2261	−166.0771
−80.0050	−26.1909	−166.7480
−79.9150	−26.1576	−167.4356
−79.8249	−26.1260	−168.1397
−79.7349	−26.0961	−168.8599
−79.6448	−26.0678	−169.5959
−79.5548	−26.0412	−170.3472
−79.4647	−26.0162	−171.1136
−79.3747	−25.9928	−171.8945
−79.2846	−25.9709	−172.6895
−79.1946	−25.9506	−173.4982
−79.1046	−25.9318	−174.3199
−79.0145	−25.9144	−175.1543
−78.9245	−25.8987	−176.0007
−78.8344	−25.8844	−176.8587
−78.7444	−25.8717	−177.7275
−78.6543	−25.8605	−178.6066
−78.5643	−25.8510	−179.4954
−78.4742	−25.8431	179.6069
−78.3842	−25.8369	178.7007
−78.2941	−25.8325	177.7870
−78.2041	−25.8300	176.8662
−78.1141	−25.8295	175.9392
−78.0240	−25.8310	175.0065
−77.9340	−25.8348	174.0691
−77.8439	−25.8409	173.1274
−77.7539	−25.8496	172.1823
−77.6638	−25.8609	171.2273
−77.5738	−25.8745	170.2300

(*Continues*)

(*Continued*)

Vertical port reading for directivity: Frequency 5.8 GHz		
Elevation (°)	**Amplitude (dB)**	**Phase (°)**
−77.4837	−25.8902	169.1861
−77.3937	−25.9075	168.0954
−77.3036	−25.9262	166.9581
−77.2136	−25.9459	165.7741
−77.1236	−25.9663	164.5437
−77.0335	−25.9869	163.2673
−76.9435	−26.0074	161.9454
−76.8534	−26.0275	160.5788
−76.7634	−26.0466	159.1684
−76.6733	−26.0645	157.7153
−76.5833	−26.0808	156.2211
−76.4932	−26.0951	154.6874
−76.4032	−26.1070	153.1160
−76.3132	−26.1162	151.5091
−76.2231	−26.1224	149.8691
−76.1331	−26.1254	148.1986
−76.0430	−26.1248	146.5005
−75.9530	−26.1205	144.7778
−75.8629	−26.1123	143.0338
−75.7729	−26.1002	141.2717
−75.6828	−26.0841	139.4950
−75.5928	−26.0639	137.7072
−75.5028	−26.0399	135.9118
−75.4127	−26.0120	134.1123
−75.3227	−25.9806	132.3121
−75.2326	−25.9458	130.5144
−75.1426	−25.9079	128.7224
−75.0525	−25.8674	126.9389
−74.9625	−25.8247	125.1666
−74.8724	−25.7802	123.4079
−74.7824	−25.7343	121.6649
−74.6923	−25.6877	119.9388
−74.6023	−25.6383	118.1906
−74.5123	−25.5840	116.3990
−74.4222	−25.5245	114.5695
−74.3322	−25.4592	112.7081
−74.2421	−25.3880	110.8207
−74.1521	−25.3107	108.9137
−74.0620	−25.2270	106.9933
−73.9720	−25.1371	105.0659
−73.8819	−25.0410	103.1378
−73.7919	−24.9390	101.2150
−73.7019	−24.8314	99.3031
−73.6118	−24.7187	97.4077
−73.5218	−24.6013	95.5337
−73.4317	−24.4797	93.6855

(*Continues*)

(*Continued*)

Vertical port reading for directivity: Frequency 5.8 GHz		
Elevation (°)	**Amplitude (dB)**	**Phase (°)**
−73.3417	−24.3548	91.8672
−73.2516	−24.2270	90.0821
−73.1616	−24.0972	88.3332
−73.0715	−23.9660	86.6229
−72.9815	−23.8344	84.9529
−72.8914	−23.7030	83.3246
−72.8014	−23.5726	81.7389
−72.7114	−23.4440	80.1962
−72.6213	−23.3180	78.6965
−72.5313	−23.1953	77.2396
−72.4412	−23.0767	75.8248
−72.3512	−22.9630	74.4512
−72.2611	−22.8548	73.1177
−72.1711	−22.7528	71.8222
−72.0810	−22.6536	70.5303
−71.9910	−22.5548	69.2251
−71.9009	−22.4563	67.9099
−71.8109	−22.3583	66.5875
−71.7209	−22.2609	65.2609
−71.6308	−22.1640	63.9330
−71.5408	−22.0680	62.6062
−71.4507	−21.9729	61.2832
−71.3607	−21.8790	59.9661
−71.2706	−21.7867	58.6572
−71.1806	−21.6961	57.3582
−71.0905	−21.6076	56.0711
−71.0005	−21.5215	54.7973
−70.9105	−21.4383	53.5382
−70.8204	−21.3583	52.2949
−70.7304	−21.2820	51.0685
−70.6403	−21.2097	49.8597
−70.5503	−21.1421	48.6691
−70.4602	−21.0796	47.4971
−70.3702	−21.0226	46.3440
−70.2801	−20.9718	45.2099
−70.1901	−20.9277	44.0946
−70.1001	−20.8910	42.9981
−70.0100	−20.8622	41.9197
−69.9200	−20.8415	40.8509
−69.8299	−20.8271	39.7540
−69.7399	−20.8182	38.6253
−69.6498	−20.8147	37.4658
−69.5598	−20.8161	36.2765
−69.4697	−20.8224	35.0586
−69.3797	−20.8332	33.8133
−69.2896	−20.8483	32.5418

(*Continues*)

(*Continued*)

Vertical port reading for directivity: Frequency 5.8 GHz		
Elevation (°)	**Amplitude (dB)**	**Phase (°)**
−69.1996	−20.8676	31.2455
−69.1096	−20.8910	29.9259
−69.0195	−20.9183	28.5843
−68.9295	−20.9496	27.2224
−68.8394	−20.9847	25.8419
−68.7494	−21.0238	24.4443
−68.6593	−21.0670	23.0314
−68.5693	−21.1142	21.6050
−68.4792	−21.1659	20.1668
−68.3892	−21.2221	18.7186
−68.2991	−21.2833	17.2621
−68.2091	−21.3499	15.7990
−68.1191	−21.4222	14.3310
−68.0290	−21.5008	12.8596
−67.9390	−21.5866	11.3810
−67.8489	−21.6820	9.8352
−67.7589	−21.7865	8.2027
−67.6688	−21.8992	6.4812
−67.5788	−22.0190	4.6688
−67.4887	−22.1445	2.7639
−67.3987	−22.2746	0.7657
−67.3087	−22.4080	−1.3259
−67.2186	−22.5433	−3.5103
−67.1286	−22.6792	−5.7853
−67.0385	−22.8142	−8.1480
−66.9485	−22.9472	−10.5938
−66.8584	−23.0767	−13.1169
−66.7684	−23.2018	−15.7097
−66.6783	−23.3215	−18.3636
−66.5883	−23.4349	−21.0682
−66.4983	−23.5417	−23.8122
−66.4082	−23.6417	−26.5835
−66.3182	−23.7351	−29.3690
−66.2281	−23.8224	−32.1558
−66.1381	−23.9046	−34.9309
−66.0480	−23.9835	−37.7555
−65.9580	−24.0583	−40.7237
−65.8679	−24.1257	−43.8272
−65.7779	−24.1825	−47.0533
−65.6878	−24.2258	−50.3855
−65.5978	−24.2531	−53.8034
−65.5078	−24.2624	−57.2836
−65.4177	−24.2526	−60.8005
−65.3277	−24.2232	−64.3267
−65.2376	−24.1747	−67.8351
−65.1476	−24.1083	−71.2995

(*Continues*)

(*Continued*)

Vertical port reading for directivity: Frequency 5.8 GHz		
Elevation (°)	**Amplitude (dB)**	**Phase (°)**
−65.0575	−24.0259	−74.6961
−64.9675	−23.9302	−78.0041
−64.8774	−23.8242	−81.2064
−64.7874	−23.7113	−84.2898
−64.6973	−23.5951	−87.2454
−64.6073	−23.4795	−90.0675
−64.5173	−23.3680	−92.7541
−64.4272	−23.2644	−95.3060
−64.3372	−23.1680	−97.7875
−64.2471	−23.0726	−100.2992
−64.1571	−22.9774	−102.8321
−64.0670	−22.8819	−105.3755
−63.9770	−22.7859	−107.9189
−63.8869	−22.6896	−110.4518
−63.7969	−22.5934	−112.9648
−63.7069	−22.4981	−115.4489
−63.6168	−22.4047	−117.8963
−63.5268	−22.3143	−120.3003
−63.4367	−22.2283	−122.6554
−63.3467	−22.1483	−124.9571
−63.2566	−22.0759	−127.2024
−63.1666	−22.0130	−129.3895
−63.0765	−21.9615	−131.5176
−62.9865	−21.9235	−133.5871
−62.8965	−21.9013	−135.5997
−62.8064	−21.8972	−137.5578
−62.7164	−21.9110	−139.5279
−62.6263	−21.9387	−141.5950
−62.5363	−21.9788	−143.7583
−62.4462	−22.0296	−146.0160
−62.3562	−22.0897	−148.3659
−62.2661	−22.1578	−150.8050
−62.1761	−22.2327	−153.3298
−62.0860	−22.3135	−155.9361
−61.9960	−22.3994	−158.6190
−61.9060	−22.4898	−161.3734
−61.8159	−22.5846	−164.1935
−61.7259	−22.6836	−167.0739
−61.6358	−22.7872	−170.0091
−61.5458	−22.8962	−172.9942
−61.4557	−23.0116	−176.0253
−61.3657	−23.1348	−179.0999
−61.2756	−23.2676	177.7825
−61.1856	−23.4109	174.5135
−61.0955	−23.5593	170.9469
−61.0055	−23.7054	167.0815

(Continues)

(*Continued*)

Vertical port reading for directivity: Frequency 5.8 GHz		
Elevation (°)	**Amplitude (dB)**	**Phase (°)**
−60.9155	−23.8411	162.9247
−60.8254	−23.9583	158.4948
−60.7354	−24.0490	153.8226
−60.6453	−24.1063	148.9511
−60.5553	−24.1249	143.9350
−60.4652	−24.1017	138.8370
−60.3752	−24.0362	133.7230
−60.2851	−23.9306	128.6566
−60.1951	−23.7894	123.6936
−60.1051	−23.6191	118.8782
−60.0150	−23.4270	114.2410
−59.9250	−23.2212	109.7989
−59.8349	−23.0096	105.5555
−59.7449	−22.7913	101.4372
−59.6548	−22.5470	97.2854
−59.5648	−22.2759	93.1474
−59.4747	−21.9805	89.0754
−59.3847	−21.6647	85.1123
−59.2946	−21.3335	81.2909
−59.2046	−20.9924	77.6331
−59.1146	−20.6470	74.1513
−59.0245	−20.3026	70.8491
−58.9345	−19.9643	67.7232
−58.8444	−19.6366	64.7652
−58.7544	−19.3234	61.9627
−58.6643	−19.0282	59.3007
−58.5743	−18.7540	56.7626
−58.4842	−18.5035	54.3302
−58.3942	−18.2783	51.9823
−58.3042	−18.0624	49.6219
−58.2141	−17.8470	47.2222
−58.1241	−17.6328	44.8016
−58.0340	−17.4204	42.3761
−57.9440	−17.2110	39.9587
−57.8539	−17.0060	37.5599
−57.7639	−16.8068	35.1874
−57.6738	−16.6151	32.8464
−57.5838	−16.4327	30.5398
−57.4937	−16.2613	28.2679
−57.4037	−16.1028	26.0292
−57.3137	−15.9588	23.8203
−57.2236	−15.8314	21.6356
−57.1336	−15.7223	19.4680
−57.0435	−15.6328	17.3041
−56.9535	−15.5520	15.0531
−56.8634	−15.4741	12.6931

(*Continues*)

(*Continued*)

Vertical port reading for directivity: Frequency 5.8 GHz		
Elevation (°)	**Amplitude (dB)**	**Phase (°)**
−56.7734	−15.3978	10.2392
−56.6833	−15.3222	7.7060
−56.5933	−15.2467	5.1077
−56.5033	−15.1711	2.4578
−56.4132	−15.0956	−0.2313
−56.3232	−15.0208	−2.9483
−56.2331	−14.9473	−5.6837
−56.1431	−14.8763	−8.4295
−56.0530	−14.8091	−11.1799
−55.9630	−14.7473	−13.9311
−55.8729	−14.6926	−16.6818
−55.7829	−14.6468	−19.4330
−55.6928	−14.6066	−22.2471
−55.6028	−14.5613	−25.2158
−55.5128	−14.5082	−28.3165
−55.4227	−14.4451	−31.5230
−55.3327	−14.3706	−34.8082
−55.2426	−14.2841	−38.1447
−55.1526	−14.1858	−41.5060
−55.0625	−14.0768	−44.8676
−54.9725	−13.9586	−48.2076
−54.8824	−13.8335	−51.5077
−54.7924	−13.7042	−54.7537
−54.7024	−13.5737	−57.9357
−54.6123	−13.4454	−61.0480
−54.5223	−13.3224	−64.0890
−54.4322	−13.1999	−67.1312
−54.3422	−13.0663	−70.2578
−54.2521	−12.9209	−73.4371
−54.1621	−12.7638	−76.6381
−54.0720	−12.5962	−79.8326
−53.9820	−12.4197	−82.9956
−53.8919	−12.2368	−86.1066
−53.8019	−12.0503	−89.1494
−53.7119	−11.8633	−92.1124
−53.6218	−11.6790	−94.9887
−53.5318	−11.5007	−97.7752
−53.4417	−11.3318	−100.4728
−53.3517	−11.1755	−103.0855
−53.2616	−11.0336	−105.6302
−53.1716	−10.8935	−108.2122
−53.0815	−10.7513	−110.8408
−52.9915	−10.6070	−113.4962
−52.9015	−10.4613	−116.1605
−52.8114	−10.3152	−118.8180
−52.7214	−10.1701	−121.4555

(*Continues*)

(*Continued*)

Vertical port reading for directivity: Frequency 5.8 GHz		
Elevation (°)	**Amplitude (dB)**	**Phase (°)**
−52.6313	−10.0277	−124.0623
−52.5413	−9.8899	−126.6307
−52.4512	−9.7587	−129.1552
−52.3612	−9.6366	−131.6332
−52.2711	−9.5257	−134.0646
−52.1811	−9.4285	−136.4515
−52.0910	−9.3472	−138.8016
−52.0010	−9.2740	−141.2078
−51.9110	−9.2035	−143.6991
−51.8209	−9.1347	−146.2603
−51.7309	−9.0669	−148.8764
−51.6408	−8.9999	−151.5332
−51.5508	−8.9338	−154.2169
−51.4607	−8.8691	−156.9154
−51.3707	−8.8067	−159.6176
−51.2806	−8.7479	−162.3144
−51.1906	−8.6943	−164.9985
−51.1006	−8.6477	−167.6648
−51.0105	−8.6101	−170.3104
−50.9205	−8.5831	−172.9481
−50.8304	−8.5599	−175.6935
−50.7404	−8.5364	−178.5593
−50.6503	−8.5104	178.4733
−50.5603	−8.4802	175.4244
−50.4702	−8.4448	172.3148
−50.3802	−8.4037	169.1653
−50.2901	−8.3572	165.9963
−50.2001	−8.3061	162.8269
−50.1101	−8.2519	159.6743
−50.0200	−8.1966	156.5529
−49.9300	−8.1427	153.4743
−49.8399	−8.0930	150.4469
−49.7499	−8.0456	147.3992
−49.6598	−7.9930	144.2364
−49.5698	−7.9327	140.9845
−49.4797	−7.8633	137.6717
−49.3897	−7.7842	134.3263
−49.2996	−7.6958	130.9758
−49.2096	−7.5991	127.6455
−49.1196	−7.4962	124.3580
−49.0295	−7.3897	121.1323
−48.9395	−7.2825	117.9831
−48.8494	−7.1782	114.9209
−48.7594	−7.0805	111.9520
−48.6693	−6.9900	109.0361
−48.5793	−6.8978	106.0535

(Continues)

(*Continued*)

Vertical port reading for directivity: Frequency 5.8 GHz		
Elevation (°)	**Amplitude (dB)**	**Phase (°)**
−48.4892	−6.8019	103.0183
−48.3992	−6.7017	99.9542
−48.3092	−6.5974	96.8838
−48.2191	−6.4897	93.8281
−48.1291	−6.3801	90.8054
−48.0390	−6.2706	87.8318
−47.9490	−6.1636	84.9196
−47.8589	−6.0618	82.0783
−47.7689	−5.9684	79.3140
−47.6788	−5.8864	76.6293
−47.5888	−5.8164	73.9809
−47.4987	−5.7508	71.2603
−47.4087	−5.6877	68.4775
−47.3187	−5.6260	65.6494
−47.2286	−5.5650	62.7929
−47.1386	−5.5049	59.9243
−47.0485	−5.4460	57.0590
−46.9585	−5.3895	54.2110
−46.8684	−5.3368	51.3925
−46.7784	−5.2899	48.6137
−46.6883	−5.2511	45.8827
−46.5983	−5.2229	43.2048
−46.5083	−5.2048	40.4954
−46.4182	−5.1920	37.6796
−46.3282	−5.1822	34.7724
−46.2381	−5.1730	31.7908
−46.1481	−5.1631	28.7530
−46.0580	−5.1515	25.6781
−45.9680	−5.1381	22.5855
−45.8779	−5.1232	19.4945
−45.7879	−5.1080	16.4231
−45.6978	−5.0943	13.3878
−45.6078	−5.0843	10.4030
−45.5178	−5.0804	7.4533
−45.4277	−5.0803	4.3983
−45.3377	−5.0804	1.2325
−45.2476	−5.0776	−2.0221
−45.1576	−5.0697	−5.3412
−45.0675	−5.0555	−8.6987
−44.9775	−5.0343	−12.0672
−44.8874	−5.0069	−15.4198
−44.7974	−4.9745	−18.7310
−44.7074	−4.9395	−21.9772
−44.6173	−4.9049	−25.1382
−44.5273	−4.8743	−28.2031
−44.4372	−4.8473	−31.3070

(*Continues*)

(*Continued*)

Vertical port reading for directivity: Frequency 5.8 GHz		
Elevation (°)	**Amplitude (dB)**	**Phase (°)**
−44.3472	−4.8201	−34.4946
−44.2571	−4.7904	−37.7425
−44.1671	−4.7565	−41.0253
−44.0770	−4.7178	−44.3166
−43.9870	−4.6744	−47.5900
−43.8969	−4.6275	−50.8203
−43.8069	−4.5791	−53.9839
−43.7169	−4.5319	−57.0604
−43.6268	−4.4893	−60.0321
−43.5368	−4.4551	−62.8870
−43.4467	−4.4305	−65.7445
−43.3567	−4.4123	−68.6682
−43.2666	−4.3983	−71.6415
−43.1766	−4.3869	−74.6465
−43.0865	−4.3772	−77.6635
−42.9965	−4.3689	−80.6724
−42.9065	−4.3626	−83.6531
−42.8164	−4.3596	−86.5858
−42.7264	−4.3617	−89.4520
−42.6363	−4.3716	−92.2347
−42.5463	−4.3927	−94.9244
−42.4562	−4.4288	−97.6496
−42.3662	−4.4779	−100.4638
−42.2761	−4.5370	−103.3571
−42.1861	−4.6033	−106.3168
−42.0960	−4.6747	−109.3269
−42.0060	−4.7498	−112.3690
−41.9160	−4.8278	−115.4221
−41.8259	−4.9087	−118.4636
−41.7359	−4.9935	−121.4699
−41.6458	−5.0839	−124.4166
−41.5558	−5.1840	−127.3090
−41.4657	−5.3017	−130.3125
−41.3757	−5.4338	−133.4540
−41.2856	−5.5755	−136.7257
−41.1956	−5.7226	−140.1137
−41.1056	−5.8713	−143.5980
−41.0155	−6.0186	−147.1527
−40.9255	−6.1626	−150.7460
−40.8354	−6.3026	−154.3416
−40.7454	−6.4394	−157.8997
−40.6553	−6.5752	−161.3788
−40.5653	−6.7194	−164.8694
−40.4752	−6.8773	−168.5385
−40.3852	−7.0422	−172.3789
−40.2951	−7.2080	−176.3724

(*Continues*)

(*Continued*)

Vertical port reading for directivity: Frequency 5.8 GHz		
Elevation (°)	**Amplitude (dB)**	**Phase (°)**
−40.2051	−7.3692	179.5088
−40.1151	−7.5216	175.3027
−40.0250	−7.6627	171.0554
−39.9350	−7.7919	166.8194
−39.8449	−7.9111	162.6506
−39.7549	−8.0244	158.6045
−39.6648	−8.1397	154.6907
−39.5748	−8.2686	150.6578
−39.4847	−8.4061	146.4642
−39.3947	−8.5458	142.1282
−39.3047	−8.6822	137.6788
−39.2146	−8.8116	133.1545
−39.1246	−8.9319	128.6018
−39.0345	−9.0436	124.0721
−38.9445	−9.1495	119.6179
−38.8544	−9.2552	115.2896
−38.7644	−9.3689	111.1254
−38.6743	−9.5043	106.8817
−38.5843	−9.6587	102.4182
−38.4942	−9.8229	97.7382
−38.4042	−9.9887	92.8586
−38.3142	−10.1489	87.8105
−38.2241	−10.2982	82.6376
−38.1341	−10.4339	77.3939
−38.0440	−10.5563	72.1392
−37.9540	−10.6694	66.9329
−37.8639	−10.7803	61.8282
−37.7739	−10.9006	56.5846
−37.6838	−11.0222	50.9644
−37.5938	−11.1276	45.0012
−37.5038	−11.2008	38.7613
−37.4137	−11.2300	32.3417
−37.3237	−11.2090	25.8610
−37.2336	−11.1385	19.4445
−37.1436	−11.0258	13.2065
−37.0535	−10.8837	7.2370
−36.9635	−10.7284	1.5930
−36.8734	−10.5640	−3.9727
−36.7834	−10.3748	−9.6736
−36.6933	−10.1539	−15.4025
−36.6033	−9.9006	−21.0520
−36.5133	−9.6193	−26.5263
−36.4232	−9.3185	−31.7502
−36.3332	−9.0094	−36.6736
−36.2431	−8.7044	−41.2709
−36.1531	−8.4159	−45.5384

(*Continues*)

(*Continued*)

Vertical port reading for directivity: Frequency 5.8 GHz		
Elevation (°)	**Amplitude (dB)**	**Phase (°)**
−36.0630	−8.1554	−49.4901
−35.9730	−7.9145	−53.3671
−35.8829	−7.6783	−57.3009
−35.7929	−7.4446	−61.2360
−35.7029	−7.2135	−65.1224
−35.6128	−6.9871	−68.9176
−35.5228	−6.7692	−72.5876
−35.4327	−6.5649	−76.1081
−35.3427	−6.3799	−79.4634
−35.2526	−6.2208	−82.6462
−35.1626	−6.0939	−85.6677
−35.0725	−5.9910	−88.7519
−34.9825	−5.9035	−91.9566
−34.8924	−5.8276	−95.2456
−34.8024	−5.7614	−98.5827
−34.7124	−5.7040	−101.9321
−34.6223	−5.6566	−105.2603
−34.5323	−5.6214	−108.5362
−34.4422	−5.6022	−111.7331
−34.3522	−5.6041	−114.8282
−34.2621	−5.6324	−117.8604
−34.1721	−5.6848	−121.0774
−34.0820	−5.7544	−124.4882
−33.9920	−5.8355	−128.0624
−33.9020	−5.9239	−131.7641
−33.8119	−6.0169	−135.5520
−33.7219	−6.1137	−139.3816
−33.6318	−6.2159	−143.2068
−33.5418	−6.3271	−146.9820
−33.4517	−6.4534	−150.6647
−33.3617	−6.6025	−154.4322
−33.2716	−6.7703	−158.5095
−33.1816	−6.9472	−162.8703
−33.0915	−7.1246	−167.4749
−33.0015	−7.2963	−172.2701
−32.9115	−7.4585	−177.1916
−32.8214	−7.6114	177.8318
−32.7314	−7.7591	172.8721
−32.6413	−7.9098	167.9949
−32.5513	−8.0750	163.1915
−32.4612	−8.2518	158.0496
−32.3712	−8.4256	152.5152
−32.2811	−8.5831	146.6449
−32.1911	−8.7145	140.5179
−32.1011	−8.8147	134.2295
−32.0110	−8.8845	127.8794

(*Continues*)

(*Continued*)

Vertical port reading for directivity: Frequency 5.8 GHz		
Elevation (°)	**Amplitude (dB)**	**Phase (°)**
−31.9210	−8.9309	121.5580
−31.8309	−8.9660	115.3326
−31.7409	−9.0062	109.2371
−31.6508	−9.0441	102.9412
−31.5608	−9.0488	96.2229
−31.4707	−9.0040	89.2027
−31.3807	−8.9006	82.0251
−31.2906	−8.7383	74.8388
−31.2006	−8.5249	67.7733
−31.1106	−8.2740	60.9215
−31.0205	−8.0028	54.3307
−30.9305	−7.7285	48.0036
−30.8404	−7.4506	41.7974
−30.7504	−7.1265	35.4640
−30.6603	−6.7523	29.1433
−30.5703	−6.3343	22.9728
−30.4802	−5.8842	17.0522
−30.3902	−5.4165	11.4394
−30.3001	−4.9460	6.1535
−30.2101	−4.4861	1.1831
−30.1201	−4.0483	−3.5052
−30.0300	−3.6356	−7.9867
−29.9400	−3.2118	−12.4545
−29.8499	−2.7730	−16.8642
−29.7599	−2.3256	−21.1491
−29.6698	−1.8774	−25.2665
−29.5798	−1.4365	−29.1945
−29.4897	−1.0104	−32.9289
−29.3997	−0.6061	−36.4787
−29.3097	−0.2298	−39.8629
−29.2196	0.1152	−43.1150
−29.1296	0.4536	−46.3735
−29.0395	0.7933	−49.6406
−28.9495	1.1311	−52.8736
−28.8594	1.4633	−56.0417
−28.7694	1.7855	−59.1247
−28.6793	2.0933	−62.1117
−28.5893	2.3824	−65.0000
−28.4993	2.6484	−67.7935
−28.4092	2.8879	−70.5045
−28.3192	3.1154	−73.2446
−28.2291	3.3396	−76.0375
−28.1391	3.5596	−78.8503
−28.0490	3.7737	−81.6553
−27.9590	3.9798	−84.4310
−27.8689	4.1752	−87.1613

(*Continues*)

(*Continued*)

Vertical port reading for directivity: Frequency 5.8 GHz		
Elevation (°)	**Amplitude (dB)**	**Phase (°)**
−27.7789	4.3569	−89.8352
−27.6888	4.5214	−92.4470
−27.5988	4.6656	−94.9975
−27.5088	4.7976	−97.5924
−27.4187	4.9243	−100.2668
−27.3287	5.0464	−102.9917
−27.2386	5.1638	−105.7402
−27.1486	5.2756	−108.4882
−27.0585	5.3802	−111.2150
−26.9685	5.4754	−113.9031
−26.8784	5.5584	−116.5388
−26.7884	5.6262	−119.1157
−26.6983	5.6831	−121.7483
−26.6083	5.7339	−124.4740
−26.5183	5.7804	−127.2652
−26.4282	5.8231	−130.0945
−26.3382	5.8621	−132.9354
−26.2481	5.8963	−135.7627
−26.1581	5.9237	−138.5536
−26.0680	5.9419	−141.2876
−25.9780	5.9476	−143.9563
−25.8879	5.9426	−146.6923
−25.7979	5.9306	−149.5248
−25.7079	5.9134	−152.4281
−25.6178	5.8923	−155.3752
−25.5278	5.8673	−158.3388
−25.4377	5.8377	−161.2914
−25.3477	5.8016	−164.2068
−25.2576	5.7562	−167.0601
−25.1676	5.6979	−169.8523
−25.0775	5.6264	−172.7343
−24.9875	5.5448	−175.7215
−24.8974	5.4554	−178.7926
−24.8074	5.3595	178.0757
−24.7174	5.2579	174.9091
−24.6273	5.1497	171.7347
−24.5373	5.0335	168.5797
−24.4472	4.9060	165.4710
−24.3572	4.7628	162.3775
−24.2671	4.6015	159.1346
−24.1771	4.4261	155.7375
−24.0870	4.2399	152.1990
−23.9970	4.0459	148.5363
−23.9070	3.8457	144.7712
−23.8169	3.6401	140.9294
−23.7269	3.4283	137.0389

(Continues)

(*Continued*)

Vertical port reading for directivity: Frequency 5.8 GHz		
Elevation (°)	**Amplitude (dB)**	**Phase (°)**
−23.6368	3.2077	133.1290
−23.5468	2.9718	129.0887
−23.4567	2.7214	124.7520
−23.3667	2.4653	120.1169
−23.2766	2.2121	115.1940
−23.1866	1.9701	110.0078
−23.0965	1.7465	104.5975
−23.0065	1.5459	99.0153
−22.9165	1.3705	93.3221
−22.8264	1.2188	87.5666
−22.7364	1.0920	81.5238
−22.6463	1.0052	75.1489
−22.5563	0.9726	68.5360
−22.4662	1.0031	61.8068
−22.3762	1.0986	55.0959
−22.2861	1.2541	48.5315
−22.1961	1.4585	42.2191
−22.1061	1.6968	36.2300
−22.0160	1.9576	30.5269
−21.9260	2.2505	24.9056
−21.8359	2.5746	19.4266
−21.7459	2.9245	14.1597
−21.6558	3.2919	9.1548
−21.5658	3.6676	4.4414
−21.4757	4.0419	0.0301
−21.3857	4.4058	−4.0839
−21.2956	4.7512	−7.9176
−21.2056	5.0846	−11.6128
−21.1156	5.4150	−15.2428
−21.0255	5.7411	−18.7777
−20.9355	6.0611	−22.1942
−20.8454	6.3721	−25.4761
−20.7554	6.6710	−28.6133
−20.6653	6.9543	−31.6016
−20.5753	7.2186	−34.4414
−20.4852	7.4624	−37.1614
−20.3952	7.6950	−39.8811
−20.3052	7.9189	−42.5948
−20.2151	8.1343	−45.2802
−20.1251	8.3408	−47.9179
−20.0350	8.5372	−50.4917
−19.9450	8.7220	−52.9883
−19.8549	8.8932	−55.3974
−19.7649	9.0486	−57.7115
−19.6748	9.1902	−60.0063
−19.5848	9.3220	−62.3340

(*Continues*)

(Continued)

Vertical port reading for directivity: Frequency 5.8 GHz		
Elevation (°)	**Amplitude (dB)**	**Phase (°)**
−19.4947	9.4454	−64.6738
−19.4047	9.5606	−67.0061
−19.3147	9.6677	−69.3127
−19.2246	9.7659	−71.5764
−19.1346	9.8541	−73.7818
−19.0445	9.9308	−75.9147
−18.9545	9.9945	−77.9887
−18.8644	10.0470	−80.0986
−18.7744	10.0897	−82.2392
−18.6843	10.1237	−84.3929
−18.5943	10.1493	−86.5421
−18.5043	10.1664	−88.6692
−18.4142	10.1744	−90.7570
−18.3242	10.1722	−92.7883
−18.2341	10.1582	−94.7489
−18.1441	10.1306	−96.7161
−18.0540	10.0903	−98.7242
−17.9640	10.0383	−100.7597
−17.8739	9.9753	−102.8081
−17.7839	9.9013	−104.8539
−17.6938	9.8162	−106.8807
−17.6038	9.7190	−108.8713
−17.5138	9.6086	−110.8076
−17.4237	9.4819	−112.7166
−17.3337	9.3361	−114.6786
−17.2436	9.1721	−116.6890
−17.1536	8.9905	−118.7394
−17.0635	8.7914	−120.8195
−16.9735	8.5747	−122.9176
−16.8834	8.3396	−125.0197
−16.7934	8.0847	−127.1099
−16.7034	7.8070	−129.1861
−16.6133	7.4978	−131.3480
−16.5233	7.1558	−133.6232
−16.4332	6.7809	−136.0194
−16.3432	6.3723	−138.5448
−16.2531	5.9290	−141.2075
−16.1631	5.4492	−144.0158
−16.0730	4.9304	−146.9788
−15.9830	4.3685	−150.1093
−15.8929	3.7480	−153.5616
−15.8029	3.0606	−157.4965
−15.7129	2.3067	−162.0421
−15.6228	1.4903	−167.3635
−15.5328	0.6232	−173.6688
−15.4427	−0.2699	178.7942

(Continues)

(*Continued*)

Vertical port reading for directivity: Frequency 5.8 GHz		
Elevation (°)	**Amplitude (dB)**	**Phase (°)**
−15.3527	−1.1424	169.7708
−15.2626	−1.9177	159.0888
−15.1726	−2.4881	146.6229
−15.0825	−2.7066	132.5941
−14.9925	−2.4717	118.1923
−14.9025	−1.8184	104.8136
−14.8124	−0.8886	93.3187
−14.7224	0.1650	83.8540
−14.6323	1.2357	76.1523
−14.5423	2.2648	69.8373
−14.4522	3.2338	64.5447
−14.3622	4.1589	59.9704
−14.2721	5.0350	55.9785
−14.1821	5.8586	52.4621
−14.0920	6.6289	49.3321
−14.0020	7.3470	46.5158
−13.9120	8.0146	43.9538
−13.8219	8.6339	41.5977
−13.7319	9.2089	39.4034
−13.6418	9.7560	37.3132
−13.5518	10.2785	35.3162
−13.4617	10.7763	33.4091
−13.3717	11.2493	31.5869
−13.2816	11.6972	29.8433
−13.1916	12.1203	28.1716
−13.1016	12.5184	26.5643
−13.0115	12.8920	25.0131
−12.9215	13.2495	23.4853
−12.8314	13.5942	21.9750
−12.7414	13.9256	20.4890
−12.6513	14.2431	19.0323
−12.5613	14.5461	17.6080
−12.4712	14.8342	16.2178
−12.3812	15.1066	14.8623
−12.2911	15.3629	13.5410
−12.2011	15.6073	12.2288
−12.1111	15.8434	10.9126
−12.0210	16.0705	9.6010
−11.9310	16.2882	8.3013
−11.8409	16.4957	7.0194
−11.7509	16.6925	5.7599
−11.6608	16.8777	4.5263
−11.5708	17.0507	3.3215
−11.4807	17.2134	2.1273
−11.3907	17.3692	0.9210
−11.3007	17.5176	−0.2885

(*Continues*)

(*Continued*)

Vertical port reading for directivity: Frequency 5.8 GHz		
Elevation (°)	**Amplitude (dB)**	**Phase (°)**
−11.2106	17.6582	−1.4936
−11.1206	17.7903	−2.6873
−11.0305	17.9133	−3.8638
−10.9405	18.0266	−5.0180
−10.8504	18.1291	−6.1453
−10.7604	18.2216	−7.2579
−10.6703	18.3070	−8.3846
−10.5803	18.3849	−9.5188
−10.4902	18.4550	−10.6531
−10.4002	18.5169	−11.7805
−10.3102	18.5699	−12.8949
−10.2201	18.6132	−13.9903
−10.1301	18.6462	−15.0614
−10.0400	18.6686	−16.1151
−9.9500	18.6821	−17.1835
−9.8599	18.6870	−18.2630
−9.7699	18.6827	−19.3473
−9.6798	18.6691	−20.4301
−9.5898	18.6455	−21.5054
−9.4998	18.6114	−22.5674
−9.4097	18.5658	−23.6102
−9.3197	18.5083	−24.6369
−9.2296	18.4390	−25.6809
−9.1396	18.3579	−26.7419
−9.0495	18.2648	−27.8154
−8.9595	18.1595	−28.8966
−8.8694	18.0415	−29.9806
−8.7794	17.9102	−31.0622
−8.6893	17.7647	−32.1360
−8.5993	17.6040	−33.2027
−8.5093	17.4260	−34.2969
−8.4192	17.2303	−35.4232
−8.3292	17.0166	−36.5804
−8.2391	16.7843	−37.7672
−8.1491	16.5326	−38.9818
−8.0590	16.2607	−40.2224
−7.9690	15.9672	−41.4867
−7.8789	15.6502	−42.7777
−7.7889	15.3041	−44.1372
−7.6988	14.9266	−45.5821
−7.6088	14.5161	−47.1238
−7.5188	14.0705	−48.7758
−7.4287	13.5874	−50.5546
−7.3387	13.0637	−52.4803
−7.2486	12.4960	−54.5784
−7.1586	11.8792	−56.8872

(*Continues*)

(*Continued*)

Vertical port reading for directivity: Frequency 5.8 GHz		
Elevation (°)	**Amplitude (dB)**	**Phase (°)**
−7.0685	11.1998	−59.5138
−6.9785	10.4507	−62.5635
−6.8884	9.6269	−66.1653
−6.7984	8.7253	−70.4975
−6.7084	7.7487	−75.8057
−6.6183	6.7122	−82.4205
−6.5283	5.6569	−90.7543
−6.4382	4.6690	−101.2369
−6.3482	3.8950	−114.2265
−6.2581	3.5485	−129.2652
−6.1681	3.7765	−144.7783
−6.0780	4.5239	−158.8735
−5.9880	5.5869	−170.5135
−5.8979	6.7685	−179.6765
−5.8079	7.9469	173.1891
−5.7179	9.0661	167.5759
−5.6278	10.1164	163.0260
−5.5378	11.0953	159.2665
−5.4477	12.0039	156.1127
−5.3577	12.8463	153.4294
−5.2676	13.6276	151.1167
−5.1776	14.3529	149.1004
−5.0875	15.0272	147.3244
−4.9975	15.6554	145.7431
−4.9075	16.2468	144.2987
−4.8174	16.8058	142.9683
−4.7274	17.3345	141.7397
−4.6373	17.8350	140.6030
−4.5473	18.3092	139.5494
−4.4572	18.7584	138.5715
−4.3672	19.1841	137.6628
−4.2771	19.5879	136.8160
−4.1871	19.9740	136.0085
−4.0970	20.3440	135.2354
−4.0070	20.6985	134.4972
−3.9170	21.0380	133.7938
−3.8269	21.3631	133.1252
−3.7369	21.6740	132.4911
−3.6468	21.9712	131.8912
−3.5568	22.2552	131.3236
−3.4667	22.5287	130.7745
−3.3767	22.7924	130.2417
−3.2866	23.0463	129.7269
−3.1966	23.2905	129.2313
−3.1066	23.5251	128.7559
−3.0165	23.7500	128.3015

(Continues)

(*Continued*)

Vertical port reading for directivity: Frequency 5.8 GHz		
Elevation (°)	**Amplitude (dB)**	**Phase (°)**
−2.9265	23.9654	127.8690
−2.8364	24.1715	127.4577
−2.7464	24.3702	127.0581
−2.6563	24.5618	126.6691
−2.5663	24.7463	126.2919
−2.4762	24.9235	125.9276
−2.3862	25.0933	125.5773
−2.2961	25.2557	125.2416
−2.2061	25.4104	124.9216
−2.1161	25.5577	124.6169
−2.0260	25.6990	124.3209
−1.9360	25.8345	124.0329
−1.8459	25.9641	123.7538
−1.7559	26.0876	123.4843
−1.6658	26.2048	123.2253
−1.5758	26.3156	122.9774
−1.4857	26.4198	122.7411
−1.3957	26.5174	122.5166
−1.3057	26.6095	122.2991
−1.2156	26.6964	122.0882
−1.1256	26.7778	121.8845
−1.0355	26.8536	121.6885
−0.9455	26.9237	121.5009
−0.8554	26.9879	121.3221
−0.7654	27.0461	121.1526
−0.6753	27.0981	120.9924
−0.5853	27.1447	120.8385
−0.4952	27.1861	120.6905
−0.4052	27.2220	120.5489
−0.3152	27.2525	120.4140
−0.2251	27.2775	120.2863
−0.1351	27.2969	120.1661
−0.0450	27.3105	120.0539
0.0450	27.3183	119.9499
0.1351	27.3205	119.8521
0.2251	27.3170	119.7603
0.3152	27.3080	119.6748
0.4052	27.2935	119.5958
0.4952	27.2734	119.5236
0.5853	27.2478	119.4586
0.6753	27.2167	119.4010
0.7654	27.1799	119.3511
0.8554	27.1369	119.3083
0.9455	27.0879	119.2726
1.0355	27.0330	119.2439
1.1256	26.9722	119.2222

(Continues)

(*Continued*)

Vertical port reading for directivity: Frequency 5.8 GHz		
Elevation (°)	**Amplitude (dB)**	**Phase (°)**
1.2156	26.9057	119.2076
1.3057	26.8336	119.2003
1.3957	26.7559	119.2003
1.4857	26.6726	119.2080
1.5758	26.5825	119.2247
1.6658	26.4856	119.2504
1.7559	26.3822	119.2845
1.8459	26.2724	119.3269
1.9360	26.1564	119.3771
2.0260	26.0344	119.4349
2.1161	25.9065	119.4998
2.2061	25.7726	119.5727
2.2961	25.6309	119.6572
2.3862	25.4815	119.7531
2.4762	25.3246	119.8597
2.5663	25.1602	119.9760
2.6563	24.9886	120.1012
2.7464	24.8101	120.2343
2.8364	24.6247	120.3742
2.9265	24.4321	120.5217
3.0165	24.2302	120.6838
3.1066	24.0189	120.8603
3.1966	23.7981	121.0499
3.2866	23.5680	121.2512
3.3767	23.3287	121.4625
3.4667	23.0801	121.6819
3.5568	22.8224	121.9074
3.6468	22.5550	122.1398
3.7369	22.2750	122.3898
3.8269	21.9821	122.6572
3.9170	21.6759	122.9406
4.0070	21.3562	123.2381
4.0970	21.0226	123.5474
4.1871	20.6748	123.8659
4.2771	20.3125	124.1902
4.3672	19.9343	124.5211
4.4572	19.5361	124.8731
4.5473	19.1164	125.2469
4.6373	18.6740	125.6416
4.7274	18.2074	126.0558
4.8174	17.7149	126.4877
4.9075	17.1946	126.9348
4.9975	16.6443	127.3934
5.0875	16.0599	127.8648
5.1776	15.4325	128.3684
5.2676	14.7564	128.9099

(Continues)

(*Continued*)

Vertical port reading for directivity: Frequency 5.8 GHz		
Elevation (°)	**Amplitude (dB)**	**Phase (°)**
5.3577	14.0249	129.4940
5.4477	13.2301	130.1265
5.5378	12.3616	130.8153
5.6278	11.4059	131.5708
5.7179	10.3452	132.4078
5.8079	9.1514	133.3553
5.8979	7.7738	134.4906
5.9880	6.1476	135.9265
6.0780	4.1677	137.8817
6.1681	1.6435	140.8578
6.2581	−1.8223	146.3033
6.3482	−7.2021	160.4921
6.4382	−13.5599	−134.9403
6.5283	−6.9746	−75.1399
6.6183	−1.8797	−61.8093
6.7084	1.3919	−56.5005
6.7984	3.7563	−53.5279
6.8884	5.5911	−51.5348
6.9785	7.0793	−50.0429
7.0685	8.3222	−48.8411
7.1586	9.3815	−47.8220
7.2486	10.2998	−46.9177
7.3387	11.1147	−46.0695
7.4287	11.8445	−45.2648
7.5188	12.5016	−44.4976
7.6088	13.0957	−43.7634
7.6988	13.6342	−43.0585
7.7889	14.1231	−42.3798
7.8789	14.5669	−41.7246
7.9690	14.9707	−41.0856
8.0590	15.3427	−40.4381
8.1491	15.6862	−39.7840
8.2391	16.0030	−39.1279
8.3292	16.2945	−38.4735
8.4192	16.5622	−37.8238
8.5093	16.8071	−37.1811
8.5993	17.0299	−36.5474
8.6893	17.2317	−35.9203
8.7794	17.4156	−35.2772
8.8694	17.5829	−34.6188
8.9595	17.7344	−33.9495
9.0495	17.8707	−33.2737
9.1396	17.9923	−32.5951
9.2296	18.0995	−31.9172
9.3197	18.1928	−31.2432
9.4097	18.2721	−30.5730

(*Continues*)

(*Continued*)

Vertical port reading for directivity: Frequency 5.8 GHz		
Elevation (°)	**Amplitude (dB)**	**Phase (°)**
9.4998	18.3378	−29.8836
9.5898	18.3907	−29.1733
9.6798	18.4315	−28.4466
9.7699	18.4607	−27.7077
9.8599	18.4789	−26.9609
9.9500	18.4866	−26.2104
10.0400	18.4840	−25.4605
10.1301	18.4712	−24.7134
10.2201	18.4466	−23.9449
10.3102	18.4104	−23.1500
10.4002	18.3635	−22.3331
10.4902	18.3064	−21.4987
10.5803	18.2400	−20.6516
10.6703	18.1647	−19.7970
10.7604	18.0812	−18.9405
10.8504	17.9897	−18.0869
10.9405	17.8875	−17.2116
11.0305	17.7738	−16.3046
11.1206	17.6495	−15.3703
11.2106	17.5154	−14.4132
11.3007	17.3724	−13.4389
11.3907	17.2210	−12.4535
11.4807	17.0620	−11.4639
11.5708	16.8960	−10.4781
11.6608	16.7197	−9.4736
11.7509	16.5312	−8.4337
11.8409	16.3313	−7.3618
11.9310	16.1207	−6.2627
12.0210	15.9001	−5.1417
12.1111	15.6702	−4.0056
12.2011	15.4317	−2.8625
12.2911	15.1852	−1.7220
12.3812	14.9276	−0.5698
12.4712	14.6552	0.6201
12.5613	14.3683	1.8458
12.6513	14.0675	3.1043
12.7414	13.7530	4.3913
12.8314	13.4254	5.7011
12.9215	13.0849	7.0261
13.0115	12.7319	8.3566
13.1016	12.3634	9.6973
13.1916	11.9725	11.0805
13.2816	11.5585	12.5103
13.3717	11.1210	13.9889
13.4617	10.6592	15.5177
13.5518	10.1724	17.0970

(*Continues*)

(*Continued*)

Vertical port reading for directivity: Frequency 5.8 GHz		
Elevation (°)	**Amplitude (dB)**	**Phase (°)**
13.6418	9.6599	18.7259
13.7319	9.1204	20.4012
13.8219	8.5499	22.1255
13.9120	7.9338	23.9396
14.0020	7.2666	25.8706
14.0920	6.5432	27.9486
14.1821	5.7576	30.2110
14.2721	4.9026	32.7065
14.3622	3.9699	35.4993
14.4522	2.9492	38.6772
14.5423	1.8264	42.3673
14.6323	0.5556	46.8351
14.7224	−0.8964	52.5196
14.8124	−2.5474	60.1490
14.9025	−4.3683	70.9514
14.9925	−6.1670	86.7173
15.0825	−7.3864	108.4998
15.1726	−7.3223	132.8843
15.2626	−6.0973	153.3681
15.3527	−4.4303	168.1473
15.4427	−2.7573	178.5822
15.5328	−1.2324	−173.8659
15.6228	0.1159	−168.1682
15.7129	1.2989	−163.6904
15.8029	2.3369	−160.0445
15.8929	3.2490	−156.9862
15.9830	4.0517	−154.3561
16.0730	4.7675	−151.9957
16.1631	5.4222	−149.7722
16.2531	6.0222	−147.6743
16.3432	6.5721	−145.6934
16.4332	7.0752	−143.8207
16.5233	7.5344	−142.0478
16.6133	7.9515	−140.3664
16.7034	8.3281	−138.7684
16.7934	8.6673	−137.2319
16.8834	8.9808	−135.6844
16.9735	9.2713	−134.1290
17.0635	9.5397	−132.5779
17.1536	9.7863	−131.0408
17.2436	10.0112	−129.5252
17.3337	10.2143	−128.0370
17.4237	10.3951	−126.5807
17.5138	10.5532	−125.1575
17.6038	10.6927	−123.7105
17.6938	10.8168	−122.2212

(*Continues*)

(*Continued*)

Vertical port reading for directivity: Frequency 5.8 GHz		
Elevation (°)	**Amplitude (dB)**	**Phase (°)**
17.7839	10.9261	−120.7024
17.8739	11.0212	−119.1661
17.9640	11.1023	−117.6229
18.0540	11.1691	−116.0827
18.1441	11.2213	−114.5546
18.2341	11.2582	−113.0468
18.3242	11.2796	−111.5327
18.4142	11.2869	−109.9553
18.5043	11.2816	−108.3253
18.5943	11.2645	−106.6549
18.6843	11.2363	−104.9564
18.7744	11.1974	−103.2426
18.8644	11.1478	−101.5262
18.9545	11.0871	−99.8202
19.0445	11.0144	−98.1278
19.1346	10.9275	−96.3718
19.2246	10.8275	−94.5414
19.3147	10.7160	−92.6481
19.4047	10.5942	−90.7047
19.4947	10.4629	−88.7254
19.5848	10.3226	−86.7260
19.6748	10.1735	−84.7234
19.7649	10.0151	−82.7365
19.8549	9.8437	−80.7194
19.9450	9.6569	−78.6060
20.0350	9.4566	−76.4040
20.1251	9.2448	−74.1239
20.2151	9.0227	−71.7786
20.3052	8.7915	−69.3841
20.3952	8.5518	−66.9594
20.4852	8.3036	−64.5266
20.5753	8.0448	−62.0934
20.6653	7.7673	−59.5541
20.7554	7.4715	−56.8855
20.8454	7.1602	−54.0892
20.9355	6.8355	−51.1705
21.0255	6.4995	−48.1391
21.1156	6.1541	−45.0097
21.2056	5.8001	−41.8022
21.2956	5.4378	−38.5424
21.3857	5.0588	−35.1701
21.4757	4.6553	−31.5648
21.5658	4.2322	−27.7038
21.6558	3.7952	−23.5703
21.7459	3.3507	−19.1546
21.8359	2.9056	−14.4576

(*Continues*)

(*Continued*)

Vertical port reading for directivity: Frequency 5.8 GHz		
Elevation (°)	**Amplitude (dB)**	**Phase (°)**
21.9260	2.4668	−9.4933
22.0160	2.0405	−4.2907
22.1061	1.6299	1.1245
22.1961	1.2237	6.9522
22.2861	0.8346	13.2757
22.3762	0.4841	20.0829
22.4662	0.1943	27.3114
22.5563	−0.0153	34.8423
22.6463	−0.1328	42.5084
22.7364	−0.1563	50.1173
22.8264	−0.0949	57.4854
22.9165	0.0338	64.5845
23.0065	0.2231	71.6063
23.0965	0.4726	78.4437
23.1866	0.7755	84.9843
23.2766	1.1200	91.1413
23.3667	1.4914	96.8605
23.4567	1.8745	102.1198
23.5468	2.2548	106.9230
23.6368	2.6198	111.2930
23.7269	2.9694	115.4239
23.8169	3.3105	119.4365
23.9070	3.6432	123.3044
23.9970	3.9661	127.0063
24.0870	4.2768	130.5269
24.1771	4.5723	133.8566
24.2671	4.8491	136.9911
24.3572	5.1034	139.9306
24.4472	5.3321	142.6902
24.5373	5.5411	145.4225
24.6273	5.7350	148.1556
24.7174	5.9154	150.8679
24.8074	6.0829	153.5394
24.8974	6.2374	156.1520
24.9875	6.3780	158.6898
25.0775	6.5032	161.1390
25.1676	6.6111	163.4881
25.2576	6.6998	165.7537
25.3477	6.7711	168.0582
25.4377	6.8278	170.4042
25.5278	6.8721	172.7746
25.6178	6.9053	175.1520
25.7079	6.9281	177.5186
25.7979	6.9408	179.8562
25.8879	6.9428	−177.8530
25.9780	6.9330	−175.6266

(*Continues*)

(*Continued*)

Vertical port reading for directivity: Frequency 5.8 GHz		
Elevation (°)	**Amplitude (dB)**	**Phase (°)**
26.0680	6.9088	−173.4401
26.1581	6.8677	−171.1873
26.2481	6.8124	−168.8718
26.3382	6.7457	−166.5068
26.4282	6.6695	−164.1075
26.5183	6.5855	−161.6916
26.6083	6.4947	−159.2788
26.6983	6.3975	−156.8909
26.7884	6.2938	−154.5512
26.8784	6.1790	−152.2289
26.9685	6.0474	−149.8254
27.0585	5.9022	−147.3438
27.1486	5.7466	−144.7940
27.2386	5.5834	−142.1897
27.3287	5.4150	−139.5490
27.4187	5.2434	−136.8939
27.5088	5.0698	−134.2508
27.5988	4.8949	−131.6499
27.6888	4.7125	−129.0616
27.7789	4.5136	−126.3879
27.8689	4.3019	−123.6287
27.9590	4.0810	−120.7911
28.0490	3.8546	−117.8876
28.1391	3.6258	−114.9369
28.2291	3.3974	−111.9641
28.3192	3.1716	−109.0006
28.4092	2.9495	−106.0843
28.4993	2.7237	−103.1982
28.5893	2.4812	−100.2409
28.6793	2.2255	−97.2068
28.7694	1.9605	−94.0993
28.8594	1.6905	−90.9288
28.9495	1.4190	−87.7135
29.0395	1.1497	−84.4801
29.1296	0.8853	−81.2639
29.2196	0.6277	−78.1088
29.3097	0.3697	−75.0210
29.3997	0.0907	−71.8957
29.4897	−0.2068	−68.7159
29.5798	−0.5189	−65.4774
29.6698	−0.8413	−62.1839
29.7599	−1.1697	−58.8483
29.8499	−1.4999	−55.4939
29.9400	−1.8280	−52.1555
30.0300	−2.1509	−48.8799
30.1201	−2.4734	−45.7010

(*Continues*)

(*Continued*)

Vertical port reading for directivity: Frequency 5.8 GHz		
Elevation (°)	**Amplitude (dB)**	**Phase (°)**
30.2101	−2.8273	−42.5185
30.3001	−3.2148	−39.2908
30.3902	−3.6325	−35.9974
30.4802	−4.0762	−32.6251
30.5703	−4.5413	−29.1706
30.6603	−5.0223	−25.6431
30.7504	−5.5137	−22.0679
30.8404	−6.0099	−18.4899
30.9305	−6.5089	−14.9696
31.0205	−7.0593	−11.4226
31.1106	−7.6800	−7.7447
31.2006	−8.3715	−3.8591
31.2906	−9.1325	0.3167
31.3807	−9.9599	4.8707
31.4707	−10.8472	9.8928
31.5608	−11.7834	15.4669
31.6508	−12.7520	21.6525
31.7409	−13.7310	28.4563
31.8309	−14.7680	36.1180
31.9210	−15.9192	45.3653
32.0110	−17.1113	56.9276
32.1011	−18.1728	71.4502
32.1911	−18.8203	88.7639
32.2811	−18.8062	107.0692
32.3712	−18.1641	123.7290
32.4612	−17.1733	137.2629
32.5513	−16.1046	147.7283
32.6413	−15.1008	155.9703
32.7314	−14.1529	163.2640
32.8214	−13.2532	169.7993
32.9115	−12.4050	175.6311
33.0015	−11.6130	−179.1798
33.0915	−10.8823	−174.5683
33.1816	−10.2172	−170.4683
33.2716	−9.6212	−166.8166
33.3617	−9.0974	−163.5546
33.4517	−8.6485	−160.6293
33.5418	−8.2554	−157.7777
33.6318	−7.8951	−154.8445
33.7219	−7.5626	−151.8873
33.8119	−7.2556	−148.9536
33.9020	−6.9733	−146.0828
33.9920	−6.7167	−143.3063
34.0820	−6.4881	−140.6483
34.1721	−6.2904	−138.1272
34.2621	−6.1276	−135.7561

(Continues)

(*Continued*)

Vertical port reading for directivity: Frequency 5.8 GHz		
Elevation (°)	**Amplitude (dB)**	**Phase (°)**
34.3522	−6.0030	−133.5146
34.4422	−5.9085	−131.1834
34.5323	−5.8380	−128.7430
34.6223	−5.7877	−126.2241
34.7124	−5.7552	−123.6577
34.8024	−5.7392	−121.0740
34.8924	−5.7397	−118.5030
34.9825	−5.7576	−115.9738
35.0725	−5.7951	−113.5147
35.1626	−5.8555	−111.1529
35.2526	−5.9440	−108.8344
35.3427	−6.0638	−106.3692
35.4327	−6.2101	−103.7573
35.5228	−6.3785	−101.0122
35.6128	−6.5654	−98.1515
35.7029	−6.7679	−95.1974
35.7929	−6.9840	−92.1770
35.8829	−7.2129	−89.1224
35.9730	−7.4553	−86.0704
36.0630	−7.7133	−83.0617
36.1531	−7.9994	−79.9861
36.2431	−8.3227	−76.6441
36.3332	−8.6758	−73.0174
36.4232	−9.0512	−69.0952
36.5133	−9.4406	−64.8764
36.6033	−9.8357	−60.3732
36.6933	−10.2287	−55.6143
36.7834	−10.6133	−50.6467
36.8734	−10.9857	−45.5349
36.9635	−11.3458	−40.3580
37.0535	−11.7088	−34.9126
37.1436	−12.0729	−28.9293
37.2336	−12.4148	−22.4082
37.3237	−12.7100	−15.3999
37.4137	−12.9363	−8.0150
37.5038	−13.0782	−0.4197
37.5938	−13.1311	7.1856
37.6838	−13.1023	14.6020
37.7739	−13.0102	21.6667
37.8639	−12.8798	28.2748
37.9540	−12.7255	34.6562
38.0440	−12.5375	41.0485
38.1341	−12.3120	47.3552
38.2241	−12.0513	53.4828
38.3142	−11.7622	59.3532
38.4042	−11.4548	64.9107

(Continues)

(*Continued*)

Vertical port reading for directivity: Frequency 5.8 GHz		
Elevation (°)	**Amplitude (dB)**	**Phase (°)**
38.4942	−11.1409	70.1247
38.5843	−10.8325	74.9877
38.6743	−10.5411	79.5107
38.7644	−10.2774	83.7190
38.8544	−10.0417	87.7657
38.9445	−9.8166	91.8775
39.0345	−9.5967	96.0181
39.1246	−9.3792	100.1452
39.2146	−9.1634	104.2203
39.3047	−8.9506	108.2098
39.3947	−8.7432	112.0867
39.4847	−8.5450	115.8315
39.5748	−8.3604	119.4324
39.6648	−8.1942	122.8847
39.7549	−8.0496	126.2313
39.8449	−7.9152	129.6871
39.9350	−7.7845	133.2481
40.0250	−7.6537	136.8730
40.1151	−7.5211	140.5209
40.2051	−7.3861	144.1535
40.2951	−7.2501	147.7357
40.3852	−7.1152	151.2377
40.4752	−6.9849	154.6352
40.5653	−6.8633	157.9105
40.6553	−6.7551	161.0527
40.7454	−6.6568	164.2110
40.8354	−6.5610	167.4696
40.9255	−6.4646	170.7928
41.0155	−6.3661	174.1447
41.1056	−6.2650	177.4909
41.1956	−6.1621	−179.2008
41.2856	−6.0593	−175.9588
41.3757	−5.9595	−172.8071
41.4657	−5.8660	−169.7653
41.5558	−5.7831	−166.8477
41.6458	−5.7140	−164.0343
41.7359	−5.6533	−161.1632
41.8259	−5.5975	−158.2287
41.9160	−5.5446	−155.2548
42.0060	−5.4937	−152.2656
42.0960	−5.4445	−149.2846
42.1861	−5.3978	−146.3341
42.2761	−5.3550	−143.4346
42.3662	−5.3180	−140.6042
42.4562	−5.2896	−137.8584
42.5463	−5.2730	−135.2101

(*Continues*)

(*Continued*)

Vertical port reading for directivity: Frequency 5.8 GHz		
Elevation (°)	**Amplitude (dB)**	**Phase (°)**
42.6363	−5.2707	−132.6100
42.7264	−5.2801	−129.9320
42.8164	−5.2985	−127.1830
42.9065	−5.3236	−124.3775
42.9965	−5.3537	−121.5310
43.0865	−5.3876	−118.6605
43.1766	−5.4249	−115.7838
43.2666	−5.4659	−112.9189
43.3567	−5.5112	−110.0835
43.4467	−5.5626	−107.2948
43.5368	−5.6220	−104.5687
43.6268	−5.6935	−101.8464
43.7169	−5.7779	−99.0124
43.8069	−5.8715	−96.0714
43.8969	−5.9709	−93.0335
43.9870	−6.0731	−89.9121
44.0770	−6.1756	−86.7241
44.1671	−6.2765	−83.4892
44.2571	−6.3747	−80.2299
44.3472	−6.4699	−76.9702
44.4372	−6.5627	−73.7354
44.5273	−6.6545	−70.5503
44.6173	−6.7498	−67.3724
44.7074	−6.8521	−64.0621
44.7974	−6.9565	−60.6232
44.8874	−7.0587	−57.0712
44.9775	−7.1547	−53.4272
45.0675	−7.2413	−49.7174
45.1576	−7.3164	−45.9726
45.2476	−7.3789	−42.2265
45.3377	−7.4291	−38.5143
45.4277	−7.4684	−34.8709
45.5178	−7.4996	−31.3290
45.6078	−7.5270	−27.8900
45.6978	−7.5551	−24.3956
45.7879	−7.5809	−20.8308
45.8779	−7.6012	−17.2165
45.9680	−7.6135	−13.5774
46.0580	−7.6164	−9.9408
46.1481	−7.6095	−6.3356
46.2381	−7.5934	−2.7905
46.3282	−7.5698	0.6671
46.4182	−7.5412	4.0121
46.5083	−7.5108	7.2227
46.5983	−7.4825	10.2807
46.6883	−7.4612	13.2730

(Continues)

(*Continued*)

Vertical port reading for directivity: Frequency 5.8 GHz		
Elevation (°)	**Amplitude (dB)**	**Phase (°)**
46.7784	−7.4468	16.2946
46.8684	−7.4369	19.3354
46.9585	−7.4295	22.3833
47.0485	−7.4233	25.4241
47.1386	−7.4176	28.4424
47.2286	−7.4120	31.4222
47.3187	−7.4070	34.3469
47.4087	−7.4036	37.2005
47.4987	−7.4031	39.9678
47.5888	−7.4075	42.6345
47.6788	−7.4196	45.1942
47.7689	−7.4450	47.7577
47.8589	−7.4831	50.3676
47.9490	−7.5313	53.0201
48.0390	−7.5870	55.7088
48.1291	−7.6482	58.4253
48.2191	−7.7132	61.1586
48.3092	−7.7807	63.8959
48.3992	−7.8499	66.6227
48.4892	−7.9205	69.3231
48.5793	−7.9926	71.9801
48.6693	−8.0669	74.5762
48.7594	−8.1454	77.1024
48.8494	−8.2368	79.6693
48.9395	−8.3410	82.3157
49.0295	−8.4547	85.0396
49.1196	−8.5748	87.8359
49.2096	−8.6984	90.6961
49.2996	−8.8228	93.6082
49.3897	−8.9461	96.5569
49.4797	−9.0663	99.5236
49.5698	−9.1824	102.4870
49.6598	−9.2940	105.4236
49.7499	−9.4012	108.3081
49.8399	−9.5052	111.1154
49.9300	−9.6150	113.9363
50.0200	−9.7335	116.8433
50.1101	−9.8573	119.8325
50.2001	−9.9831	122.8961
50.2901	−10.1079	126.0224
50.3802	−10.2291	129.1955
50.4702	−10.3445	132.3959
50.5603	−10.4528	135.6008
50.6503	−10.5532	138.7848
50.7404	−10.6459	141.9208
50.8304	−10.7316	144.9809

(*Continues*)

(*Continued*)

Vertical port reading for directivity: Frequency 5.8 GHz		
Elevation (°)	**Amplitude (dB)**	**Phase (°)**
50.9205	−10.8123	147.9370
51.0105	−10.8923	150.7969
51.1006	−10.9806	153.6889
51.1906	−11.0756	156.6265
51.2806	−11.1748	159.6049
51.3707	−11.2759	162.6161
51.4607	−11.3771	165.6494
51.5508	−11.4768	168.6913
51.6408	−11.5740	171.7257
51.7309	−11.6682	174.7348
51.8209	−11.7595	177.6992
51.9110	−11.8485	−179.4010
52.0010	−11.9366	−176.5866
52.0910	−12.0256	−173.8779
52.1811	−12.1218	−171.2428
52.2711	−12.2334	−168.5728
52.3612	−12.3585	−165.8586
52.4512	−12.4945	−163.0971
52.5413	−12.6392	−160.2881
52.6313	−12.7904	−157.4339
52.7214	−12.9459	−154.5399
52.8114	−13.1042	−151.6145
52.9015	−13.2636	−148.6688
52.9915	−13.4230	−145.7167
53.0815	−13.5817	−142.7744
53.1716	−13.7398	−139.8606
53.2616	−13.8975	−136.9953
53.3517	−14.0598	−134.1653
53.4417	−14.2403	−131.2472
53.5318	−14.4367	−128.2161
53.6218	−14.6452	−125.0618
53.7119	−14.8616	−121.7779
53.8019	−15.0816	−118.3627
53.8919	−15.3006	−114.8200
53.9820	−15.5144	−111.1595
54.0720	−15.7187	−107.3981
54.1621	−15.9101	−103.5591
54.2521	−16.0856	−99.6725
54.3422	−16.2434	−95.7737
54.4322	−16.3831	−91.9020
54.5223	−16.5056	−88.0973
54.6123	−16.6231	−84.2468
54.7024	−16.7387	−80.2566
54.7924	−16.8463	−76.1357
54.8824	−16.9402	−71.9016
54.9725	−17.0151	−67.5810

(*Continues*)

(*Continued*)

Vertical port reading for directivity: Frequency 5.8 GHz		
Elevation (°)	**Amplitude (dB)**	**Phase (°)**
55.0625	−17.0670	−63.2084
55.1526	−17.0931	−58.8247
55.2426	−17.0925	−54.4745
55.3327	−17.0657	−50.2034
55.4227	−17.0148	−46.0547
55.5128	−16.9436	−42.0669
55.6028	−16.8568	−38.2723
55.6928	−16.7598	−34.6955
55.7829	−16.6586	−31.3515
55.8729	−16.5591	−28.1046
55.9630	−16.4615	−24.8799
56.0530	−16.3639	−21.6878
56.1431	−16.2653	−18.5403
56.2331	−16.1651	−15.4499
56.3232	−16.0633	−12.4295
56.4132	−15.9604	−9.4913
56.5033	−15.8570	−6.6470
56.5933	−15.7546	−3.9069
56.6833	−15.6544	−1.2801
56.7734	−15.5583	1.2260
56.8634	−15.4682	3.6050
56.9535	−15.3860	5.8520
57.0435	−15.3139	7.9633
57.1336	−15.2546	9.9750
57.2236	−15.2098	11.9852
57.3137	−15.1778	13.9982
57.4037	−15.1569	16.0123
57.4937	−15.1458	18.0249
57.5838	−15.1434	20.0326
57.6738	−15.1485	22.0310
57.7639	−15.1605	24.0148
57.8539	−15.1788	25.9783
57.9440	−15.2029	27.9152
58.0340	−15.2328	29.8186
58.1241	−15.2685	31.6814
58.2141	−15.3101	33.4960
58.3042	−15.3581	35.2546
58.3942	−15.4131	36.9491
58.4842	−15.4789	38.6079
58.5743	−15.5614	40.3020
58.6643	−15.6589	42.0364
58.7544	−15.7698	43.8127
58.8444	−15.8925	45.6317
58.9345	−16.0252	47.4931
59.0245	−16.1667	49.3954
59.1146	−16.3153	51.3357

(*Continues*)

(*Continued*)

Vertical port reading for directivity: Frequency 5.8 GHz		
Elevation (°)	**Amplitude (dB)**	**Phase (°)**
59.2046	−16.4697	53.3101
59.2946	−16.6287	55.3130
59.3847	−16.7910	57.3372
59.4747	−16.9557	59.3743
59.5648	−17.1218	61.4139
59.6548	−17.2888	63.4446
59.7449	−17.4563	65.4530
59.8349	−17.6247	67.4295
59.9250	−17.8050	69.4510
60.0150	−17.9997	71.5540
60.1051	−18.2068	73.7444
60.1951	−18.4240	76.0268
60.2851	−18.6492	78.4042
60.3752	−18.8800	80.8777
60.4652	−19.1139	83.4462
60.5553	−19.3484	86.1060
60.6453	−19.5811	88.8503
60.7354	−19.8095	91.6690
60.8254	−20.0314	94.5485
60.9155	−20.2450	97.4717
61.0055	−20.4488	100.4183
61.0955	−20.6418	103.3649
61.1856	−20.8238	106.2858
61.2756	−20.9950	109.1539
61.3657	−21.1649	112.0398
61.4557	−21.3379	115.0119
61.5458	−21.5115	118.0714
61.6358	−21.6830	121.2164
61.7259	−21.8498	124.4421
61.8159	−22.0091	127.7403
61.9060	−22.1588	131.0993
61.9960	−22.2968	134.5041
62.0860	−22.4215	137.9368
62.1761	−22.5318	141.3770
62.2661	−22.6273	144.8028
62.3562	−22.7083	148.1912
62.4462	−22.7756	151.5194
62.5363	−22.8307	154.7655
62.6263	−22.8759	157.9095
62.7164	−22.9137	160.9332
62.8064	−22.9472	163.8216
62.8965	−22.9838	166.6478
62.9865	−23.0268	169.4765
63.0765	−23.0745	172.3116
63.1666	−23.1255	175.1548
63.2566	−23.1784	178.0063

(*Continues*)

(Continued)

Vertical port reading for directivity: Frequency 5.8 GHz		
Elevation (°)	**Amplitude (dB)**	**Phase (°)**
63.3467	−23.2320	−179.1353
63.4367	−23.2851	−176.2731
63.5268	−23.3366	−173.4117
63.6168	−23.3858	−170.5569
63.7069	−23.4320	−167.7156
63.7969	−23.4748	−164.8955
63.8869	−23.5141	−162.1050
63.9770	−23.5501	−159.3529
64.0670	−23.5830	−156.6479
64.1571	−23.6135	−153.9990
64.2471	−23.6424	−151.4146
64.3372	−23.6708	−148.9026
64.4272	−23.7000	−146.4691
64.5173	−23.7377	−144.0390
64.6073	−23.7863	−141.5617
64.6973	−23.8431	−139.0331
64.7874	−23.9060	−136.4509
64.8774	−23.9724	−133.8149
64.9675	−24.0401	−131.1270
65.0575	−24.1070	−128.3912
65.1476	−24.1710	−125.6136
65.2376	−24.2304	−122.8022
65.3277	−24.2835	−119.9672
65.4177	−24.3291	−117.1203
65.5078	−24.3661	−114.2747
65.5978	−24.3941	−111.4443
65.6878	−24.4127	−108.6441
65.7779	−24.4222	−105.8888
65.8679	−24.4231	−103.1930
65.9580	−24.4163	−100.5706
66.0480	−24.4031	−98.0344
66.1381	−24.3847	−95.5956
66.2281	−24.3667	−93.1763
66.3182	−24.3505	−90.7201
66.4082	−24.3342	−88.2327
66.4983	−24.3163	−85.7214
66.5883	−24.2953	−83.1948
66.6783	−24.2701	−80.6623
66.7684	−24.2398	−78.1342
66.8584	−24.2038	−75.6211
66.9485	−24.1619	−73.1341
67.0385	−24.1141	−70.6840
67.1286	−24.0605	−68.2814
67.2186	−24.0017	−65.9363
67.3087	−23.9384	−63.6578
67.3987	−23.8715	−61.4544

(Continues)

(Continued)

Vertical port reading for directivity: Frequency 5.8 GHz		
Elevation (°)	**Amplitude (dB)**	**Phase (°)**
67.4887	−23.8020	−59.3333
67.5788	−23.7310	−57.3009
67.6688	−23.6598	−55.3625
67.7589	−23.5895	−53.5224
67.8489	−23.5216	−51.7843
67.9390	−23.4571	−50.1507
68.0290	−23.3975	−48.6035
68.1191	−23.3434	−47.0723
68.2091	−23.2943	−45.5525
68.2991	−23.2498	−44.0458
68.3892	−23.2093	−42.5541
68.4792	−23.1727	−41.0795
68.5693	−23.1396	−39.6240
68.6593	−23.1098	−38.1898
68.7494	−23.0833	−36.7792
68.8394	−23.0598	−35.3945
68.9295	−23.0395	−34.0380
69.0195	−23.0223	−32.7119
69.1096	−23.0084	−31.4184
69.1996	−22.9977	−30.1598
69.2896	−22.9906	−28.9383
69.3797	−22.9871	−27.7560
69.4697	−22.9875	−26.6149
69.5598	−22.9920	−25.5172
69.6498	−23.0010	−24.4650
69.7399	−23.0148	−23.4601
69.8299	−23.0336	−22.5048
69.9200	−23.0579	−21.6009
70.0100	−23.0882	−20.7468
70.1001	−23.1274	−19.9114
70.1901	−23.1756	−19.0863
70.2801	−23.2323	−18.2702
70.3702	−23.2974	−17.4619
70.4602	−23.3703	−16.6603
70.5503	−23.4509	−15.8646
70.6403	−23.5387	−15.0740
70.7304	−23.6335	−14.2882
70.8204	−23.7351	−13.5067
70.9105	−23.8432	−12.7293
71.0005	−23.9576	−11.9559
71.0905	−24.0780	−11.1866
71.1806	−24.2042	−10.4215
71.2706	−24.3360	−9.6611
71.3607	−24.4732	−8.9059
71.4507	−24.6157	−8.1565
71.5408	−24.7633	−7.4138

(Continues)

(*Continued*)

Vertical port reading for directivity: Frequency 5.8 GHz		
Elevation (°)	**Amplitude (dB)**	**Phase (°)**
71.6308	–24.9157	–6.6788
71.7209	–25.0729	–5.9526
71.8109	–25.2347	–5.2368
71.9009	–25.4010	–4.5327
71.9910	–25.5716	–3.8423
72.0810	–25.7464	–3.1676
72.1711	–25.9254	–2.5107
72.2611	–26.1120	–1.8686
72.3512	–26.3127	–1.2310
72.4412	–26.5277	–0.5948
72.5313	–26.7569	0.0428
72.6213	–27.0005	0.6848
72.7114	–27.2586	1.3341
72.8014	–27.5314	1.9938
72.8914	–27.8191	2.6670
72.9815	–28.1219	3.3573
73.0715	–28.4402	4.0681
73.1616	–28.7742	4.8036
73.2516	–29.1244	5.5680
73.3417	–29.4911	6.3660
73.4317	–29.8747	7.2029
73.5218	–30.2758	8.0846
73.6118	–30.6949	9.0176
73.7019	–31.1324	10.0095
73.7919	–31.5889	11.0688
73.8819	–32.0649	12.2050
73.9720	–32.5610	13.4292
74.0620	–33.0775	14.7542
74.1521	–33.6151	16.1944
74.2421	–34.1737	17.7668
74.3322	–34.7536	19.4907
74.4222	–35.3544	21.3880
74.5123	–35.9755	23.4840
74.6023	–36.6154	25.8071
74.6923	–37.2720	28.3886
74.7824	–37.9557	31.2914
74.8724	–38.6874	34.6353
74.9625	–39.4662	38.5419
75.0525	–40.2863	43.1649
75.1426	–41.1343	48.6928
75.2326	–41.9832	55.3389
75.3227	–42.7856	63.3028
75.4127	–43.4690	72.6809
75.5028	–43.9420	83.3244
75.5928	–44.1215	94.7268
75.6828	–43.9740	106.0949

(*Continues*)

(*Continued*)

Vertical port reading for directivity: Frequency 5.8 GHz		
Elevation (°)	**Amplitude (dB)**	**Phase (°)**
75.7729	−43.5361	116.6443
75.8629	−42.8920	125.8856
75.9530	−42.1325	133.6853
76.0430	−41.3289	140.1472
76.1331	−40.5270	145.4725
76.2231	−39.7529	149.8734
76.3132	−39.0196	153.5359
76.4032	−38.3322	156.6110
76.4932	−37.6917	159.2173
76.5833	−37.0966	161.4465
76.6733	−36.5446	163.3696
76.7634	−36.0330	165.0415
76.8534	−35.5589	166.5052
76.9435	−35.1196	167.7946
77.0335	−34.7126	168.9366
77.1236	−34.3355	169.9527
77.2136	−33.9862	170.8603
77.3036	−33.6630	171.6736
77.3937	−33.3640	172.4043
77.4837	−33.0879	173.0620
77.5738	−32.8333	173.6546
77.6638	−32.5990	174.1890
77.7539	−32.3831	174.6749
77.8439	−32.1776	175.1490
77.9340	−31.9807	175.6176
78.0240	−31.7922	176.0797
78.1141	−31.6117	176.5345
78.2041	−31.4391	176.9811
78.2941	−31.2741	177.4191
78.3842	−31.1166	177.8479
78.4742	−30.9664	178.2672
78.5643	−30.8233	178.6768
78.6543	−30.6871	179.0763
78.7444	−30.5577	179.4658
78.8344	−30.4349	179.8450
78.9245	−30.3187	−179.7861
79.0145	−30.2089	−179.4274
79.1046	−30.1054	−179.0790
79.1946	−30.0081	−178.7408
79.2846	−29.9168	−178.4128
79.3747	−29.8315	−178.0948
79.4647	−29.7521	−177.7869
79.5548	−29.6785	−177.4889
79.6448	−29.6105	−177.2007
79.7349	−29.5483	−176.9222
79.8249	−29.4916	−176.6533

(*Continues*)

(*Continued*)

Vertical port reading for directivity: Frequency 5.8 GHz		
Elevation (°)	**Amplitude (dB)**	**Phase (°)**
79.9150	−29.4404	−176.3938
80.0050	−29.3946	−176.1437
80.0950	−29.3542	−175.9029
80.1851	−29.3192	−175.6712
80.2751	−29.2894	−175.4486
80.3652	−29.2648	−175.2349
80.4552	−29.2454	−175.0300
80.5453	−29.2312	−174.8339
80.6353	−29.2221	−174.6465
80.7254	−29.2180	−174.4678
80.8154	−29.2190	−174.2975
80.9054	−29.2250	−174.1358
80.9955	−29.2360	−173.9825
81.0855	−29.2519	−173.8376
81.1756	−29.2729	−173.7011
81.2656	−29.2987	−173.5730
81.3557	−29.3295	−173.4532
81.4457	−29.3653	−173.3417
81.5358	−29.4060	−173.2387
81.6258	−29.4515	−173.1440
81.7159	−29.5021	−173.0578
81.8059	−29.5588	−172.9793
81.8960	−29.6249	−172.9061
81.9860	−29.7002	−172.8378
82.0760	−29.7842	−172.7738
82.1661	−29.8765	−172.7140
82.2561	−29.9769	−172.6580
82.3462	−30.0849	−172.6054
82.4362	−30.2003	−172.5562
82.5263	−30.3229	−172.5099
82.6163	−30.4523	−172.4665
82.7064	−30.5883	−172.4258
82.7964	−30.7307	−172.3875
82.8864	−30.8792	−172.3516
82.9765	−31.0338	−172.3179
83.0665	−31.1941	−172.2863
83.1566	−31.3600	−172.2566
83.2466	−31.5314	−172.2288
83.3367	−31.7080	−172.2027
83.4267	−31.8898	−172.1783
83.5168	−32.0765	−172.1555
83.6068	−32.2680	−172.1341
83.6968	−32.4642	−172.1143
83.7869	−32.6650	−172.0957
83.8769	−32.8702	−172.0785
83.9670	−33.0796	−172.0625

(*Continues*)

(*Continued*)

Vertical port reading for directivity: Frequency 5.8 GHz		
Elevation (°)	**Amplitude (dB)**	**Phase (°)**
84.0570	−33.2932	−172.0478
84.1471	−33.5109	−172.0341
84.2371	−33.7324	−172.0216
84.3272	−33.9577	−172.0101
84.4172	−34.1867	−171.9997
84.5073	−34.4193	−171.9902
84.5973	−34.6552	−171.9816
84.6873	−34.8945	−171.9740
84.7774	−35.1369	−171.9672
84.8674	−35.3823	−171.9612
84.9575	−35.6307	−171.9561
85.0475	−35.8818	−171.9517
85.1376	−36.1356	−171.9481
85.2276	−36.3919	−171.9452
85.3177	−36.6506	−171.9430
85.4077	−36.9115	−171.9414
85.4977	−37.1745	−171.9405
85.5878	−37.4393	−171.9402
85.6778	−37.7060	−171.9406
85.7679	−37.9742	−171.9415
85.8579	−38.2438	−171.9429
85.9480	−38.5147	−171.9448
86.0380	−38.7866	−171.9473
86.1281	−39.0594	−171.9502
86.2181	−39.3328	−171.9536
86.3082	−39.6067	−171.9574
86.3982	−39.8807	−171.9616
86.4882	−40.1548	−171.9662
86.5783	−40.4285	−171.9712
86.6683	−40.7018	−171.9765
86.7584	−40.9743	−171.9821
86.8484	−41.2458	−171.9880
86.9385	−41.5159	−171.9942
87.0285	−41.7844	−172.0005
87.1186	−42.0510	−172.0071
87.2086	−42.3153	−172.0139
87.2987	−42.5770	−172.0208
87.3887	−42.8358	−172.0279
87.4787	−43.0914	−172.0351
87.5688	−43.3433	−172.0423
87.6588	−43.5912	−172.0496
87.7489	−43.8347	−172.0569
87.8389	−44.0734	−172.0642
87.9290	−44.3068	−172.0714
88.0190	−44.5347	−172.0786
88.1091	−44.7565	−172.0857

(Continues)

(*Continued*)

Vertical port reading for directivity: Frequency 5.8 GHz		
Elevation (°)	**Amplitude (dB)**	**Phase (°)**
88.1991	−44.9719	−172.0927
88.2891	−45.1803	−172.0995
88.3792	−45.3814	−172.1062
88.4692	−45.5748	−172.1126
88.5593	−45.7600	−172.1188
88.6493	−45.9365	−172.1248
88.7394	−46.1040	−172.1305
88.8294	−46.2620	−172.1358
88.9195	−46.4102	−172.1409
89.0095	−46.5481	−172.1456
89.0995	−46.6754	−172.1500
89.1896	−46.7917	−172.1539
89.2796	−46.8968	−172.1575
89.3697	−46.9902	−172.1606
89.4597	−47.0718	−172.1633
89.5498	−47.1412	−172.1656
89.6398	−47.1983	−172.1674
89.7299	−47.2429	−172.1687
89.8199	−47.2748	−172.1696
89.9100	−47.2941	−172.1700
90.0000	−47.3005	−172.1699
Horizontal port measured directivity	Frequency 5.8 GHz	
Azimuth (°)	Amplitude (dB)	Phase (°)
−90.0000	−36.6228	−98.5090
−89.9100	−36.6165	−98.5079
−89.8199	−36.5974	−98.5049
−89.7299	−36.5657	−98.4999
−89.6398	−36.5215	−98.4928
−89.5498	−36.4648	−98.4838
−89.4597	−36.3959	−98.4729
−89.3697	−36.3150	−98.4601
−89.2796	−36.2223	−98.4455
−89.1896	−36.1181	−98.4291
−89.0995	−36.0027	−98.4110
−89.0095	−35.8763	−98.3913
−88.9195	−35.7395	−98.3701
−88.8294	−35.5924	−98.3474
−88.7394	−35.4355	−98.3233
−88.6493	−35.2693	−98.2979
−88.5593	−35.0940	−98.2713
−88.4692	−34.9102	−98.2437
−88.3792	−34.7182	−98.2151
−88.2891	−34.5185	−98.1856
−88.1991	−34.3115	−98.1554
−88.1091	−34.0976	−98.1246

(Continues)

(*Continued*)

Vertical port reading for directivity: Frequency 5.8 GHz		
Elevation (°)	**Amplitude (dB)**	**Phase (°)**
−88.0190	−33.8773	−98.0932
−87.9290	−33.6509	−98.0614
−87.8389	−33.4190	−98.0294
−87.7489	−33.1818	−97.9972
−87.6588	−32.9399	−97.9649
−87.5688	−32.6935	−97.9327
−87.4787	−32.4431	−97.9007
−87.3887	−32.1891	−97.8690
−87.2987	−31.9317	−97.8377
−87.2086	−31.6714	−97.8070
−87.1186	−31.4085	−97.7769
−87.0285	−31.1433	−97.7476
−86.9385	−30.8762	−97.7192
−86.8484	−30.6074	−97.6917
−86.7584	−30.3372	−97.6654
−86.6683	−30.0659	−97.6402
−86.5783	−29.7937	−97.6164
−86.4882	−29.5211	−97.5940
−86.3982	−29.2480	−97.5731
−86.3082	−28.9749	−97.5538
−86.2181	−28.7020	−97.5363
−86.1281	−28.4294	−97.5205
−86.0380	−28.1574	−97.5067
−85.9480	−27.8861	−97.4950
−85.8579	−27.6158	−97.4853
−85.7679	−27.3466	−97.4779
−85.6778	−27.0788	−97.4729
−85.5878	−26.8125	−97.4702
−85.4977	−26.5478	−97.4701
−85.4077	−26.2849	−97.4726
−85.3177	−26.0240	−97.4778
−85.2276	−25.7652	−97.4859
−85.1376	−25.5086	−97.4969
−85.0475	−25.2545	−97.5110
−84.9575	−25.0028	−97.5283
−84.8674	−24.7538	−97.5488
−84.7774	−24.5076	−97.5728
−84.6873	−24.2643	−97.6003
−84.5973	−24.0240	−97.6315
−84.5073	−23.7869	−97.6665
−84.4172	−23.5530	−97.7055
−84.3272	−23.3224	−97.7485
−84.2371	−23.0954	−97.7959
−84.1471	−22.8720	−97.8477
−84.0570	−22.6523	−97.9041
−83.9670	−22.4364	−97.9653

(Continues)

(*Continued*)

Vertical port reading for directivity: Frequency 5.8 GHz		
Elevation (°)	**Amplitude (dB)**	**Phase (°)**
−83.8769	−22.2244	−98.0316
−83.7869	−22.0166	−98.1031
−83.6968	−21.8129	−98.1800
−83.6068	−21.6134	−98.2627
−83.5168	−21.4185	−98.3513
−83.4267	−21.2280	−98.4462
−83.3367	−21.0422	−98.5477
−83.2466	−20.8612	−98.6561
−83.1566	−20.6852	−98.7717
−83.0665	−20.5142	−98.8950
−82.9765	−20.3485	−99.0263
−82.8864	−20.1881	−99.1661
−82.7964	−20.0333	−99.3149
−82.7064	−19.8841	−99.4733
−82.6163	−19.7409	−99.6418
−82.5263	−19.6036	−99.8210
−82.4362	−19.4727	−100.0116
−82.3462	−19.3482	−100.2145
−82.2561	−19.2304	−100.4304
−82.1661	−19.1194	−100.6603
−82.0760	−19.0157	−100.9051
−81.9860	−18.9193	−101.1660
−81.8960	−18.8307	−101.4443
−81.8059	−18.7500	−101.7413
−81.7159	−18.6774	−102.0587
−81.6258	−18.6093	−102.4012
−81.5358	−18.5443	−102.7697
−81.4457	−18.4823	−103.1639
−81.3557	−18.4233	−103.5836
−81.2656	−18.3671	−104.0285
−81.1756	−18.3139	−104.4983
−81.0855	−18.2633	−104.9927
−80.9955	−18.2155	−105.5115
−80.9054	−18.1703	−106.0542
−80.8154	−18.1276	−106.6207
−80.7254	−18.0874	−107.2106
−80.6353	−18.0496	−107.8237
−80.5453	−18.0141	−108.4595
−80.4552	−17.9809	−109.1177
−80.3652	−17.9499	−109.7980
−80.2751	−17.9209	−110.4999
−80.1851	−17.8941	−111.2229
−80.0950	−17.8691	−111.9667
−80.0050	−17.8461	−112.7307
−79.9150	−17.8250	−113.5144
−79.8249	−17.8057	−114.3172

(*Continues*)

(*Continued*)

Vertical port reading for directivity: Frequency 5.8 GHz		
Elevation (°)	**Amplitude (dB)**	**Phase (°)**
−79.7349	−17.7881	−115.1385
−79.6448	−17.7722	−115.9777
−79.5548	−17.7580	−116.8340
−79.4647	−17.7455	−117.7068
−79.3747	−17.7346	−118.5952
−79.2846	−17.7253	−119.4985
−79.1946	−17.7177	−120.4158
−79.1046	−17.7117	−121.3462
−79.0145	−17.7074	−122.2889
−78.9245	−17.7048	−123.2427
−78.8344	−17.7040	−124.2068
−78.7444	−17.7051	−125.1800
−78.6543	−17.7082	−126.1614
−78.5643	−17.7133	−127.1499
−78.4742	−17.7206	−128.1443
−78.3842	−17.7303	−129.1436
−78.2941	−17.7426	−130.1466
−78.2041	−17.7577	−131.1521
−78.1141	−17.7757	−132.1591
−78.0240	−17.7971	−133.1664
−77.9340	−17.8220	−134.1729
−77.8439	−17.8509	−135.1773
−77.7539	−17.8840	−136.1786
−77.6638	−17.9218	−137.1854
−77.5738	−17.9641	−138.2407
−77.4837	−18.0107	−139.3508
−77.3937	−18.0612	−140.5171
−77.3036	−18.1152	−141.7408
−77.2136	−18.1721	−143.0230
−77.1236	−18.2316	−144.3650
−77.0335	−18.2932	−145.7674
−76.9435	−18.3563	−147.2312
−76.8534	−18.4204	−148.7569
−76.7634	−18.4848	−150.3447
−76.6733	−18.5490	−151.9945
−76.5833	−18.6122	−153.7059
−76.4932	−18.6739	−155.4782
−76.4032	−18.7334	−157.3100
−76.3132	−18.7899	−159.1995
−76.2231	−18.8429	−161.1445
−76.1331	−18.8917	−163.1420
−76.0430	−18.9356	−165.1888
−75.9530	−18.9742	−167.2809
−75.8629	−19.0069	−169.4138
−75.7729	−19.0334	−171.5825
−75.6828	−19.0532	−173.7818

(*Continues*)

(*Continued*)

Vertical port reading for directivity: Frequency 5.8 GHz		
Elevation (°)	**Amplitude (dB)**	**Phase (°)**
−75.5928	−19.0663	−176.0058
−75.5028	−19.0724	−178.2488
−75.4127	−19.0716	179.4954
−75.3227	−19.0640	177.2328
−75.2326	−19.0499	174.9695
−75.1426	−19.0297	172.7112
−75.0525	−19.0039	170.4634
−74.9625	−18.9731	168.2311
−74.8724	−18.9379	166.0188
−74.7824	−18.8993	163.8306
−74.6923	−18.8581	161.6688
−74.6023	−18.8121	159.4785
−74.5123	−18.7588	157.2306
−74.4222	−18.6974	154.9334
−74.3322	−18.6271	152.5957
−74.2421	−18.5472	150.2270
−74.1521	−18.4575	147.8373
−74.0620	−18.3576	145.4368
−73.9720	−18.2477	143.0356
−73.8819	−18.1279	140.6436
−73.7919	−17.9987	138.2703
−73.7019	−17.8607	135.9246
−73.6118	−17.7146	133.6146
−73.5218	−17.5614	131.3473
−73.4317	−17.4020	129.1290
−73.3417	−17.2377	126.9646
−73.2516	−17.0694	124.8584
−73.1616	−16.8984	122.8133
−73.0715	−16.7259	120.8314
−72.9815	−16.5530	118.9141
−72.8914	−16.3808	117.0617
−72.8014	−16.2106	115.2739
−72.7114	−16.0434	113.5500
−72.6213	−15.8801	111.8884
−72.5313	−15.7218	110.2875
−72.4412	−15.5695	108.7450
−72.3512	−15.4240	107.2585
−72.2611	−15.2863	105.8253
−72.1711	−15.1570	104.4417
−72.0810	−15.0318	103.0653
−71.9910	−14.9078	101.6758
−71.9009	−14.7851	100.2768
−71.8109	−14.6635	98.8720
−71.7209	−14.5431	97.4646
−71.6308	−14.4241	96.0578
−71.5408	−14.3065	94.6547

(*Continues*)

(*Continued*)

Vertical port reading for directivity: Frequency 5.8 GHz		
Elevation (°)	**Amplitude (dB)**	**Phase (°)**
−71.4507	−14.1906	93.2580
−71.3607	−14.0766	91.8702
−71.2706	−13.9648	90.4937
−71.1806	−13.8554	89.1307
−71.0905	−13.7489	87.7829
−71.0005	−13.6456	86.4520
−70.9105	−13.5459	85.1394
−70.8204	−13.4503	83.8464
−70.7304	−13.3591	82.5738
−70.6403	−13.2730	81.3225
−70.5503	−13.1925	80.0929
−70.4602	−13.1180	78.8855
−70.3702	−13.0501	77.7003
−70.2801	−12.9895	76.5374
−70.1901	−12.9368	75.3965
−70.1001	−12.8926	74.2772
−70.0100	−12.8576	73.1792
−69.9200	−12.8321	72.0918
−69.8299	−12.8137	70.9712
−69.7399	−12.8019	69.8130
−69.6498	−12.7963	68.6186
−69.5598	−12.7964	67.3892
−69.4697	−12.8019	66.1262
−69.3797	−12.8126	64.8310
−69.2896	−12.8282	63.5051
−69.1996	−12.8485	62.1500
−69.1096	−12.8733	60.7676
−69.0195	−12.9025	59.3596
−68.9295	−12.9359	57.9279
−68.8394	−12.9736	56.4745
−68.7494	−13.0156	55.0013
−68.6593	−13.0619	53.5104
−68.5693	−13.1128	52.0041
−68.4792	−13.1684	50.4843
−68.3892	−13.2291	48.9532
−68.2991	−13.2951	47.4130
−68.2091	−13.3671	45.8656
−68.1191	−13.4455	44.3130
−68.0290	−13.5310	42.7572
−67.9390	−13.6246	41.1934
−67.8489	−13.7287	39.5515
−67.7589	−13.8427	37.8089
−67.6688	−13.9654	35.9625
−67.5788	−14.0955	34.0097
−67.4887	−14.2314	31.9485
−67.3987	−14.3718	29.7775

(Continues)

(*Continued*)

Vertical port reading for directivity: Frequency 5.8 GHz		
Elevation (°)	**Amplitude (dB)**	**Phase (°)**
−67.3087	−14.5149	27.4965
−67.2186	−14.6592	25.1062
−67.1286	−14.8029	22.6088
−67.0385	−14.9444	20.0084
−66.9485	−15.0819	17.3105
−66.8584	−15.2140	14.5225
−66.7684	−15.3393	11.6540
−66.6783	−15.4567	8.7162
−66.5883	−15.5653	5.7217
−66.4983	−15.6647	2.6848
−66.4082	−15.7547	−0.3796
−66.3182	−15.8359	−3.4562
−66.2281	−15.9091	−6.5299
−66.1381	−15.9757	−9.5863
−66.0480	−16.0366	−12.6940
−65.9580	−16.0892	−15.9545
−65.8679	−16.1296	−19.3542
−65.7779	−16.1544	−22.8743
−65.6878	−16.1607	−26.4917
−65.5978	−16.1462	−30.1793
−65.5078	−16.1095	−33.9076
−65.4177	−16.0501	−37.6455
−65.3277	−15.9685	−41.3621
−65.2376	−15.8662	−45.0282
−65.1476	−15.7454	−48.6176
−65.0575	−15.6091	−52.1083
−64.9675	−15.4608	−55.4826
−64.8774	−15.3043	−58.7279
−64.7874	−15.1434	−61.8363
−64.6973	−14.9820	−64.8041
−64.6073	−14.8241	−67.6314
−64.5173	−14.6732	−70.3216
−64.4272	−14.5329	−72.8802
−64.3372	−14.4010	−75.3720
−64.2471	−14.2695	−77.8891
−64.1571	−14.1375	−80.4217
−64.0670	−14.0048	−82.9588
−63.9770	−13.8716	−85.4898
−63.8869	−13.7380	−88.0050
−63.7969	−13.6048	−90.4954
−63.7069	−13.4726	−92.9535
−63.6168	−13.3427	−95.3727
−63.5268	−13.2161	−97.7477
−63.4367	−13.0941	−100.0747
−63.3467	−12.9783	−102.3509
−63.2566	−12.8701	−104.5750

(*Continues*)

(*Continued*)

Vertical port reading for directivity: Frequency 5.8 GHz		
Elevation (°)	**Amplitude (dB)**	**Phase (°)**
−63.1666	−12.7712	−106.7468
−63.0765	−12.6832	−108.8674
−62.9865	−12.6079	−110.9390
−62.8965	−12.5471	−112.9647
−62.8064	−12.5027	−114.9489
−62.7164	−12.4730	−116.9513
−62.6263	−12.4526	−119.0435
−62.5363	−12.4400	−121.2213
−62.4462	−12.4337	−123.4798
−62.3562	−12.4324	−125.8136
−62.2661	−12.4350	−128.2168
−62.1761	−12.4405	−130.6830
−62.0860	−12.4483	−133.2054
−61.9960	−12.4579	−135.7769
−61.9060	−12.4689	−138.3904
−61.8159	−12.4814	−141.0387
−61.7259	−12.4956	−143.7150
−61.6358	−12.5120	−146.4131
−61.5458	−12.5314	−149.1276
−61.4557	−12.5548	−151.8537
−61.3657	−12.5835	−154.5885
−61.2756	−12.6190	−157.3300
−61.1856	−12.6605	−160.1571
−61.0955	−12.7030	−163.1688
−61.0055	−12.7421	−166.3551
−60.9155	−12.7737	−169.7014
−60.8254	−12.7942	−173.1890
−60.7354	−12.8003	−176.7950
−60.6453	−12.7895	179.5071
−60.5553	−12.7600	175.7467
−60.4652	−12.7113	171.9545
−60.3752	−12.6436	168.1611
−60.2851	−12.5582	164.3955
−60.1951	−12.4572	160.6837
−60.1051	−12.3436	157.0476
−60.0150	−12.2208	153.5042
−59.9250	−12.0927	150.0653
−59.8349	−11.9634	146.7378
−59.7449	−11.8324	143.4700
−59.6548	−11.6887	140.1335
−59.5648	−11.5305	136.7533
−59.4747	−11.3574	133.3620
−59.3847	−11.1702	129.9903
−59.2946	−10.9702	126.6658
−59.2046	−10.7598	123.4123
−59.1146	−10.5417	120.2488

(*Continues*)

(*Continued*)

Vertical port reading for directivity: Frequency 5.8 GHz		
Elevation (°)	**Amplitude (dB)**	**Phase (°)**
−59.0245	−10.3190	117.1896
−58.9345	−10.0950	114.2444
−58.8444	−9.8731	111.4186
−58.7544	−9.6566	108.7137
−58.6643	−9.4488	106.1279
−58.5743	−9.2528	103.6570
−58.4842	−9.0714	101.2942
−58.3942	−8.9072	99.0287
−58.3042	−8.7505	96.7624
−58.2141	−8.5957	94.4617
−58.1241	−8.4425	92.1413
−58.0340	−8.2911	89.8145
−57.9440	−8.1421	87.4934
−57.8539	−7.9962	85.1885
−57.7639	−7.8543	82.9087
−57.6738	−7.7179	80.6613
−57.5838	−7.5881	78.4519
−57.4937	−7.4666	76.2844
−57.4037	−7.3551	74.1613
−57.3137	−7.2553	72.0834
−57.2236	−7.1691	70.0503
−57.1336	−7.0985	68.0600
−57.0435	−7.0451	66.1048
−56.9535	−7.0035	64.0936
−56.8634	−6.9698	61.9983
−56.7734	−6.9428	59.8278
−56.6833	−6.9213	57.5908
−56.5933	−6.9046	55.2964
−56.5033	−6.8922	52.9538
−56.4132	−6.8839	50.5718
−56.3232	−6.8796	48.1593
−56.2331	−6.8798	45.7246
−56.1431	−6.8850	43.2753
−56.0530	−6.8962	40.8183
−55.9630	−6.9146	38.3593
−55.8729	−6.9417	35.9030
−55.7829	−6.9793	33.4524
−55.6928	−7.0277	30.9469
−55.6028	−7.0828	28.2830
−55.5128	−7.1413	25.4672
−55.4227	−7.2001	22.5104
−55.3327	−7.2565	19.4263
−55.2426	−7.3079	16.2311
−55.1526	−7.3525	12.9433
−55.0625	−7.3888	9.5837
−54.9725	−7.4163	6.1740

(Continues)

(*Continued*)

Vertical port reading for directivity: Frequency 5.8 GHz		
Elevation (°)	**Amplitude (dB)**	**Phase (°)**
−54.8824	−7.4350	2.7364
−54.7924	−7.4458	−0.7073
−54.7024	−7.4502	−4.1370
−54.6123	−7.4506	−7.5351
−54.5223	−7.4498	−10.8871
−54.4322	−7.4471	−14.2768
−54.3422	−7.4351	−17.8204
−54.2521	−7.4100	−21.4892
−54.1621	−7.3689	−25.2496
−54.0720	−7.3103	−29.0655
−53.9820	−7.2336	−32.9000
−53.8919	−7.1397	−36.7175
−53.8019	−7.0306	−40.4852
−53.7119	−6.9093	−44.1751
−53.6218	−6.7796	−47.7645
−53.5318	−6.6459	−51.2370
−53.4417	−6.5128	−54.5824
−53.3517	−6.3852	−57.7961
−53.2616	−6.2669	−60.8935
−53.1716	−6.1480	−64.0208
−53.0815	−6.0241	−67.1949
−52.9915	−5.8948	−70.3910
−52.9015	−5.7601	−73.5856
−52.8114	−5.6211	−76.7572
−52.7214	−5.4793	−79.8871
−52.6313	−5.3367	−82.9596
−52.5413	−5.1959	−85.9628
−52.4512	−5.0596	−88.8881
−52.3612	−4.9308	−91.7307
−52.2711	−4.8127	−94.4889
−52.1811	−4.7086	−97.1642
−52.0910	−4.6215	−99.7654
−52.0010	−4.5446	−102.4079
−51.9110	−4.4722	−105.1311
−51.8209	−4.4029	−107.9198
−51.7309	−4.3356	−110.7586
−51.6408	−4.2698	−113.6324
−51.5508	−4.2054	−116.5269
−51.4607	−4.1428	−119.4284
−51.3707	−4.0831	−122.3249
−51.2806	−4.0274	−125.2061
−51.1906	−3.9776	−128.0638
−51.1006	−3.9357	−130.8921
−51.0105	−3.9040	−133.6875
−50.9205	−3.8846	−136.4645
−50.8304	−3.8719	−139.3565

(*Continues*)

(*Continued*)

Vertical port reading for directivity: Frequency 5.8 GHz		
Elevation (°)	**Amplitude (dB)**	**Phase (°)**
−50.7404	−3.8613	−142.3821
−50.6503	−3.8499	−145.5236
−50.5603	−3.8351	−148.7614
−50.4702	−3.8152	−152.0742
−50.3802	−3.7891	−155.4397
−50.2901	−3.7566	−158.8351
−50.2001	−3.7180	−162.2386
−50.1101	−3.6746	−165.6297
−50.0200	−3.6282	−168.9906
−49.9300	−3.5815	−172.3065
−49.8399	−3.5374	−175.5660
−49.7499	−3.4948	−178.8496
−49.6598	−3.4461	177.7345
−49.5698	−3.3880	174.2150
−49.4797	−3.3182	170.6244
−49.3897	−3.2357	166.9960
−49.2996	−3.1405	163.3630
−49.2096	−3.0336	159.7563
−49.1196	−2.9171	156.2038
−49.0295	−2.7939	152.7289
−48.9395	−2.6676	149.3501
−48.8494	−2.5423	146.0804
−48.7594	−2.4223	142.9275
−48.6693	−2.3086	139.8473
−48.5793	−2.1918	136.7086
−48.4892	−2.0699	133.5281
−48.3992	−1.9425	130.3324
−48.3092	−1.8098	127.1465
−48.2191	−1.6731	123.9933
−48.1291	−1.5340	120.8927
−48.0390	−1.3950	117.8609
−47.9490	−1.2588	114.9107
−47.8589	−1.1284	112.0509
−47.7689	−1.0071	109.2869
−47.6788	−0.8986	106.6205
−47.5888	−0.8029	104.0065
−47.4987	−0.7125	101.3348
−47.4087	−0.6253	98.6154
−47.3187	−0.5406	95.8649
−47.2286	−0.4580	93.0995
−47.1386	−0.3777	90.3347
−47.0485	−0.3002	87.5844
−46.9585	−0.2268	84.8616
−46.8684	−0.1590	82.1771
−46.7784	−0.0987	79.5399
−46.6883	−0.0482	76.9569

(*Continues*)

(*Continued*)

Vertical port reading for directivity: Frequency 5.8 GHz		
Elevation (°)	**Amplitude (dB)**	**Phase (°)**
−46.5983	−0.0101	74.4326
−46.5083	0.0164	71.8860
−46.4182	0.0352	69.2453
−46.3282	0.0488	66.5232
−46.2381	0.0590	63.7340
−46.1481	0.0674	60.8929
−46.0580	0.0747	58.0162
−45.9680	0.0813	55.1203
−45.8779	0.0870	52.2218
−45.7879	0.0910	49.3364
−45.6978	0.0918	46.4790
−45.6078	0.0877	43.6631
−45.5178	0.0763	40.8744
−45.4277	0.0586	37.9804
−45.3377	0.0379	34.9729
−45.2476	0.0175	31.8694
−45.1576	−0.0003	28.6899
−45.0675	−0.0136	25.4570
−44.9775	−0.0213	22.1950
−44.8874	−0.0233	18.9292
−44.7974	−0.0202	15.6846
−44.7074	−0.0136	12.4853
−44.6173	−0.0058	9.3536
−44.5273	0.0002	6.3031
−44.4372	0.0034	3.2048
−44.3472	0.0068	0.0140
−44.2571	0.0132	−3.2474
−44.1671	0.0248	−6.5550
−44.0770	0.0427	−9.8824
−43.9870	0.0671	−13.2025
−43.8969	0.0974	−16.4884
−43.8069	0.1320	−19.7142
−43.7169	0.1686	−22.8566
−43.6268	0.2039	−25.8948
−43.5368	0.2343	−28.8134
−43.4467	0.2578	−31.7264
−43.3567	0.2768	−34.6948
−43.2666	0.2934	−37.7027
−43.1766	0.3091	−40.7320
−43.0865	0.3248	−43.7634
−42.9965	0.3406	−46.7769
−42.9065	0.3560	−49.7524
−42.8164	0.3697	−52.6705
−42.7264	0.3799	−55.5126
−42.6363	0.3840	−58.2618
−42.5463	0.3788	−60.9080

(*Continues*)

(*Continued*)

Vertical port reading for directivity: Frequency 5.8 GHz		
Elevation (°)	**Amplitude (dB)**	**Phase (°)**
−42.4562	0.3604	−63.5672
−42.3662	0.3301	−66.2877
−42.2761	0.2905	−69.0613
−42.1861	0.2438	−71.8771
−42.0960	0.1919	−74.7216
−42.0060	0.1360	−77.5792
−41.9160	0.0766	−80.4324
−41.8259	0.0137	−83.2621
−41.7359	−0.0532	−86.0482
−41.6458	−0.1256	−88.7700
−41.5558	−0.2070	−91.4306
−41.4657	−0.3056	−94.1655
−41.3757	−0.4192	−96.9983
−41.2856	−0.5441	−99.9244
−41.1956	−0.6766	−102.9347
−41.1056	−0.8137	−106.0155
−41.0155	−0.9526	−109.1482
−40.9255	−1.0913	−112.3095
−40.8354	−1.2287	−115.4717
−40.7454	−1.3645	−118.6036
−40.6553	−1.4993	−121.6708
−40.5653	−1.6424	−124.7397
−40.4752	−1.8012	−127.9401
−40.3852	−1.9706	−131.2708
−40.2951	−2.1455	−134.7224
−40.2051	−2.3211	−138.2783
−40.1151	−2.4932	−141.9134
−40.0250	−2.6585	−145.5947
−39.9350	−2.8150	−149.2819
−39.8449	−2.9619	−152.9285
−39.7549	−3.1004	−156.4841
−39.6648	−3.2356	−159.9289
−39.5748	−3.3834	−163.4461
−39.4847	−3.5416	−167.0702
−39.3947	−3.7050	−170.7911
−39.3047	−3.8690	−174.5902
−39.2146	−4.0298	−178.4401
−39.1246	−4.1846	177.6948
−39.0345	−4.3322	173.8572
−38.9445	−4.4728	170.0953
−38.8544	−4.6087	166.4607
−38.7644	−4.7445	163.0016
−38.6743	−4.8992	159.5658
−38.5843	−5.0776	156.0523
−38.4942	−5.2748	152.4554
−38.4042	−5.4864	148.7775

(*Continues*)

(*Continued*)

Vertical port reading for directivity: Frequency 5.8 GHz		
Elevation (°)	**Amplitude (dB)**	**Phase (°)**
−38.3142	−5.7081	145.0296
−38.2241	−5.9362	141.2321
−38.1341	−6.1683	137.4147
−38.0440	−6.4029	133.6162
−37.9540	−6.6403	129.8838
−37.8639	−6.8829	126.2716
−37.7739	−7.1550	122.6572
−37.6838	−7.4721	118.8653
−37.5938	−7.8271	114.8553
−37.5038	−8.2121	110.5952
−37.4137	−8.6181	106.0640
−37.3237	−9.0350	101.2575
−37.2336	−9.4524	96.1924
−37.1436	−9.8604	90.9122
−37.0535	−10.2515	85.4895
−36.9635	−10.6217	80.0247
−36.8734	−11.0011	74.3317
−36.7834	−11.4050	68.0768
−36.6933	−11.8051	61.2016
−36.6033	−12.1678	53.7099
−36.5133	−12.4584	45.6948
−36.4232	−12.6476	37.3515
−36.3332	−12.7204	28.9581
−36.2431	−12.6812	20.8207
−36.1531	−12.5535	13.2068
−36.0630	−12.3734	6.2952
−35.9730	−12.1739	−0.3090
−35.8829	−11.9519	−6.8921
−35.7929	−11.7003	−13.3484
−35.7029	−11.4196	−19.5683
−35.6128	−11.1169	−25.4553
−35.5228	−10.8039	−30.9370
−35.4327	−10.4951	−35.9690
−35.3427	−10.2059	−40.5329
−35.2526	−9.9515	−44.6303
−35.1626	−9.7463	−48.2959
−35.0725	−9.5886	−51.9311
−34.9825	−9.4665	−55.6616
−34.8924	−9.3720	−59.4502
−34.8024	−9.2999	−63.2551
−34.7124	−9.2480	−67.0315
−34.6223	−9.2166	−70.7337
−34.5323	−9.2086	−74.3169
−34.4422	−9.2295	−77.7387
−34.3522	−9.2870	−80.9596
−34.2621	−9.3932	−84.0320

(*Continues*)

(*Continued*)

Vertical port reading for directivity: Frequency 5.8 GHz		
Elevation (°)	**Amplitude (dB)**	**Phase (°)**
−34.1721	−9.5592	−87.3526
−34.0820	−9.7758	−90.9728
−33.9920	−10.0328	−94.8842
−33.9020	−10.3216	−99.0672
−33.8119	−10.6349	−103.4882
−33.7219	−10.9681	−108.0979
−33.6318	−11.3203	−112.8319
−33.5418	−11.6952	−117.6134
−33.4517	−12.1029	−122.3577
−33.3617	−12.5721	−127.4295
−33.2716	−13.1061	−133.3951
−33.1816	−13.6721	−140.3633
−33.0915	−14.2280	−148.3931
−33.0015	−14.7242	−157.4426
−32.9115	−15.1130	−167.3226
−32.8214	−15.3630	−177.6966
−32.7314	−15.4726	171.8408
−32.6413	−15.4730	161.6351
−32.5513	−15.4036	151.7324
−32.4612	−15.1912	141.3890
−32.3712	−14.7979	130.9537
−32.2811	−14.2427	120.9580
−32.1911	−13.5763	111.7526
−32.1011	−12.8584	103.4633
−32.0110	−12.1404	96.0414
−31.9210	−11.4598	89.3419
−31.8309	−10.8400	83.1820
−31.7409	−10.2943	77.3721
−31.6508	−9.7656	71.6097
−31.5608	−9.2056	65.8576
−31.4707	−8.6243	60.2468
−31.3807	−8.0350	54.8556
−31.2906	−7.4511	49.7162
−31.2006	−6.8847	44.8245
−31.1106	−6.3458	40.1514
−31.0205	−5.8416	35.6513
−30.9305	−5.3772	31.2686
−30.8404	−4.9377	26.9021
−30.7504	−4.4762	22.4673
−30.6603	−3.9960	18.0586
−30.5703	−3.5056	13.7493
−30.4802	−3.0139	9.5839
−30.3902	−2.5299	5.5823
−30.3001	−2.0617	1.7451
−30.2101	−1.6161	−1.9418
−30.1201	−1.1986	−5.5027

(*Continues*)

(*Continued*)

Vertical port reading for directivity: Frequency 5.8 GHz		
Elevation (°)	**Amplitude (dB)**	**Phase (°)**
–30.0300	–0.8082	–8.9852
–29.9400	–0.4106	–12.4998
–29.8499	–0.0034	–16.0048
–29.7599	0.4072	–19.4488
–29.6698	0.8149	–22.7983
–29.5798	1.2131	–26.0348
–29.4897	1.5958	–29.1528
–29.3997	1.9576	–32.1566
–29.3097	2.2937	–35.0583
–29.2196	2.6013	–37.8820
–29.1296	2.9036	–40.7327
–29.0395	3.2073	–43.6081
–28.9495	3.5090	–46.4715
–28.8594	3.8050	–49.2962
–28.7694	4.0913	–52.0642
–28.6793	4.3638	–54.7654
–28.5893	4.6184	–57.3965
–28.4993	4.8513	–59.9601
–28.4092	5.0592	–62.4662
–28.3192	5.2560	–65.0116
–28.2291	5.4494	–67.6167
–28.1391	5.6384	–70.2521
–28.0490	5.8215	–72.8931
–27.9590	5.9966	–75.5203
–27.8689	6.1610	–78.1189
–27.7789	6.3120	–80.6788
–27.6888	6.4463	–83.1946
–27.5988	6.5608	–85.6670
–27.5088	6.6634	–88.1948
–27.4187	6.7606	–90.8119
–27.3287	6.8531	–93.4911
–27.2386	6.9408	–96.2077
–27.1486	7.0231	–98.9389
–27.0585	7.0985	–101.6649
–26.9685	7.1650	–104.3688
–26.8784	7.2199	–107.0370
–26.7884	7.2604	–109.6632
–26.6983	7.2905	–112.3608
–26.6083	7.3154	–115.1673
–26.5183	7.3367	–118.0558
–26.4282	7.3555	–120.9989
–26.3382	7.3718	–123.9701
–26.2481	7.3847	–126.9437
–26.1581	7.3926	–129.8962
–26.0680	7.3927	–132.8065
–25.9780	7.3820	–135.6660

(*Continues*)

(*Continued*)

Vertical port reading for directivity: Frequency 5.8 GHz		
Elevation (°)	**Amplitude (dB)**	**Phase (°)**
−25.8879	7.3628	−138.6107
−25.7979	7.3392	−141.6697
−25.7079	7.3133	−144.8158
−25.6178	7.2866	−148.0200
−25.5278	7.2592	−151.2529
−25.4377	7.2303	−154.4853
−25.3477	7.1978	−157.6893
−25.2576	7.1587	−160.8395
−25.1676	7.1094	−163.9383
−25.0775	7.0515	−167.1410
−24.9875	6.9885	−170.4606
−24.8974	6.9232	−173.8723
−24.8074	6.8568	−177.3485
−24.7174	6.7899	179.1394
−24.6273	6.7216	175.6206
−24.5373	6.6497	172.1234
−24.4472	6.5709	168.6736
−24.3572	6.4812	165.2368
−24.2671	6.3820	161.6514
−24.1771	6.2780	157.9209
−24.0870	6.1727	154.0667
−23.9970	6.0687	150.1146
−23.9070	5.9674	146.0940
−23.8169	5.8687	142.0359
−23.7269	5.7712	137.9711
−23.6368	5.6717	133.9285
−23.5468	5.5669	129.8076
−23.4567	5.4609	125.4743
−23.3667	5.3609	120.9517
−23.2766	5.2729	116.2734
−23.1866	5.2011	111.4816
−23.0965	5.1479	106.6245
−23.0065	5.1133	101.7527
−22.9165	5.0951	96.9151
−22.8264	5.0887	92.1434
−22.7364	5.0949	87.2584
−22.6463	5.1215	82.2270
−22.5563	5.1740	77.1099
−22.4662	5.2551	71.9741
−22.3762	5.3643	66.8870
−22.2861	5.4983	61.9107
−22.1961	5.6511	57.0966
−22.1061	5.8150	52.4828
−22.0160	5.9844	48.0341
−21.9260	6.1658	43.5788
−21.8359	6.3615	39.1506

(*Continues*)

(*Continued*)

Vertical port reading for directivity: Frequency 5.8 GHz		
Elevation (°)	**Amplitude (dB)**	**Phase (°)**
−21.7459	6.5709	34.7984
−21.6558	6.7913	30.5650
−21.5658	7.0186	26.4852
−21.4757	7.2476	22.5843
−21.3857	7.4725	18.8781
−21.2956	7.6871	15.3732
−21.2056	7.8937	11.9512
−21.1156	8.0980	8.5423
−21.0255	8.3006	5.1754
−20.9355	8.5007	1.8764
−20.8454	8.6968	−1.3323
−20.7554	8.8866	−4.4328
−20.6653	9.0671	−7.4111
−20.5753	9.2352	−10.2575
−20.4852	9.3883	−12.9916
−20.3952	9.5309	−15.7351
−20.3052	9.6654	−18.4868
−20.2151	9.7928	−21.2259
−20.1251	9.9133	−23.9327
−20.0350	10.0265	−26.5888
−19.9450	10.1312	−29.1773
−19.8549	10.2258	−31.6833
−19.7649	10.3082	−34.0938
−19.6748	10.3774	−36.4854
−19.5848	10.4355	−38.9166
−19.4947	10.4846	−41.3697
−19.4047	10.5257	−43.8264
−19.3147	10.5594	−46.2678
−19.2246	10.5858	−48.6751
−19.1346	10.6041	−51.0295
−19.0445	10.6132	−53.3125
−18.9545	10.6108	−55.5349
−18.8644	10.5952	−57.7998
−18.7744	10.5682	−60.1060
−18.6843	10.5314	−62.4372
−18.5943	10.4861	−64.7760
−18.5043	10.4325	−67.1032
−18.4142	10.3708	−69.3985
−18.3242	10.3002	−71.6408
−18.2341	10.2193	−73.8106
−18.1441	10.1233	−75.9922
−18.0540	10.0117	−78.2277
−17.9640	9.8861	−80.5058
−17.8739	9.7478	−82.8127
−17.7839	9.5973	−85.1326
−17.6938	9.4349	−87.4471

(Continues)

(*Continued*)

Vertical port reading for directivity: Frequency 5.8 GHz		
Elevation (°)	**Amplitude (dB)**	**Phase (°)**
−17.6038	9.2601	−89.7356
−17.5138	9.0718	−91.9752
−17.4237	8.8651	−94.1946
−17.3337	8.6340	−96.4899
−17.2436	8.3795	−98.8629
−17.1536	8.1024	−101.3102
−17.0635	7.8030	−103.8262
−16.9735	7.4813	−106.4025
−16.8834	7.1366	−109.0280
−16.7934	6.7672	−111.6884
−16.7034	6.3689	−114.3856
−16.6133	5.9279	−117.2543
−16.5233	5.4414	−120.3530
−16.4332	4.9093	−123.7225
−16.3432	4.3316	−127.4104
−16.2531	3.7085	−131.4727
−16.1631	3.0414	−135.9759
−16.0730	2.3324	−140.9985
−15.9830	1.5858	−146.6369
−15.8929	0.7963	−153.2271
−15.8029	−0.0123	−161.1394
−15.7129	−0.7789	−170.6367
−15.6228	−1.4070	178.1703
−15.5328	−1.7778	165.5371
−15.4427	−1.7957	152.2199
−15.3527	−1.4476	139.2720
−15.2626	−0.8112	127.5627
−15.1726	0.0125	117.3489
−15.0825	0.9510	108.5433
−14.9925	1.9317	101.0907
−14.9025	2.9044	94.8117
−14.8124	3.8397	89.4967
−14.7224	4.7223	84.9526
−14.6323	5.5456	81.0183
−14.5423	6.3079	77.5641
−14.4522	7.0154	74.4702
−14.3622	7.6881	71.6157
−14.2721	8.3274	68.9732
−14.1821	8.9329	66.5220
−14.0920	9.5048	64.2418
−14.0020	10.0431	62.1133
−13.9120	10.5484	60.1184
−13.8219	11.0211	58.2402
−13.7319	11.4630	56.4586
−13.6418	11.8854	54.7263
−13.5518	12.2907	53.0395

(*Continues*)

(*Continued*)

Vertical port reading for directivity: Frequency 5.8 GHz		
Elevation (°)	**Amplitude (dB)**	**Phase (°)**
−13.4617	12.6785	51.4022
−13.3717	13.0485	49.8167
−13.2816	13.4002	48.2832
−13.1916	13.7334	46.8012
−13.1016	14.0476	45.3688
−13.0115	14.3427	43.9824
−12.9215	14.6246	42.6098
−12.8314	14.8959	41.2440
−12.7414	15.1563	39.8924
−12.6513	15.4054	38.5607
−12.5613	15.6428	37.2536
−12.4712	15.8679	35.9745
−12.3812	16.0802	34.7259
−12.2911	16.2793	33.5095
−12.2011	16.4678	32.3004
−12.1111	16.6484	31.0844
−12.0210	16.8209	29.8695
−11.9310	16.9849	28.6629
−11.8409	17.1401	27.4708
−11.7509	17.2859	26.2985
−11.6608	17.4218	25.1507
−11.5708	17.5473	24.0314
−11.4807	17.6634	22.9232
−11.3907	17.7723	21.8029
−11.3007	17.8741	20.6785
−11.2106	17.9683	19.5575
−11.1206	18.0547	18.4470
−11.0305	18.1327	17.3532
−10.9405	18.2018	16.2822
−10.8504	18.2614	15.2395
−10.7604	18.3116	14.2140
−10.6703	18.3541	13.1765
−10.5803	18.3889	12.1332
−10.4902	18.4157	11.0913
−10.4002	18.4342	10.0578
−10.3102	18.4439	9.0394
−10.2201	18.4443	8.0427
−10.1301	18.4346	7.0743
−10.0400	18.4144	6.1285
−9.9500	18.3845	5.1733
−9.8599	18.3448	4.2120
−9.7699	18.2950	3.2509
−9.6798	18.2350	2.2963
−9.5898	18.1641	1.3548
−9.4998	18.0817	0.4330
−9.4097	17.9872	−0.4620

(*Continues*)

(*Continued*)

Vertical port reading for directivity: Frequency 5.8 GHz		
Elevation (°)	**Amplitude (dB)**	**Phase (°)**
−9.3197	17.8796	−1.3317
−9.2296	17.7579	−2.2077
−9.1396	17.6220	−3.0901
−9.0495	17.4715	−3.9740
−8.9595	17.3059	−4.8544
−8.8694	17.1246	−5.7255
−8.7794	16.9267	−6.5809
−8.6893	16.7111	−7.4135
−8.5993	16.4763	−8.2214
−8.5093	16.2185	−9.0346
−8.4192	15.9362	−9.8565
−8.3292	15.6283	−10.6847
−8.2391	15.2931	−11.5161
−8.1491	14.9290	−12.3468
−8.0590	14.5335	−13.1718
−7.9690	14.1038	−13.9845
−7.8789	13.6356	−14.7815
−7.7889	13.1179	−15.5925
−7.6988	12.5434	−16.4279
−7.6088	11.9050	−17.2925
−7.5188	11.1930	−18.1924
−7.4287	10.3951	−19.1356
−7.3387	9.4946	−20.1323
−7.2486	8.4685	−21.1978
−7.1586	7.2818	−22.3601
−7.0685	5.8632	−23.7196
−6.9785	4.1156	−25.4360
−6.8884	1.8712	−27.8438
−6.7984	−1.2142	−31.8555
−6.7084	−6.0132	−41.0918
−6.6183	−14.4186	−85.8166
−6.5283	−9.0604	−173.2518
−6.4382	−2.8210	170.7911
−6.3482	0.9591	165.3124
−6.2581	3.6487	162.3738
−6.1681	5.7338	160.4300
−6.0780	7.4347	158.9802
−5.9880	8.8689	157.8135
−5.8979	10.1063	156.8253
−5.8079	11.1919	155.9574
−5.7179	12.1573	155.1730
−5.6278	13.0339	154.4355
−5.5378	13.8378	153.7362
−5.4477	14.5787	153.0724
−5.3577	15.2644	152.4418
−5.2676	15.9010	151.8420

(Continues)

(*Continued*)

Vertical port reading for directivity: Frequency 5.8 GHz		
Elevation (°)	**Amplitude (dB)**	**Phase (°)**
−5.1776	16.4937	151.2708
−5.0875	17.0466	150.7262
−4.9975	17.5637	150.2049
−4.9075	18.0527	149.6909
−4.8174	18.5169	149.1838
−4.7274	18.9579	148.6871
−4.6373	19.3768	148.2034
−4.5473	19.7749	147.7347
−4.4572	20.1532	147.2823
−4.3672	20.5126	146.8474
−4.2771	20.8541	146.4293
−4.1871	21.1814	146.0153
−4.0970	21.4958	145.6048
−4.0070	21.7974	145.2010
−3.9170	22.0868	144.8065
−3.8269	22.3641	144.4235
−3.7369	22.6297	144.0538
−3.6468	22.8836	143.6991
−3.5568	23.1264	143.3593
−3.4667	23.3602	143.0239
−3.3767	23.5857	142.6924
−3.2866	23.8027	142.3671
−3.1966	24.0114	142.0502
−3.1066	24.2118	141.7436
−3.0165	24.4039	141.4489
−2.9265	24.5877	141.1675
−2.8364	24.7633	140.8999
−2.7464	24.9323	140.6377
−2.6563	25.0951	140.3803
−2.5663	25.2515	140.1293
−2.4762	25.4014	139.8865
−2.3862	25.5448	139.6531
−2.2961	25.6816	139.4305
−2.2061	25.8115	139.2199
−2.1161	25.9348	139.0217
−2.0260	26.0526	138.8297
−1.9360	26.1650	138.6435
−1.8459	26.2721	138.4641
−1.7559	26.3736	138.2924
−1.6658	26.4693	138.1297
−1.5758	26.5592	137.9766
−1.4857	26.6431	137.8342
−1.3957	26.7210	137.7026
−1.3057	26.7938	137.5780
−1.2156	26.8616	137.4599
−1.1256	26.9242	137.3489

(*Continues*)

(*Continued*)

Vertical port reading for directivity: Frequency 5.8 GHz		
Elevation (°)	**Amplitude (dB)**	**Phase (°)**
−1.0355	26.9817	137.2455
−0.9455	27.0338	137.1504
−0.8554	27.0804	137.0641
−0.7654	27.1213	136.9871
−0.6753	27.1566	136.9198
−0.5853	27.1867	136.8600
−0.4952	27.2118	136.8075
−0.4052	27.2316	136.7625
−0.3152	27.2463	136.7253
−0.2251	27.2557	136.6959
−0.1351	27.2598	136.6746
−0.0450	27.2584	136.6618
0.0450	27.2515	136.6578
0.1351	27.2391	136.6622
0.2251	27.2212	136.6750
0.3152	27.1980	136.6959
0.4052	27.1693	136.7250
0.4952	27.1353	136.7621
0.5853	27.0959	136.8071
0.6753	27.0512	136.8602
0.7654	27.0010	136.9215
0.8554	26.9447	136.9929
0.9455	26.8825	137.0741
1.0355	26.8144	137.1648
1.1256	26.7406	137.2646
1.2156	26.6611	137.3732
1.3057	26.5761	137.4901
1.3957	26.4856	137.6150
1.4857	26.3895	137.7484
1.5758	26.2867	137.8940
1.6658	26.1772	138.0519
1.7559	26.0611	138.2213
1.8459	25.9386	138.4016
1.9360	25.8100	138.5921
2.0260	25.6753	138.7920
2.1161	25.5346	139.0007
2.2061	25.3879	139.2187
2.2961	25.2335	139.4522
2.3862	25.0712	139.7014
2.4762	24.9013	139.9654
2.5663	24.7239	140.2433
2.6563	24.5392	140.5341
2.7464	24.3473	140.8365
2.8364	24.1483	141.1495
2.9265	23.9421	141.4739
3.0165	23.7263	141.8190

(*Continues*)

(*Continued*)

Vertical port reading for directivity: Frequency 5.8 GHz		
Elevation (°)	**Amplitude (dB)**	**Phase (°)**
3.1066	23.5008	142.1851
3.1966	23.2657	142.5715
3.2866	23.0209	142.9772
3.3767	22.7666	143.4012
3.4667	22.5026	143.8421
3.5568	22.2291	144.2982
3.6468	21.9454	144.7714
3.7369	21.6487	145.2746
3.8269	21.3385	145.8093
3.9170	21.0145	146.3759
4.0070	20.6762	146.9744
4.0970	20.3235	147.6047
4.1871	19.9557	148.2664
4.2771	19.5725	148.9590
4.3672	19.1724	149.6865
4.4572	18.7513	150.4690
4.5473	18.3077	151.3128
4.6373	17.8402	152.2234
4.7274	17.3473	153.2069
4.8174	16.8273	154.2707
4.9075	16.2783	155.4232
4.9975	15.6981	156.6743
5.0875	15.0826	158.0434
5.1776	14.4234	159.5784
5.2676	13.7154	161.3170
5.3577	12.9536	163.3055
5.4477	12.1323	165.6051
5.5378	11.2451	168.2983
5.6278	10.2857	171.4981
5.7179	9.2488	175.3606
5.8079	8.1297	−179.8782
5.8979	6.9251	−173.8162
5.9880	5.6638	−165.9488
6.0780	4.4275	−155.6791
6.1681	3.3845	−142.5731
6.2581	2.7841	−127.0005
6.3482	2.8180	−110.6641
6.4382	3.4396	−95.7988
6.5283	4.4157	−83.6157
6.6183	5.5321	−74.0426
6.7084	6.6540	−66.6142
6.7984	7.7170	−60.8004
6.8884	8.6971	−56.1666
6.9785	9.5899	−52.3947
7.0685	10.3990	−49.2596
7.1586	11.1308	−46.6025

(*Continues*)

(*Continued*)

Vertical port reading for directivity: Frequency 5.8 GHz		
Elevation (°)	**Amplitude (dB)**	**Phase (°)**
7.2486	11.7939	−44.3011
7.3387	12.4025	−42.2434
7.4287	12.9625	−40.3815
7.5188	13.4780	−38.6824
7.6088	13.9527	−37.1204
7.6988	14.3895	−35.6750
7.7889	14.7910	−34.3295
7.8789	15.1594	−33.0704
7.9690	15.4973	−31.8806
8.0590	15.8106	−30.7248
8.1491	16.1012	−29.5986
8.2391	16.3705	−28.5020
8.3292	16.6194	−27.4349
8.4192	16.8488	−26.3971
8.5093	17.0592	−25.3885
8.5993	17.2511	−24.4088
8.6893	17.4251	−23.4531
8.7794	17.5836	−22.4941
8.8694	17.7276	−21.5305
8.9595	17.8578	−20.5663
9.0495	17.9748	−19.6051
9.1396	18.0789	−18.6503
9.2296	18.1705	−17.7050
9.3197	18.2498	−16.7724
9.4097	18.3168	−15.8518
9.4998	18.3715	−14.9158
9.5898	18.4147	−13.9620
9.6798	18.4471	−12.9951
9.7699	18.4692	−12.0197
9.8599	18.4815	−11.0406
9.9500	18.4843	−10.0626
10.0400	18.4780	−9.0907
10.1301	18.4624	−8.1275
10.2201	18.4361	−7.1445
10.3102	18.3994	−6.1359
10.4002	18.3529	−5.1066
10.4902	18.2974	−4.0618
10.5803	18.2334	−3.0071
10.6703	18.1614	−1.9485
10.7604	18.0818	−0.8923
10.8504	17.9948	0.1557
10.9405	17.8979	1.2232
11.0305	17.7905	2.3214
11.1206	17.6734	3.4456
11.2106	17.5474	4.5907
11.3007	17.4131	5.7508

(*Continues*)

(*Continued*)

Vertical port reading for directivity: Frequency 5.8 GHz		
Elevation (°)	**Amplitude (dB)**	**Phase (°)**
11.3907	17.2710	6.9194
11.4807	17.1217	8.0887
11.5708	16.9655	9.2508
11.6608	16.7994	10.4288
11.7509	16.6217	11.6411
11.8409	16.4331	12.8842
11.9310	16.2344	14.1539
12.0210	16.0261	15.4450
12.1111	15.8088	16.7512
12.2011	15.5831	18.0648
12.2911	15.3493	19.3768
12.3812	15.1043	20.7019
12.4712	14.8446	22.0670
12.5613	14.5706	23.4715
12.6513	14.2829	24.9138
12.7414	13.9819	26.3914
12.8314	13.6681	27.9002
12.9215	13.3420	29.4348
13.0115	13.0037	30.9876
13.1016	12.6504	32.5657
13.1916	12.2750	34.2042
13.2816	11.8771	35.9108
13.3717	11.4569	37.6917
13.4617	11.0144	39.5526
13.5518	10.5496	41.4985
13.6418	10.0627	43.5334
13.7319	9.5536	45.6601
13.8219	9.0195	47.8894
13.9120	8.4466	50.2802
14.0020	7.8320	52.8727
14.0920	7.1746	55.7099
14.1821	6.4741	58.8426
14.2721	5.7308	62.3311
14.3622	4.9465	66.2467
14.4522	4.1257	70.6714
14.5423	3.2741	75.7061
14.6323	2.3796	81.5925
14.7224	1.4583	88.5904
14.8124	0.5537	96.9457
14.9025	−0.2622	106.8277
14.9925	−0.8904	118.1726
15.0825	−1.2305	130.5161
15.1726	−1.2300	143.0093
15.2626	−0.9181	154.7391
15.3527	−0.3778	165.3447
15.4427	0.3107	174.6894

(*Continues*)

(*Continued*)

Vertical port reading for directivity: Frequency 5.8 GHz		
Elevation (°)	**Amplitude (dB)**	**Phase (°)**
15.5328	1.0739	−177.2767
15.6228	1.8556	−170.4407
15.7129	2.6189	−164.6257
15.8029	3.3422	−159.6502
15.8929	4.0139	−155.3532
15.9830	4.6286	−151.6011
16.0730	5.1913	−148.2284
16.1631	5.7160	−145.0772
16.2531	6.2046	−142.1253
16.3432	6.6583	−139.3548
16.4332	7.0780	−136.7495
16.5233	7.4642	−134.2946
16.6133	7.8171	−131.9759
16.7034	8.1367	−129.7805
16.7934	8.4244	−127.6791
16.8834	8.6894	−125.5811
16.9735	8.9344	−123.4861
17.0635	9.1605	−121.4052
17.1536	9.3683	−119.3479
17.2436	9.5577	−117.3220
17.3337	9.7287	−115.3338
17.4237	9.8807	−113.3881
17.5138	10.0130	−111.4863
17.6038	10.1287	−109.5565
17.6938	10.2311	−107.5745
17.7839	10.3213	−105.5558
17.8739	10.4000	−103.5150
17.9640	10.4676	−101.4656
18.0540	10.5239	−99.4204
18.1441	10.5684	−97.3912
18.2341	10.6002	−95.3888
18.3242	10.6190	−93.3795
18.4142	10.6271	−91.2917
18.5043	10.6260	−89.1394
18.5943	10.6168	−86.9391
18.6843	10.6002	−84.7075
18.7744	10.5762	−82.4612
18.8644	10.5446	−80.2169
18.9545	10.5044	−77.9905
19.0445	10.4543	−75.7857
19.1346	10.3941	−73.5080
19.2246	10.3256	−71.1463
19.3147	10.2503	−68.7169
19.4047	10.1692	−66.2371
19.4947	10.0827	−63.7253
19.5848	9.9908	−61.2004

(*Continues*)

(*Continued*)

Vertical port reading for directivity: Frequency 5.8 GHz		
Elevation (°)	**Amplitude (dB)**	**Phase (°)**
19.6748	9.8925	−58.6820
19.7649	9.7866	−56.1895
19.8549	9.6707	−53.6679
19.9450	9.5450	−51.0431
20.0350	9.4118	−48.3274
20.1251	9.2727	−45.5357
20.2151	9.1287	−42.6849
20.3052	8.9804	−39.7939
20.3952	8.8275	−36.8827
20.4852	8.6689	−33.9725
20.5753	8.5023	−31.0659
20.6653	8.3239	−28.0456
20.7554	8.1362	−24.8928
20.8454	7.9423	−21.6167
20.9355	7.7448	−18.2307
21.0255	7.5458	−14.7522
21.1156	7.3466	−11.2021
21.2056	7.1476	−7.6045
21.2956	6.9481	−3.9855
21.3857	6.7433	−0.2779
21.4757	6.5321	3.6299
21.5658	6.3207	7.7304
21.6558	6.1150	12.0073
21.7459	5.9201	16.4344
21.8359	5.7400	20.9771
21.9260	5.5773	25.5928
22.0160	5.4325	30.2345
22.1061	5.3036	34.8704
22.1961	5.1858	39.6613
22.2861	5.0855	44.6314
22.3762	5.0097	49.7339
22.4662	4.9635	54.9103
22.5563	4.9490	60.0941
22.6463	4.9654	65.2170
22.7364	5.0087	70.2149
22.8264	5.0725	75.0333
22.9165	5.1497	79.7079
23.0065	5.2412	84.4034
23.0965	5.3496	89.0827
23.1866	5.4755	93.6969
23.2766	5.6169	98.1996
23.3667	5.7702	102.5504
23.4567	5.9303	106.7172
23.5468	6.0910	110.6770
23.6368	6.2458	114.4164
23.7269	6.3930	118.0586

(*Continues*)

(*Continued*)

Vertical port reading for directivity: Frequency 5.8 GHz		
Elevation (°)	**Amplitude (dB)**	**Phase (°)**
23.8169	6.5365	121.6907
23.9070	6.6779	125.2870
23.9970	6.8173	128.8225
24.0870	6.9540	132.2740
24.1771	7.0860	135.6211
24.2671	7.2108	138.8468
24.3572	7.3253	141.9378
24.4472	7.4258	144.8950
24.5373	7.5126	147.8532
24.6273	7.5890	150.8418
24.7174	7.6574	153.8419
24.8074	7.7190	156.8336
24.8974	7.7747	159.7968
24.9875	7.8241	162.7114
25.0775	7.8663	165.5583
25.1676	7.8995	168.3196
25.2576	7.9210	171.0072
25.3477	7.9289	173.7494
25.4377	7.9262	176.5510
25.5278	7.9159	179.3945
25.6178	7.9001	−177.7403
25.7079	7.8803	−174.8754
25.7979	7.8570	−172.0345
25.8879	7.8300	−169.2418
25.9780	7.7983	−166.5221
26.0680	7.7583	−163.8511
26.1581	7.7060	−161.1079
26.2481	7.6447	−158.2983
26.3382	7.5775	−155.4397
26.4282	7.5068	−152.5534
26.5183	7.4342	−149.6636
26.6083	7.3608	−146.7972
26.6983	7.2863	−143.9831
26.7884	7.2099	−141.2513
26.8784	7.1266	−138.5689
26.9685	7.0307	−135.8249
27.0585	6.9256	−133.0266
27.1486	6.8141	−130.1891
27.2386	6.6985	−127.3315
27.3287	6.5806	−124.4767
27.4187	6.4613	−121.6513
27.5088	6.3409	−118.8845
27.5988	6.2185	−116.2080
27.6888	6.0874	−113.5904
27.7789	5.9396	−110.9303
27.8689	5.7780	−108.2289

(*Continues*)

(*Continued*)

Vertical port reading for directivity: Frequency 5.8 GHz		
Elevation (°)	**Amplitude (dB)**	**Phase (°)**
27.9590	5.6054	−105.4940
28.0490	5.4245	−102.7381
28.1391	5.2375	−99.9784
28.2291	5.0460	−97.2372
28.3192	4.8510	−94.5414
28.4092	4.6530	−91.9225
28.4993	4.4437	−89.3619
28.5893	4.2094	−86.7633
28.6793	3.9528	−84.1163
28.7694	3.6773	−81.4181
28.8594	3.3861	−78.6712
28.9495	3.0828	−75.8846
29.0395	2.7706	−73.0747
29.1296	2.4525	−70.2658
29.2196	2.1312	−67.4915
29.3097	1.7989	−64.7539
29.3997	1.4305	−61.9515
29.4897	1.0276	−59.0544
29.5798	0.5941	−56.0425
29.6698	0.1349	−52.9017
29.7599	−0.3449	−49.6264
29.8499	−0.8387	−46.2221
29.9400	−1.3400	−42.7099
30.0300	−1.8414	−39.1299
30.1201	−2.3458	−35.5148
30.2101	−2.8960	−31.7295
30.3001	−3.4948	−27.6911
30.3902	−4.1356	−23.3342
30.4802	−4.8091	−18.5985
30.5703	−5.5021	−13.4372
30.6603	−6.1975	−7.8304
30.7504	−6.8749	−1.8033
30.8404	−7.5131	4.5545
30.9305	−8.0986	11.0968
31.0205	−8.6859	18.0388
31.1106	−9.2768	25.5845
31.2006	−9.8393	33.7993
31.2906	−10.3345	42.6601
31.3807	−10.7228	52.0185
31.4707	−10.9754	61.5933
31.5608	−11.0855	71.0174
31.6508	−11.0728	79.9290
31.7409	−10.9783	88.0570
31.8309	−10.8638	95.6834
31.9210	−10.7482	103.2889
32.0110	−10.6147	110.8482

(*Continues*)

(*Continued*)

Vertical port reading for directivity: Frequency 5.8 GHz		
Elevation (°)	**Amplitude (dB)**	**Phase (°)**
32.1011	−10.4522	118.2896
32.1911	−10.2569	125.5165
32.2811	−10.0330	132.4290
32.3712	−9.7908	138.9429
32.4612	−9.5455	145.0011
32.5513	−9.3146	150.5768
32.6413	−9.1153	155.7810
32.7314	−8.9443	161.1645
32.8214	−8.7813	166.7419
32.9115	−8.6115	172.4248
33.0015	−8.4264	178.1112
33.0915	−8.2240	−176.3027
33.1816	−8.0083	−170.9113
33.2716	−7.7879	−165.7898
33.3617	−7.5745	−160.9902
33.4517	−7.3818	−156.5403
33.5418	−7.2090	−152.1309
33.6318	−7.0393	−147.5443
33.7219	−6.8646	−142.8685
33.8119	−6.6813	−138.1916
33.9020	−6.4901	−133.5957
33.9920	−6.2953	−129.1514
34.0820	−6.1040	−124.9138
34.1721	−5.9256	−120.9219
34.2621	−5.7702	−117.1985
34.3522	−5.6469	−113.7108
34.4422	−5.5437	−110.1400
34.5323	−5.4522	−106.4599
34.6223	−5.3686	−102.7192
34.7124	−5.2915	−98.9659
34.8024	−5.2215	−95.2449
34.8924	−5.1614	−91.5968
34.9825	−5.1156	−88.0556
35.0725	−5.0901	−84.6485
35.1626	−5.0923	−81.3954
35.2526	−5.1280	−78.2070
35.3427	−5.1912	−74.8311
35.4327	−5.2750	−71.2725
35.5228	−5.3739	−67.5546
35.6128	−5.4837	−63.7053
35.7029	−5.6017	−59.7567
35.7929	−5.7273	−55.7436
35.8829	−5.8613	−51.7017
35.9730	−6.0071	−47.6652
36.0630	−6.1701	−43.6644
36.1531	−6.3578	−39.5284

(Continues)

(*Continued*)

Vertical port reading for directivity: Frequency 5.8 GHz		
Elevation (°)	**Amplitude (dB)**	**Phase (°)**
36.2431	−6.5653	−35.0053
36.3332	−6.7789	−30.1016
36.4232	−6.9852	−24.8440
36.5133	−7.1717	−19.2812
36.6033	−7.3288	−13.4840
36.6933	−7.4504	−7.5408
36.7834	−7.5355	−1.5489
36.8734	−7.5884	4.3971
36.9635	−7.6183	10.2175
37.0535	−7.6261	16.1416
37.1436	−7.5935	22.3601
37.2336	−7.5083	28.7461
37.3237	−7.3655	35.1569
37.4137	−7.1678	41.4531
37.5038	−6.9246	47.5168
37.5938	−6.6500	53.2641
37.6838	−6.3603	58.6483
37.7739	−6.0719	63.6574
37.8639	−5.8000	68.3063
37.9540	−5.5387	72.7997
38.0440	−5.2691	77.2988
38.1341	−4.9912	81.7432
38.2241	−4.7077	86.0828
38.3142	−4.4228	90.2796
38.4042	−4.1420	94.3084
38.4942	−3.8715	98.1562
38.5843	−3.6173	101.8207
38.6743	−3.3858	105.3089
38.7644	−3.1826	108.6350
38.8544	−3.0049	111.9002
38.9445	−2.8356	115.2554
39.0345	−2.6715	118.6703
39.1246	−2.5114	122.1124
39.2146	−2.3549	125.5520
39.3047	−2.2031	128.9641
39.3947	−2.0577	132.3277
39.4847	−1.9214	135.6275
39.5748	−1.7973	138.8531
39.6648	−1.6889	141.9994
39.7549	−1.5983	145.0987
39.8449	−1.5153	148.3197
39.9350	−1.4344	151.6582
40.0250	−1.3525	155.0806
40.1151	−1.2676	158.5526
40.2051	−1.1792	162.0411
40.2951	−1.0878	165.5153

(*Continues*)

(*Continued*)

Vertical port reading for directivity: Frequency 5.8 GHz		
Elevation (°)	**Amplitude (dB)**	**Phase (°)**
40.3852	−0.9951	168.9476
40.4752	−0.9036	172.3147
40.5653	−0.8165	175.5983
40.6553	−0.7379	178.7863
40.7454	−0.6642	−177.9910
40.8354	−0.5887	−174.6612
40.9255	−0.5084	−171.2601
41.0155	−0.4218	−167.8242
41.1056	−0.3288	−164.3886
41.1956	−0.2301	−160.9862
41.2856	−0.1279	−157.6461
41.3757	−0.0248	−154.3931
41.4657	0.0754	−151.2466
41.5558	0.1687	−148.2209
41.6458	0.2520	−145.2977
41.7359	0.3317	−142.3272
41.8259	0.4109	−139.3082
41.9160	0.4908	−136.2668
42.0060	0.5717	−133.2282
42.0960	0.6529	−130.2161
42.1861	0.7332	−127.2518
42.2761	0.8108	−124.3542
42.3662	0.8830	−121.5386
42.4562	0.9471	−118.8175
42.5463	0.9995	−116.2001
42.6363	1.0388	−113.6397
42.7264	1.0692	−111.0233
42.8164	1.0928	−108.3600
42.9065	1.1110	−105.6646
42.9965	1.1248	−102.9524
43.0865	1.1345	−100.2388
43.1766	1.1399	−97.5390
43.2666	1.1403	−94.8673
43.3567	1.1346	−92.2373
43.4467	1.1211	−89.6614
43.5368	1.0975	−87.1502
43.6268	1.0612	−84.6512
43.7169	1.0128	−82.0657
43.8069	0.9550	−79.3996
43.8969	0.8905	−76.6626
43.9870	0.8212	−73.8666
44.0770	0.7491	−71.0252
44.1671	0.6752	−68.1538
44.2571	0.6002	−65.2694
44.3472	0.5242	−62.3896
44.4372	0.4466	−59.5325

(*Continues*)

(*Continued*)

Vertical port reading for directivity: Frequency 5.8 GHz		
Elevation (°)	**Amplitude (dB)**	**Phase (°)**
44.5273	0.3662	−56.7162
44.6173	0.2794	−53.9019
44.7074	0.1833	−50.9703
44.7974	0.0817	−47.9240
44.8874	−0.0217	−44.7744
44.9775	−0.1235	−41.5372
45.0675	−0.2209	−38.2322
45.1576	−0.3119	−34.8826
45.2476	−0.3952	−31.5145
45.3377	−0.4704	−28.1558
45.4277	−0.5380	−24.8356
45.5178	−0.5997	−21.5822
45.6078	−0.6585	−18.3974
45.6978	−0.7193	−15.1386
45.7879	−0.7792	−11.7911
45.8779	−0.8349	−8.3734
45.9680	−0.8837	−4.9079
46.0580	−0.9237	−1.4200
46.1481	−0.9539	2.0624
46.2381	−0.9745	5.5106
46.3282	−0.9864	8.8961
46.4182	−0.9916	12.1918
46.5083	−0.9928	15.3730
46.5983	−0.9935	18.4181
46.6883	−0.9984	21.4072
46.7784	−1.0080	24.4309
46.8684	−1.0199	27.4783
46.9585	−1.0325	30.5361
47.0485	−1.0447	33.5895
47.1386	−1.0560	36.6222
47.2286	−1.0662	39.6177
47.3187	−1.0760	42.5591
47.4087	−1.0866	45.4301
47.4987	−1.0996	48.2153
47.5888	−1.1170	50.9005
47.6788	−1.1418	53.4788
47.7689	−1.1794	56.0556
47.8589	−1.2293	58.6723
47.9490	−1.2892	61.3267
48.0390	−1.3569	64.0141
48.1291	−1.4306	66.7278
48.2191	−1.5088	69.4590
48.3092	−1.5905	72.1971
48.3992	−1.6749	74.9300
48.4892	−1.7617	77.6442
48.5793	−1.8512	80.3250

(*Continues*)

(*Continued*)

Vertical port reading for directivity: Frequency 5.8 GHz		
Elevation (°)	**Amplitude (dB)**	**Phase (°)**
48.6693	−1.9438	82.9575
48.7594	−2.0415	85.5340
48.8494	−2.1534	88.1601
48.9395	−2.2793	90.8760
49.0295	−2.4158	93.6831
49.1196	−2.5596	96.5794
49.2096	−2.7074	99.5598
49.2996	−2.8562	102.6148
49.3897	−3.0032	105.7314
49.4797	−3.1463	108.8925
49.5698	−3.2836	112.0774
49.6598	−3.4140	115.2625
49.7499	−3.5373	118.4215
49.8399	−3.6540	121.5276
49.9300	−3.7734	124.6758
50.0200	−3.8981	127.9419
50.1101	−4.0234	131.3203
50.2001	−4.1452	134.7998
50.2901	−4.2593	138.3632
50.3802	−4.3623	141.9880
50.4702	−4.4517	145.6464
50.5603	−4.5258	149.3068
50.6503	−4.5841	152.9349
50.7404	−4.6270	156.4956
50.8304	−4.6565	159.9543
50.9205	−4.6752	163.2787
51.0105	−4.6882	166.4778
51.1006	−4.7021	169.6926
51.1906	−4.7153	172.9297
51.2806	−4.7254	176.1754
51.3707	−4.7311	179.4135
51.4607	−4.7313	−177.3739
51.5508	−4.7256	−174.2059
51.6408	−4.7145	−171.1021
51.7309	−4.6989	−168.0813
51.8209	−4.6802	−165.1618
51.9110	−4.6603	−162.3603
52.0010	−4.6418	−159.6918
52.0910	−4.6273	−157.1697
52.1811	−4.6215	−154.7569
52.2711	−4.6284	−152.3511
52.3612	−4.6463	−149.9481
52.4512	−4.6738	−147.5495
52.5413	−4.7095	−145.1585
52.6313	−4.7523	−142.7796
52.7214	−4.8013	−140.4189

(*Continues*)

(*Continued*)

Vertical port reading for directivity: Frequency 5.8 GHz		
Elevation (°)	**Amplitude (dB)**	**Phase (°)**
52.8114	−4.8560	−138.0836
52.9015	−4.9159	−135.7819
52.9915	−4.9811	−133.5232
53.0815	−5.0517	−131.3172
53.1716	−5.1282	−129.1746
53.2616	−5.2116	−127.1065
53.3517	−5.3053	−125.1004
53.4417	−5.4188	−123.0701
53.5318	−5.5511	−121.0003
53.6218	−5.7000	−118.8855
53.7119	−5.8635	−116.7216
53.8019	−6.0397	−114.5066
53.8919	−6.2265	−112.2402
53.9820	−6.4219	−109.9242
54.0720	−6.6242	−107.5630
54.1621	−6.8315	−105.1630
54.2521	−7.0421	−102.7339
54.3422	−7.2547	−100.2877
54.4322	−7.4680	−97.8395
54.5223	−7.6814	−95.4067
54.6123	−7.9078	−92.9201
54.7024	−8.1544	−90.3120
54.7924	−8.4176	−87.5681
54.8824	−8.6938	−84.6763
54.9725	−8.9789	−81.6276
55.0625	−9.2681	−78.4172
55.1526	−9.5566	−75.0458
55.2426	−9.8392	−71.5208
55.3327	−10.1107	−67.8576
55.4227	−10.3662	−64.0803
55.5128	−10.6014	−60.2222
55.6028	−10.8130	−56.3249
55.6928	−10.9994	−52.4370
55.7829	−11.1604	−48.6089
55.8729	−11.3100	−44.7328
55.9630	−11.4518	−40.7338
56.0530	−11.5804	−36.6256
56.1431	−11.6910	−32.4301
56.2331	−11.7794	−28.1772
56.3232	−11.8428	−23.9031
56.4132	−11.8795	−19.6491
56.5033	−11.8893	−15.4586
56.5933	−11.8739	−11.3744
56.6833	−11.8359	−7.4362
56.7734	−11.7794	−3.6789
56.8634	−11.7094	−0.1310

(*Continues*)

(*Continued*)

Vertical port reading for directivity: Frequency 5.8 GHz		
Elevation (°)	**Amplitude (dB)**	**Phase (°)**
56.9535	−11.6311	3.1859
57.0435	−11.5502	6.2561
57.1336	−11.4736	9.1237
57.2236	−11.4064	11.9248
57.3137	−11.3476	14.6680
57.4037	−11.2963	17.3535
57.4937	−11.2516	19.9797
57.5838	−11.2131	22.5439
57.6738	−11.1805	25.0423
57.7639	−11.1537	27.4702
57.8539	−11.1328	29.8226
57.9440	−11.1184	32.0938
58.0340	−11.1108	34.2780
58.1241	−11.1110	36.3692
58.2141	−11.1199	38.3611
58.3042	−11.1386	40.2474
58.3942	−11.1685	42.0215
58.4842	−11.2146	43.7142
58.5743	−11.2842	45.4034
58.6643	−11.3758	47.0990
58.7544	−11.4878	48.8071
58.8444	−11.6186	50.5325
58.9345	−11.7668	52.2792
59.0245	−11.9310	54.0495
59.1146	−12.1098	55.8448
59.2046	−12.3020	57.6650
59.2946	−12.5063	59.5086
59.3847	−12.7217	61.3727
59.4747	−12.9469	63.2525
59.5648	−13.1812	65.1414
59.6548	−13.4236	67.0310
59.7449	−13.6734	68.9104
59.8349	−13.9312	70.7712
59.9250	−14.2159	72.6912
60.0150	−14.5335	74.7191
60.1051	−14.8827	76.8771
60.1951	−15.2619	79.1889
60.2851	−15.6689	81.6798
60.3752	−16.1012	84.3766
60.4652	−16.5555	87.3068
60.5553	−17.0274	90.4980
60.6453	−17.5113	93.9762
60.7354	−18.0004	97.7627
60.8254	−18.4861	101.8711
60.9155	−18.9586	106.3023
61.0055	−19.4070	111.0394

(*Continues*)

(*Continued*)

Vertical port reading for directivity: Frequency 5.8 GHz		
Elevation (°)	**Amplitude (dB)**	**Phase (°)**
61.0955	−19.8206	116.0432
61.1856	−20.1893	121.2496
61.2756	−20.5062	126.5719
61.3657	−20.7840	132.1382
61.4557	−21.0228	138.0788
61.5458	−21.2055	144.3446
61.6358	−21.3173	150.8481
61.7259	−21.3472	157.4684
61.8159	−21.2912	164.0638
61.9060	−21.1521	170.4909
61.9960	−20.9395	176.6243
62.0860	−20.6677	−177.6295
62.1761	−20.3530	−172.3273
62.2661	−20.0116	−167.4919
62.3562	−19.6581	−163.1193
62.4462	−19.3049	−159.1876
62.5363	−18.9618	−155.6647
62.6263	−18.6367	−152.5139
62.7164	−18.3355	−149.6980
62.8064	−18.0627	−147.1813
62.8965	−17.8141	−144.8271
62.9865	−17.5838	−142.5442
63.0765	−17.3701	−140.3287
63.1666	−17.1718	−138.1780
63.2566	−16.9881	−136.0905
63.3467	−16.8183	−134.0646
63.4367	−16.6618	−132.0997
63.5268	−16.5184	−130.1950
63.6168	−16.3879	−128.3499
63.7069	−16.2703	−126.5643
63.7969	−16.1657	−124.8375
63.8869	−16.0743	−123.1694
63.9770	−15.9964	−121.5596
64.0670	−15.9326	−120.0076
64.1571	−15.8832	−118.5132
64.2471	−15.8489	−117.0760
64.3372	−15.8304	−115.6956
64.4272	−15.8286	−114.3707
64.5173	−15.8465	−113.0396
64.6073	−15.8840	−111.6634
64.6973	−15.9395	−110.2397
64.7874	−16.0116	−108.7664
64.8774	−16.0987	−107.2414
64.9675	−16.1995	−105.6633
65.0575	−16.3128	−104.0310
65.1476	−16.4372	−102.3436

(*Continues*)

(*Continued*)

Vertical port reading for directivity: Frequency 5.8 GHz		
Elevation (°)	**Amplitude (dB)**	**Phase (°)**
65.2376	−16.5717	−100.6009
65.3277	−16.7151	−98.8034
65.4177	−16.8662	−96.9518
65.5078	−17.0240	−95.0479
65.5978	−17.1876	−93.0940
65.6878	−17.3561	−91.0933
65.7779	−17.5286	−89.0499
65.8679	−17.7044	−86.9688
65.9580	−17.8830	−84.8556
66.0480	−18.0638	−82.7170
66.1381	−18.2468	−80.5599
66.2281	−18.4400	−78.3055
66.3182	−18.6470	−75.8847
66.4082	−18.8644	−73.2867
66.4983	−19.0883	−70.5024
66.5883	−19.3145	−67.5252
66.6783	−19.5382	−64.3518
66.7684	−19.7544	−60.9836
66.8584	−19.9578	−57.4282
66.9485	−20.1429	−53.6997
67.0385	−20.3048	−49.8204
67.1286	−20.4388	−45.8202
67.2186	−20.5413	−41.7358
67.3087	−20.6100	−37.6098
67.3987	−20.6439	−33.4874
67.4887	−20.6437	−29.4146
67.5788	−20.6115	−25.4343
67.6688	−20.5509	−21.5848
67.7589	−20.4662	−17.8974
67.8489	−20.3625	−14.3959
67.9390	−20.2451	−11.0965
68.0290	−20.1180	−7.9689
68.1191	−19.9797	−4.8832
68.2091	−19.8299	−1.8441
68.2991	−19.6694	1.1327
68.3892	−19.4990	4.0328
68.4792	−19.3201	6.8441
68.5693	−19.1342	9.5568
68.6593	−18.9432	12.1634
68.7494	−18.7487	14.6584
68.8394	−18.5524	17.0387
68.9295	−18.3563	19.3027
69.0195	−18.1618	21.4505
69.1096	−17.9706	23.4834
69.1996	−17.7841	25.4038
69.2896	−17.6035	27.2147

(Continues)

(*Continued*)

Vertical port reading for directivity: Frequency 5.8 GHz		
Elevation (°)	**Amplitude (dB)**	**Phase (°)**
69.3797	−17.4302	28.9197
69.4697	−17.2651	30.5228
69.5598	−17.1092	32.0281
69.6498	−16.9633	33.4399
69.7399	−16.8283	34.7621
69.8299	−16.7049	35.9989
69.9200	−16.5937	37.1541
70.0100	−16.4950	38.2369
70.1001	−16.4059	39.2961
70.1901	−16.3253	40.3424
70.2801	−16.2528	41.3758
70.3702	−16.1879	42.3961
70.4602	−16.1304	43.4034
70.5503	−16.0798	44.3973
70.6403	−16.0359	45.3775
70.7304	−15.9984	46.3438
70.8204	−15.9673	47.2956
70.9105	−15.9422	48.2325
71.0005	−15.9232	49.1541
71.0905	−15.9100	50.0599
71.1806	−15.9027	50.9493
71.2706	−15.9011	51.8217
71.3607	−15.9053	52.6766
71.4507	−15.9152	53.5133
71.5408	−15.9309	54.3312
71.6308	−15.9524	55.1297
71.7209	−15.9798	55.9080
71.8109	−16.0132	56.6654
71.9009	−16.0527	57.4012
71.9910	−16.0984	58.1145
72.0810	−16.1505	58.8045
72.1711	−16.2092	59.4704
72.2611	−16.2761	60.1227
72.3512	−16.3540	60.7822
72.4412	−16.4423	61.4507
72.5313	−16.5408	62.1295
72.6213	−16.6489	62.8197
72.7114	−16.7664	63.5224
72.8014	−16.8929	64.2386
72.8914	−17.0281	64.9695
72.9815	−17.1717	65.7160
73.0715	−17.3233	66.4790
73.1616	−17.4827	67.2594
73.2516	−17.6496	68.0580
73.3417	−17.8238	68.8757
73.4317	−18.0049	69.7131

(*Continues*)

(*Continued*)

Vertical port reading for directivity: Frequency 5.8 GHz		
Elevation (°)	**Amplitude (dB)**	**Phase (°)**
73.5218	−18.1926	70.5708
73.6118	−18.3868	71.4495
73.7019	−18.5872	72.3496
73.7919	−18.7934	73.2715
73.8819	−19.0052	74.2155
73.9720	−19.2222	75.1815
74.0620	−19.4443	76.1697
74.1521	−19.6711	77.1796
74.2421	−19.9023	78.2110
74.3322	−20.1376	79.2632
74.4222	−20.3766	80.3352
74.5123	−20.6192	81.4258
74.6023	−20.8649	82.5336
74.6923	−21.1135	83.6566
74.7824	−21.3697	84.8111
74.8724	−21.6420	86.0311
74.9625	−21.9300	87.3254
75.0525	−22.2337	88.7032
75.1426	−22.5527	90.1748
75.2326	−22.8867	91.7517
75.3227	−23.2351	93.4462
75.4127	−23.5971	95.2722
75.5028	−23.9719	97.2447
75.5928	−24.3580	99.3800
75.6828	−24.7538	101.6956
75.7729	−25.1569	104.2096
75.8629	−25.5644	106.9407
75.9530	−25.9727	109.9066
76.0430	−26.3771	113.1236
76.1331	−26.7721	116.6043
76.2231	−27.1512	120.3556
76.3132	−27.5071	124.3761
76.4032	−27.8317	128.6535
76.4932	−28.1170	133.1621
76.5833	−28.3555	137.8621
76.6733	−28.5409	142.6993
76.7634	−28.6690	147.6084
76.8534	−28.7382	152.5176
76.9435	−28.7497	157.3546
77.0335	−28.7074	162.0526
77.1236	−28.6174	166.5558
77.2136	−28.4870	170.8219
77.3036	−28.3244	174.8231
77.3937	−28.1375	178.5452
77.4837	−27.9339	−178.0149
77.5738	−27.7202	−174.8517

(*Continues*)

(*Continued*)

Vertical port reading for directivity: Frequency 5.8 GHz		
Elevation (°)	**Amplitude (dB)**	**Phase (°)**
77.6638	−27.5021	−171.9535
77.7539	−27.2835	−169.2904
77.8439	−27.0606	−166.7317
77.9340	−26.8330	−164.2563
78.0240	−26.6017	−161.8699
78.1141	−26.3678	−159.5767
78.2041	−26.1327	−157.3792
78.2941	−25.8973	−155.2785
78.3842	−25.6628	−153.2746
78.4742	−25.4302	−151.3665
78.5643	−25.2004	−149.5523
78.6543	−24.9743	−147.8296
78.7444	−24.7525	−146.1955
78.8344	−24.5359	−144.6466
78.9245	−24.3249	−143.1795
79.0145	−24.1199	−141.7905
79.1046	−23.9215	−140.4760
79.1946	−23.7299	−139.2323
79.2846	−23.5455	−138.0558
79.3747	−23.3683	−136.9431
79.4647	−23.1987	−135.8907
79.5548	−23.0367	−134.8955
79.6448	−22.8824	−133.9544
79.7349	−22.7359	−133.0646
79.8249	−22.5973	−132.2231
79.9150	−22.4666	−131.4276
80.0050	−22.3437	−130.6756
80.0950	−22.2286	−129.9648
80.1851	−22.1215	−129.2931
80.2751	−22.0221	−128.6586
80.3652	−21.9305	−128.0595
80.4552	−21.8466	−127.4941
80.5453	−21.7705	−126.9607
80.6353	−21.7020	−126.4581
80.7254	−21.6411	−125.9848
80.8154	−21.5878	−125.5397
80.9054	−21.5420	−125.1216
80.9955	−21.5037	−124.7295
81.0855	−21.4728	−124.3624
81.1756	−21.4494	−124.0196
81.2656	−21.4334	−123.7002
81.3557	−21.4248	−123.4037
81.4457	−21.4235	−123.1293
81.5358	−21.4295	−122.8766
81.6258	−21.4429	−122.6451
81.7159	−21.4636	−122.4345

(*Continues*)

(*Continued*)

Vertical port reading for directivity: Frequency 5.8 GHz		
Elevation (°)	**Amplitude (dB)**	**Phase (°)**
81.8059	−21.4928	−122.2429
81.8960	−21.5332	−122.0654
81.9860	−21.5845	−121.9006
82.0760	−21.6460	−121.7474
82.1661	−21.7173	−121.6047
82.2561	−21.7980	−121.4717
82.3462	−21.8876	−121.3477
82.4362	−21.9857	−121.2318
82.5263	−22.0920	−121.1236
82.6163	−22.2062	−121.0224
82.7064	−22.3279	−120.9278
82.7964	−22.4569	−120.8392
82.8864	−22.5929	−120.7564
82.9765	−22.7356	−120.6789
83.0665	−22.8848	−120.6064
83.1566	−23.0404	−120.5386
83.2466	−23.2020	−120.4753
83.3367	−23.3695	−120.4160
83.4267	−23.5427	−120.3608
83.5168	−23.7214	−120.3092
83.6068	−23.9054	−120.2612
83.6968	−24.0947	−120.2165
83.7869	−24.2889	−120.1750
83.8769	−24.4881	−120.1366
83.9670	−24.6919	−120.1010
84.0570	−24.9003	−120.0681
84.1471	−25.1132	−120.0379
84.2371	−25.3303	−120.0103
84.3272	−25.5516	−119.9850
84.4172	−25.7769	−119.9620
84.5073	−26.0061	−119.9413
84.5973	−26.2390	−119.9227
84.6873	−26.4755	−119.9061
84.7774	−26.7155	−119.8915
84.8674	−26.9588	−119.8788
84.9575	−27.2053	−119.8679
85.0475	−27.4549	−119.8587
85.1376	−27.7073	−119.8513
85.2276	−27.9626	−119.8454
85.3177	−28.2204	−119.8412
85.4077	−28.4807	−119.8384
85.4977	−28.7433	−119.8371
85.5878	−29.0080	−119.8371
85.6778	−29.2746	−119.8385
85.7679	−29.5431	−119.8412
85.8579	−29.8131	−119.8452

(*Continues*)

(*Continued*)

Vertical port reading for directivity: Frequency 5.8 GHz		
Elevation (°)	**Amplitude (dB)**	**Phase (°)**
85.9480	−30.0846	−119.8503
86.0380	−30.3573	−119.8565
86.1281	−30.6310	−119.8638
86.2181	−30.9054	−119.8721
86.3082	−31.1805	−119.8814
86.3982	−31.4559	−119.8917
86.4882	−31.7315	−119.9027
86.5783	−32.0069	−119.9146
86.6683	−32.2819	−119.9273
86.7584	−32.5562	−119.9407
86.8484	−32.8296	−119.9547
86.9385	−33.1018	−119.9693
87.0285	−33.3724	−119.9845
87.1186	−33.6411	−120.0001
87.2086	−33.9077	−120.0162
87.2987	−34.1717	−120.0326
87.3887	−34.4328	−120.0494
87.4787	−34.6908	−120.0663
87.5688	−34.9451	−120.0835
87.6588	−35.1954	−120.1007
87.7489	−35.4413	−120.1181
87.8389	−35.6824	−120.1354
87.9290	−35.9182	−120.1526
88.0190	−36.1485	−120.1697
88.1091	−36.3726	−120.1866
88.1991	−36.5903	−120.2032
88.2891	−36.8010	−120.2195
88.3792	−37.0044	−120.2355
88.4692	−37.1999	−120.2509
88.5593	−37.3871	−120.2659
88.6493	−37.5656	−120.2803
88.7394	−37.7350	−120.2940
88.8294	−37.8948	−120.3071
88.9195	−38.0447	−120.3194
89.0095	−38.1842	−120.3309
89.0995	−38.3130	−120.3416
89.1896	−38.4306	−120.3515
89.2796	−38.5369	−120.3604
89.3697	−38.6314	−120.3683
89.4597	−38.7139	−120.3753
89.5498	−38.7841	−120.3812
89.6398	−38.8419	−120.3861
89.7299	−38.8870	−120.3900
89.8199	−38.9193	−120.3927
89.9100	−38.9388	−120.3944
90.0000	−38.9453	−120.3949

Appendix D

Dielectric materials and resources

Absorber: The quality factor is $Q = (\omega W_e)/(P_{\text{rad}})$. Hence, absorbers are dielectric resonator antennas (DRAs) with standing wave and if these standing waves are converted into travelling waves, these absorbers are converted into antennas using excitation into DRAs. Absorbers have high-Q factor and high-energy storage; if accelerating charges are created in a DR absorber, they behave like DRAs due to energy leakage into air. And thus quality factor of an absorber is reduced to act as antennas. Hence, the Q factor of an absorber is high and that of a DRA is low. The eigen state at resonance is absorber cavity and excited state of cavity is an antenna due to accelerating charge careers and simultaneously decreasing the quality factor too. The feed used at THz fDRA is Gaussian beam and feed is wave port for an absorber. For making an absorber, imaginary part must be high and second derivate image impedance w.r.t. frequency must be negative.

S. no.	Materials	Permittivity	Supplier or manufacturer
1	$MgO–SiO_2$ (CD-6)	6.3	**Countis Laboratories** 12295 Charles Dr, Grass Valley, CA 95945, United States Tel: +1 530-272-8334 Email: tcountis@countis.com
2	$MgO–SiO_2–TiO_2$ (CD-9)	9.5	
3	$MgO–TiO_2–SiO_2$ (CD-13)	13.0	
4	$MgO–TiO_2$ (CD-15)	15.0	
5	$MgO–TiO_2$ (CD-16)	16.0	
6	$MgO–CaO–TiO_2$ (CD-18)	18.0	
7	$MgO–CaO–TiO_2$ (CD-20)	20.0	
8	$MgO–CaO–TiO_2$ (CD-30)	30.0	
9	$MgO–CaO–TiO_2$ (CD-50)	50.0	
10	$MgO–CaO–TiO_2$ (CD-100)	100.0	
11	$MgO–CaO–TiO_2$ (CD-140)	140.0	
12	Boron nitride (ECCOSTOCK@)	4.0	**Emerson & Cuming Microwave Products N.V.** A unit of Laird Technologies Hong Kong Holdings (4) Ltd. Unit 2507-8, 25/F, Office Tower, Langham Place, 8 Argyle Street, Mongkok, Kowloon, Hong Kong Tel: +852-2923 0600, +852-2923 0605 Email: sales@hk.eccosorb.com
13	Beryllium oxide (ECCOSTOCK@)	6.0	
14	Magnesium oxide (ECCOSTOCK@)	9.0	
15	Magnesium titanate (ECCOSTOCK@)	10.0	
16	Zirconia (ECCOSTOCK@)	20.0	
17	Titanium dioxide (rutile) (ECCOSTOCK@)	50.0	
18	Strontium titanate (ECCOSTOCK@)	>100.0	

19	Magnesium manganese aluminum iron ferrite	9.2	**Hiltek Microwave Limited** 15200 Shady Grove Road Suite 350, Rockville, MD 20850, United States Tel: +1 (301) 670-2833 Fax: +1 (301) 670-2831 Email: www.hiltek.com
20	Magnesium titanate	16.0	
21	Lithium ferrite	20.0	
22	Zirconium tin titanate	37.0	
23	Titania ceramic	80–100	
24	MgSi (steatite) (D6)	6.0	**Morgan Advanced Materials** 150 Kampong Ampat, 05-06A KA Centre, Singapore 368324, Singapore Tel: +65 6595 0000 Fax: +65 6595 0005 Email: asia.mc@morganplc.com
25	CaMgTi (Mg, Ca, Ti) (D20)	20.0	
26	ZrTiSn (Zr, Sn, Ti) (D36)	37.0	
27	BaSmTi (Ba, Sm, Ti) (D37)	76.5	
28	Titanate with other ingredients (PD-6)	6.5	**Pacific Ceramics, Inc.** Advanced Microwave Ceramic Materials 824 San Aleso Ave Sunnyvale, CA 94085, USA Tel: +1 (408) 747-4600 Email: info@pceramics.com
29	Titanate with other ingredients (PD-9)	9.5	
30	Titanate with other ingredients (PD-12)	12.0	
31	Titanate with other ingredients (PD-13)	13.0	
32	Titanate with other ingredients (PD-15)	15.0	
33	Titanate with other ingredients (PD-16)	16.0	
34	Titanate with other ingredients (PD-18)	18.0	
35	Titanate with other ingredients (PD-25)	25.0	

(Continues)

(*Continued*)

S. no.	Materials	Permittivity	Supplier or manufacturer
36	Titanate with other ingredients (PD-38)	38.0	
37	Titanate with other ingredients (PD-50)	50.0	
38	Titanate with other ingredients (PD-100)	98.0	
39	Titanate with other ingredients (PD-160)	160.0	
40	Titanate with other ingredients (PD-270)	270.0	
41	Zr, Sn, Ti oxide (E2000)	37.0	**Temex Components & Temex Telecom, USA**
42	E3000	34.0	**Supplier 1** SM CREATIVE No 845, 2nd Cross, 7th Main HAL 2nd Stage, Indiranagar, Bangalore, 560 038, India Tel: +91 (80) 25210268, +91 (80) 41255492 Mobile: +91 (98) 45410417 Email: sundar@smcel.com http://www.smcel.com S M Creative Electronics Ltd. #10, Electronic City, Sector-18 Gurgaon 122 015, Haryana Tel: +91 124-4909850 Fax: +91 124-2455 212 Email: smcel@smcel.com **Supplier 2** SIMAL # 60 & 60/1, 18th Cross, 4th Main, Malleswaram, Bangalore, 560 055, India Tel: +91 (80) 41532079, +91 (80) 23444410 Mobile: +91 (99721) 24165 Email: agencies@simal.com.sg http://www.simal.com.sg
43	Ba Zn Ta oxide (E4000)	30.0	
44	Ba, Sm, Ti oxide (E5000)	78.0	
45	Ti, Zr, Nb, Zn Oxide (E6000)	45.0	

46	Cordierite (Mg, Al, Si) (D-4)	4.5	**Trans-Tech Skyworks Solutions, Inc.** 5520 Adamstown Road
47	Forsterite (Mg, Si, oxide) (D-6)	6.3	Adamstown, MD 21710 **Supplier**
48	Mg–Ti (D-15)	15.0	SM Electronic Technologies Pvt. Ltd. #1790, 5th Main, 9th Cross, RPC Layout, Vijayanagar 2nd Stage, Bangalore 560 040,
49	Mg–Ti (D-16)	16.0	India Mr. Manjunath
50	Ba, Zn, Ta oxide (D-29)	29.0–30.7	Tel: +91-80-23301030 Email: smgroup@vsnl.com
51	Ba, Zn, Ta oxide (perovskite) (D-87)	29.5–31.0	
52	Ba, Zn, Co, Nb (D-83)	35.0 – 36.5	
53	Zirconium titanate based (D-43)	44.7–46.2	
54	E-11	11.0	**T-CERAM, RF & Microwave**
55	E-20	20.0	Okružní 1144, 500 03 Hradec Králové, Czech Republic, EU
56	E-37	37.0	Tel: +420 774 406 438CZ 42196078 Email: sales@t-ceram.com www.t-ceram.com
57	TE-21	21.0	**Token Electronics Industry Co., Ltd.**
58	TE-30	30.0	No. 137, Sec. 1, Chung Shin Rd., Wu Ku Hsiang, Taipei Hsien, Taiwan, R.O.C.
59	TE-36	36.0	Tel: +886-2-2981 0109
60	TE-45	45.0	Fax: +886-2-2988 7487
61	TE-80	80.0	http://www.token.com.tw
62	TE-90	90.0	Email: rfq@token.com.tw

(Continues)

(Continued)

S. no.	Materials	Permittivity	Supplier or manufacturer
63	Mg-Ca-Ti (MDR20)	20.0	**MCV Microwave** 6640 Lusk Blvd, Suite A102 San Diego, CA 92121, United States Tel: +1 858-450-0468 Fax: +1 858-869-8404 www.mcv-microwave.com
64	Ta with other ingredients (MDR24)	24.0	
65	Ta with other ingredients (MDR30)	30.0	
66	Zn–Sn–Ti (MDR38)	38.0	
67	La-Ba-Ti (MDR45)	45.0	
68	DR-30	30.0	**TCI Ceramics, Inc.** 18450 Showalter Rd., Hagerstown, MD 21742, United States Tel: +1 301-766-0560 Fax: +1 301-766-0566 Email: sales@tciceramics.com www.tciceramics.com
69	DR-36	36.0	
70	DR-45	45.0	
71	DR-80	80.0	
72	RT6010, RT-6002	10.2	
73	MCT-25	25	**Trans-Tech**
	SMAT	27	
	BaTiO3	14	
74	ECCOSTOCK'SHIK	10,20,30,40	

Appendix E
Dual-band graphene antenna design and implementation

Dual-band graphene antenna design and implementation are described in Figures E.1–E3 and Table E.1.

Theory: Consider a cavity resonator of one angstrom size, i.e. a cube with each side of length $a = 10^{-10}$ m. The Maxwell equations in such a cube have solutions of the form:

$$A_r(t, x, y, z) = \sum_{mnp} c(mnp, t)u_r, mnp(x, y, z), r = 1, 2, 3$$

where u_r, mnp are spatial functions obtained by integrating the electric field w.r.t.

$$\left\{\cos\left(\frac{m\pi x}{a}\right), \sin\left(\frac{m\pi x}{a}\right)\right\} \otimes \left\{\cos\left(\frac{n\pi y}{a}\right), \sin\left(\frac{n\pi y}{a}\right)\right\}$$
$$\otimes \left\{\cos\left(\frac{p\pi y}{a}\right), \sin\left(\frac{p\pi z}{a}\right)\right\}$$

Multiplied by some constants depending on the indices (m, n, p). We may, without loss of generality, assume those are normalized so that

$$\int_C u_r, mnp(r)\bar{u}_s, m'n'p'(r)d^3r = \delta_{rs}\delta_{mm'}\delta_{nn'}\delta_{pp'}$$

The dependence of $c(mnp, t)$ on t is $\exp(i\omega(mnp)t)$ where $\omega(mnp)$ are the characteristic frequencies of oscillation:

$$\omega(mnp) = \left(\frac{\pi c}{a}\right)\sqrt{m^2 + n^2 + p^2}, \quad m, n, p = 1, 2, \ldots$$

which are of the order of magnitude

$$\omega = \frac{\pi c}{a}$$

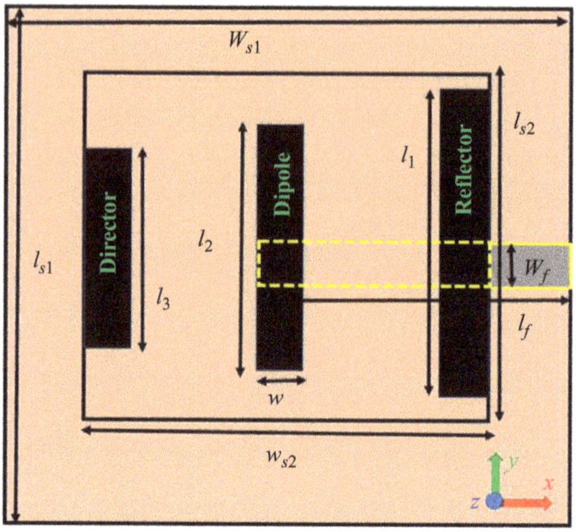

Figure E.1 Graphene antenna

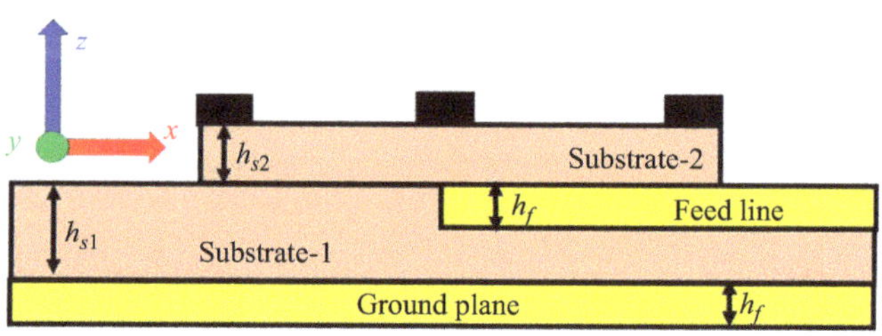

Figure E.2 The geometry of a graphene antenna

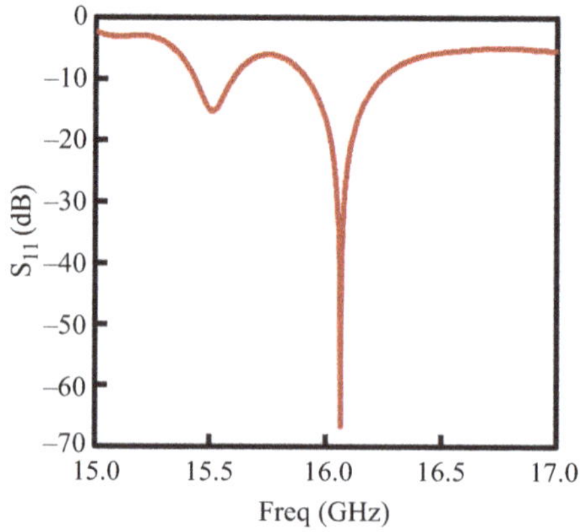

Figure E.3 Simulated response of the S_{11} parameter

Table E.1 The dimensions of the proposed antenna geometry

Parameters	Dimensions (μm)	Materials name
Substrate-1 width (w_{s1})	110	Silicon dioxide (SiO_2)
Substrate-1 length (l_{s1})	110	Silicon dioxide (SiO_2)
Substrate-1 height (h_{s1})	3.2	Silicon dioxide (SiO_2)
Substrate-2 width (w_{s2})	70	Silicon dioxide (SiO_2)
Substrate-2 length (w_{s2})	70	Silicon dioxide (SiO_2)
Substrate-2 height (h_{s2})	1.2	Silicon dioxide (SiO_2)
Feed-line length (l_f)	37	Silver
Feed-line width (w_f)	5	Silver
Feed-line thickness (t_f)	0.035	Silver
Ground-plane width (w_{s1})	110	PEC
Ground-plane length (l_{s1})	110	PEC
Ground-plane height	0.035	PEC
Reflector length (l_1)	71	Graphene
Dipole length (l_2)	65	Graphene
Director length (l_3)	56	Graphene
Width of reflector = width of dipole = width of director (w)	4	Graphene
Thickness of reflector = thickness of dipole = thickness of director (t)	1 nm	Graphene

The electric field is

$$E_r = \partial_t A_r = \sum_{mnp} c(mnp, t) i\omega(mnp) u_r, mnp(r)$$

The magnetic field is

$$B = \text{curl} A$$

which is of the order of magnitude $\frac{|c(mnp,t)|}{a}$, where by $c(mnp, t)$, we actually mean its average in a coherent state. The total electric field energy within the cavity C is

$$U_E = \left(\frac{\epsilon_0}{2}\right) \int_C |E|^2 d^3 r$$

which has components of the order of magnitude

$$\epsilon_0/\omega(mnp)c(mnp, t)|^2 a^3 = \epsilon_0 \omega(mnp)^2 a^3 |c(mnp, t)|^2$$

The total magnetic field energy within the cavity is

$$U_B = (2\mu_0)^{-1} \int_c |B|^2 \, d^3 r$$

which has components of the order of magnitude

$$\left|\frac{c(mnp, t)}{a}\right|^2 \frac{a^3}{\mu_0} = |c(mnp, t)|^2 \frac{a}{\mu_0}$$

The relation of the orders of magnitude of the electric field energy and the magnetic field energy within the cavity therefore has the order of magnitude

$$\frac{U_E}{U_B} \approx \mu_0 \epsilon_0 \omega (mnp)^2 a^2 \approx \frac{\omega^2 a^2}{c^2} \approx 1$$

As expected, the canonical commutation relations are

$$[A_r(t,r), \partial_t A_s(t,r')] = \left(\frac{ih}{2\pi}\right) \delta^3(r - r')$$

These fields are

$$\left[c_r(mnp,t), \omega(mnp) c_s(m'n'p', t^*)\right] = \left(\frac{h}{2\pi}\right) \delta_{rs} \delta_{mm'} \delta_{nn'} \delta_{pp'}$$

So that the eigenvalues of $c_r(mnp,t)^* c_r(mnp,t)$ are positive integer multiples of $h/2\pi\omega(mnp)$. This means that the field energy within the cavity when a finite number of modes are excited assumes eigenvalues that are of the same order of magnitude as positive integer multiples of $h\omega/2\pi$ as expected by Planck's quantum theory of radiation. This fact also fulfills the result that $|c(mnp,t)|$ is of the order of magnitude of $\sqrt{h/(2\pi\omega)}$.

Now we come to the question of computing the order of magnitude of the Poynting vector power flux at a given radial distance R from the quantum cavity antenna caused by the surface current density induced by the magnetic field on the antenna surface. The magnetic field on the surface and hence the corresponding induced surface current density both have the order of magnitudes of $|c(mnp,t)|/a$ that is of the order $a^{-1}\sqrt{h/\omega}$. Therefore, the far-field magnetic vector potential at a distance R from the cavity is of the order of magnitude (use the retarded potential formula) $(a/R)\sqrt{h/\omega}$ and hence the corresponding far-field-radiated magnetic field is of the order of magnitude $(\omega/c)(a/R)\sqrt{h/\omega}$ while the near-field magnetic field of the order of magnitude $(a/R^2)\sqrt{h/\omega}$. Actually, these expressions for the magnetic field must be multiplied by $\sqrt{N}$ where N is a positive integer corresponding to the largest modal eigenvalue of the operators $(2\pi\omega(mnp)/h)c(mnp,t)^* c(mnp,t)$.

The far-field Poynting vector has the order of magnitude of $B^2 c/2\mu_0$ that is of the order

$$\left(\frac{c}{2\mu_0}\right)\left(\sqrt{N} \cdot \left(\frac{\omega}{c}\right)\left(\frac{a}{R}\right)\sqrt{\frac{h}{\omega}}\right)^2 = \left(\frac{h}{2\mu_0}\right) N \cdot \left(\frac{\omega}{c}\right)\left(\frac{a^2}{R^2}\right)$$

And the total power radiated outward by this quantum antenna in the far-field zone is thus of the order of magnitude

$$P = N = \left(\frac{h}{2\mu_0}\right)\left(\frac{a^2\omega}{c}\right)$$

Now we look at the order of magnitude of the power radiated in the far-field zone by the Dirac field of electrons and positrons within the cavity. The Dirac equation is

$$[i\gamma^{\mu}\partial_{\mu} - m]\psi(x) = 0$$

Or more precisely in arbitrary units,

$$\left[\left(\frac{ih}{2\pi}\right)\partial_t - c\left(\alpha,\left(-\frac{ih}{2\pi}\right)\nabla\right) - \beta mc^2\right]\psi(x) = 0$$

Here, the appearances of the constants h, m, c are explicitly shown. Now $|\psi(x)|^2 c$ is the probability density of the electron that must integrate to unity over the cavity volume. Thus $\psi(x)$ is of the order of magnitude $a^{-3/2}$. The Dirac current density $J^{\mu} = e\psi^{*}\gamma^{0}\gamma^{\mu}\psi$ has the same order of magnitude as $e|\psi(x)|^2 c$, which is $ec/a^3/^2$. Therefore, the far-field magnetic vector potential at a radial distance of R from the cavity is, in accordance with the retarded potential theory of the order

$$\left(\frac{ec}{a^{3/2}}\right)\cdot\left(\frac{a^3}{R}\right) = \frac{eca^{3/2}}{R}$$

The electric field in the far-field zone is then of the order

$$E \approx \omega \cdot \frac{eca^{3/2}}{R}$$

where ω is the characteristic oscillation frequency of the Dirac current. The magnetic field is of the order

$$B \approx a^{-1} \cdot \frac{eca^{3/2}}{R} = ec\sqrt{\frac{a}{R}}$$

If P is the characteristic momentum of the electrons and positrons in a given state, for example P may be the average momentum of an electron in a given state, then according to de Broglie, P is of the order h/a since a is the order of the electron wavelength. Then the electron energy is of the order

$$E_e = c\sqrt{m^2c^2 + P^2} \approx c\sqrt{m^2c^2 + \frac{h^2}{a^2}}$$

And the characteristic frequency of oscillation of the Dirac wave field is then

$$\omega = \frac{E_e}{h}$$

The Poynting vector corresponding to the power radiated by the Dirac field in the far-field zone then has the order of magnitude

$$S \approx c\left(\epsilon_0 E^2 + \frac{B^2}{\mu_0}\right) = c^3\epsilon_0\omega^2 ea^3 + \frac{e^2c^3a}{\mu_0}$$

Appendix F

Miniaturization design techniques

Abstract

A novel technique is presented for a miniaturization rectangular dielectric resonator antenna (DRA). The miniaturized DRA is built by applying the metallic strips on the sidewalls of the DRA. The specific format of the applied metallic strips excites the generation of fundamental and higher order modes. The frequency ratio and aspect-ratio can be manipulated to obtain the right choice of aspect-ratio and then destined resonant modes. The frequency ratio is tuned by the right size of applied metallic strips at right locations in destined geometries. A small frequency ratio is optimized design and implementation requirements.

F.1 Introduction

A dielectric resonator antenna (DRA) is efficient technology and extends deign flexibilities with many other advantages as compared to metal or patch antennas. The rectangular DRA is simple in design and has been used to show miniaturization and compactness (Figure F.1). Here, metallic strips have been used to show the size reduction of DRA. High-permittivity materials of DRAs can also be used in size reductions.

The use of high-permittivity materials is one of the methods used for miniaturization. The limitation of using high-permittivity materials is that it reduces bandwidth. The techniques like increasing the aspect-ratio for the enhancement of bandwidth using the high-permittivity DR by merging of modes, applying the metallic coating, the fundamental and third-order orthogonal degenerate modes were excited. The simulated and measured results on miniaturization techniques have been validated and found closely matching with expectations and simulated ones. Radiation pattern, S_{11}, fields, higher order modes and tuned results are presented in this appendix (Figures F.2–F.8 and Table F.1).

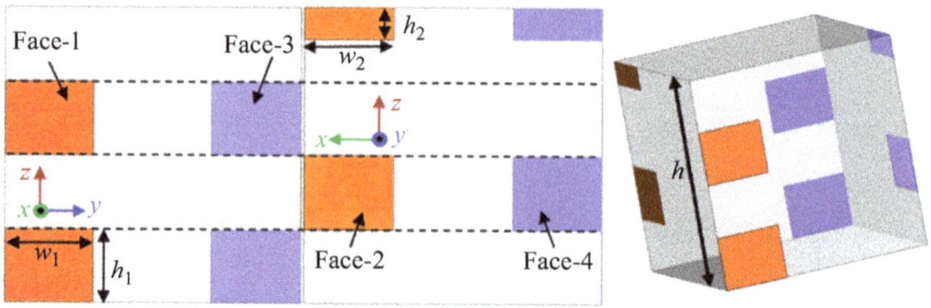

Figure F.1 *Rectangular DRA with metallic strips painted on its side walls. Dimensions (mm), face: 1—front, 2—right, 3—back and 4—left, copper color strips—at front face and blue color strips—at back face,* $w_g = l_g = 50, a = b = h = 12, l_m = 25, s = 5, w_m = 2.2, l_s = 9,$ $w_s = 1,\ w_1 = w_2 = 3.6, h_1 = 3, h_2 = 1.3, h = 12$

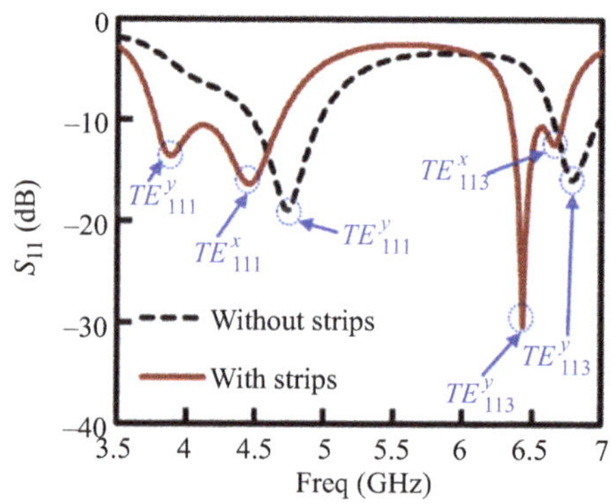

Figure F.2 *Three resonant modes*

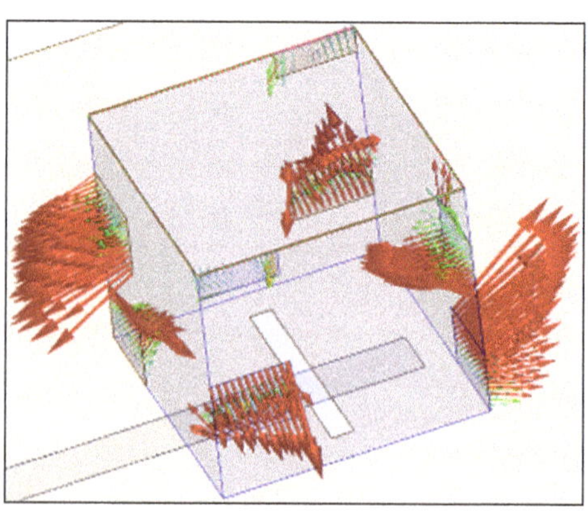

Figure F.3 *Electric fields inside DRA*

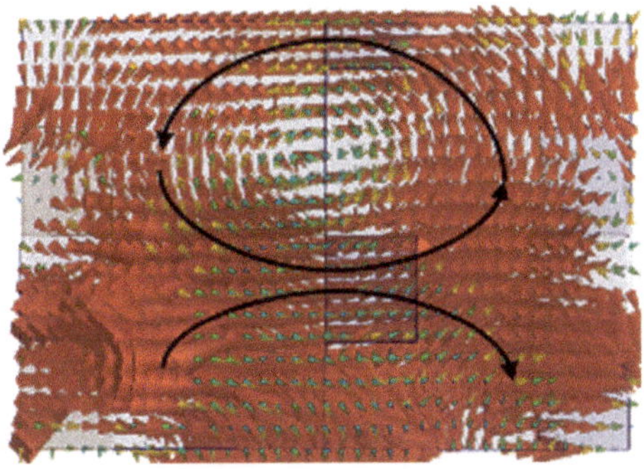

Figure F.4 E Fields, three half-wave variations

Figure F.5 Prototype model of rectangular DRA

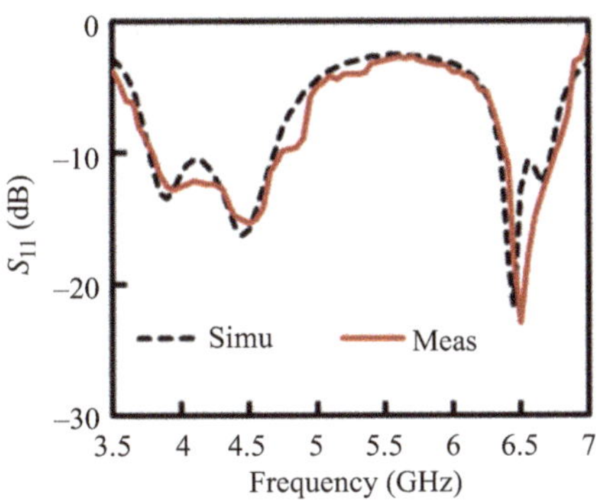

Figure F.6 Measured vs simulated results of DRA

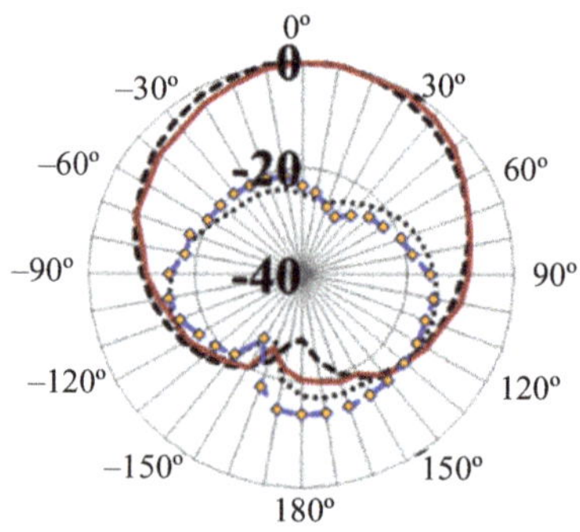

Figure F.7 Radiation pattern of rectangular DRA

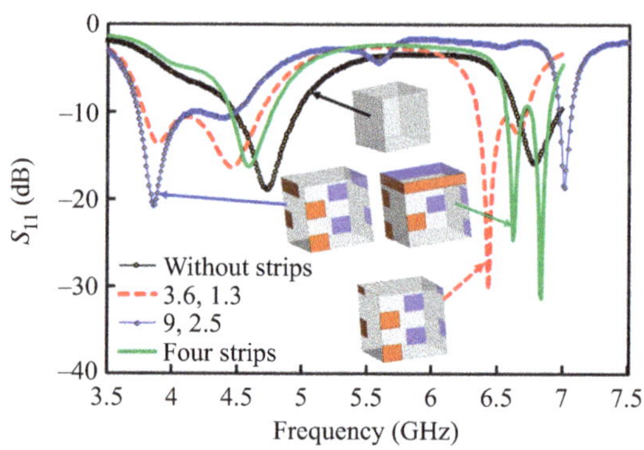

Figure F.8 Tuning of DRA using different lengths of strip lines

F.2 Conclusion

The metallic strips are applied in a specific format to get the miniaturization in the rectangular DRA. The sidewalls of the DR are divided into four layers. Each layer is having the height $h/4$. The height of strips at the lower three layers is selected as $h_1 = h/4$. The height of strips at the top layer near to the top edges of the DR is initially selected as $h_2 = h/8$.

Table F.1 Results with metallic strips

Antenna	Operating modes		f_r (GHz)		3-dB AR bandwidth (MHz)				10-dB impedance bandwidth (MHz)			
	LB	UB	LB	UB	LB		UB		LB		UB	
					S	M	S	M	S	M	S	M
DR without strips	TE^y_{111}	TE^y_{113}	4.75	6.80	0	–	0	–	490 (4.47–4.96)	*	330 (6.64–6.97)	–
DR with strips	TE^y_{111} TE^x_{111}	TE^y_{113} TE^x_{113}	3.88 4.46	6.43 6.66	130 (4.34–4.47)	160 (4.33–4.49)	60 (6.55–6.61)	90 (6.54–6.63)	910 (3.77–4.68)	1 030 (3.78–4.81)	390 (6.33–6.72)	410 (6.37–6.78)

Appendix G

Gaussian beam feed process

How to apply Gaussian beam into the feed to terahertz dielectric resonator antennas (TDRAs) (silver nano waveguide) and TDRAs. The steps have been presented bit-by-bit with screenshots. This will be helpful to beginners for ease of CST modeling.

The following figure is a CDRA model. Now we need to apply "Gaussian Beam":

So, follow the steps shown in the following figure:

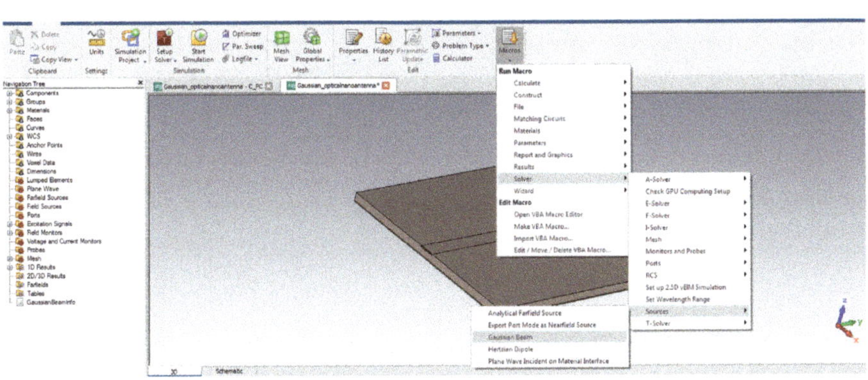

Step 1: Follow the steps and finally click on the "Gaussian Beam."

After clicking a new window will open as shown in the following:

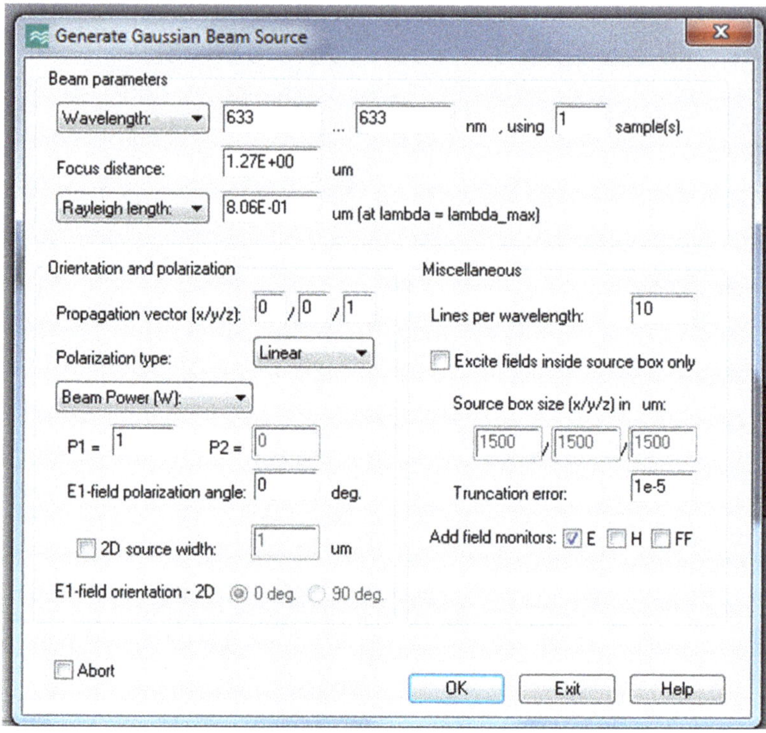

Step 2: Modify the propagation vector (*x*/*y*/*z*): 0/1/0:

Press OK.

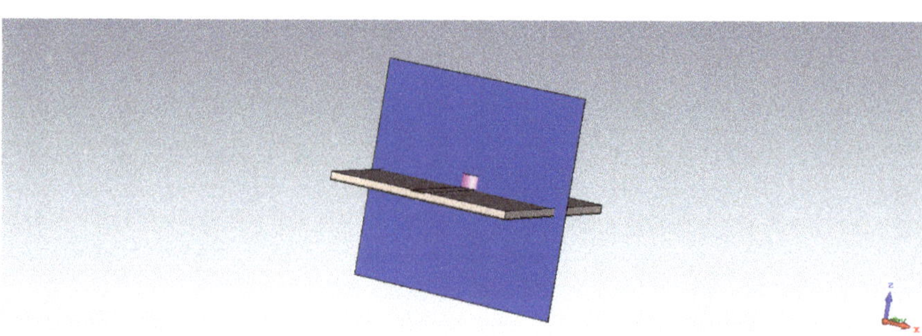

Step3: Perform Transform operation as shown in the following:

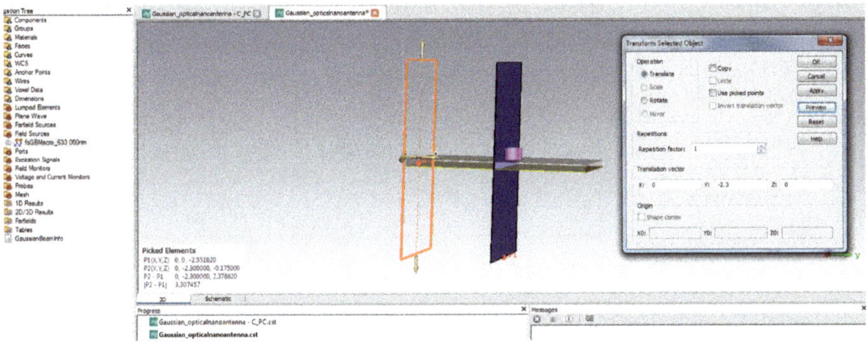

This is the final model with Gaussian Beam applied, and now simulation can be started:

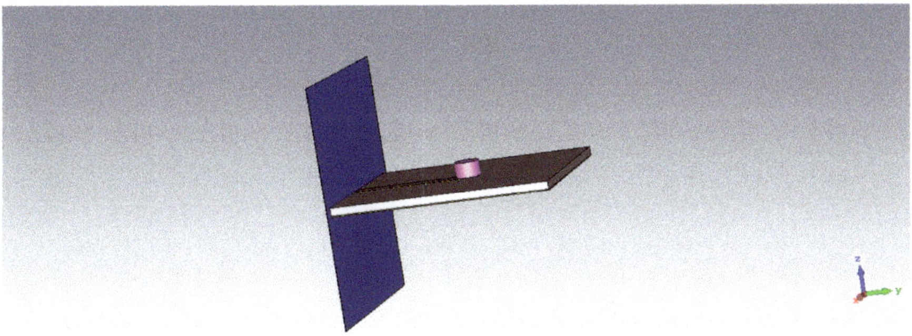

CST Model simulations with Gaussian Beam.

With Gaussian Beam:

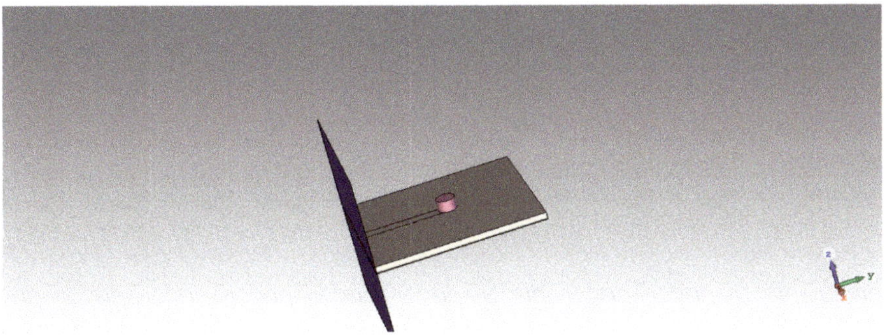

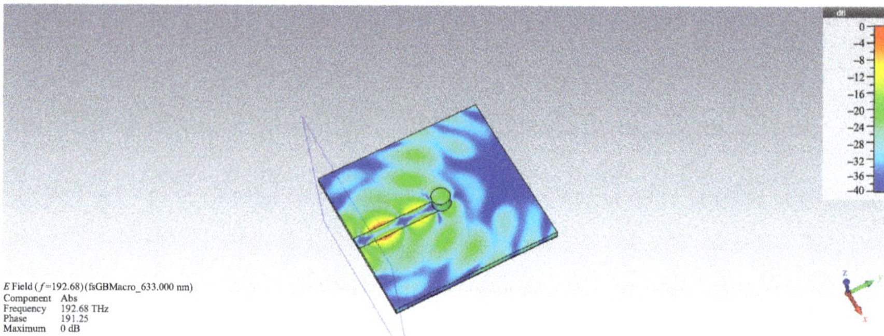
dB
0
−4
−8
−12
−16
−20
−24
−28
−32
−36
−40
E Field (f=192.68) (fsGBMacro_633.000 nm)
Component Abs
Frequency 192.68 THz
Phase 191.25
Maximum 0 dB

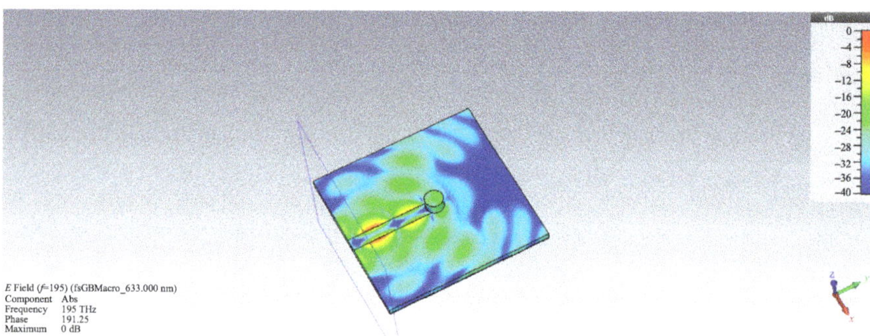
dB
0
−4
−8
−12
−16
−20
−24
−28
−32
−36
−40
E Field (f=195) (fsGBMacro_633.000 nm)
Component Abs
Frequency 195 THz
Phase 191.25
Maximum 0 dB

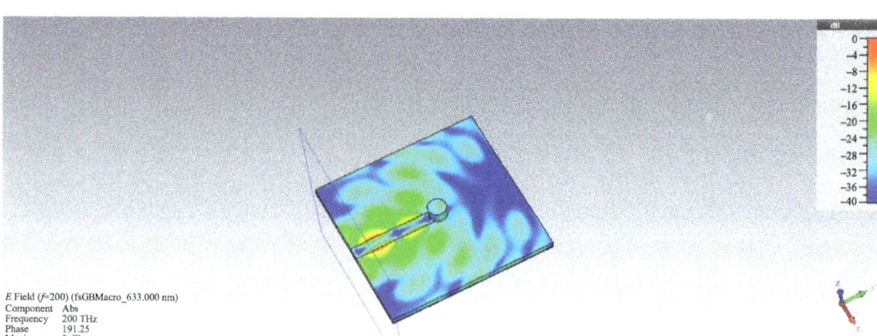
dB
0
−4
−8
−12
−16
−20
−24
−28
−32
−36
−40
E Field (f=200) (fsGBMacro_633.000 nm)
Component Abs
Frequency 200 THz
Phase 191.25
Maximum 0 dB

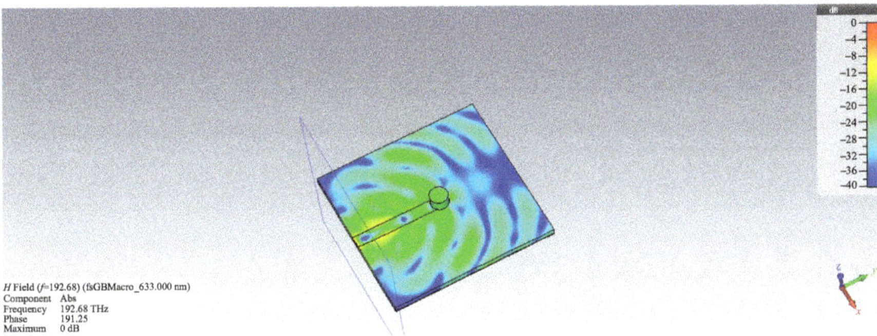
dB
0
−4
−8
−12
−16
−20
−24
−28
−32
−36
−40
H Field (f=192.68) (fsGBMacro_633.000 nm)
Component Abs
Frequency 192.68 THz
Phase 191.25
Maximum 0 dB

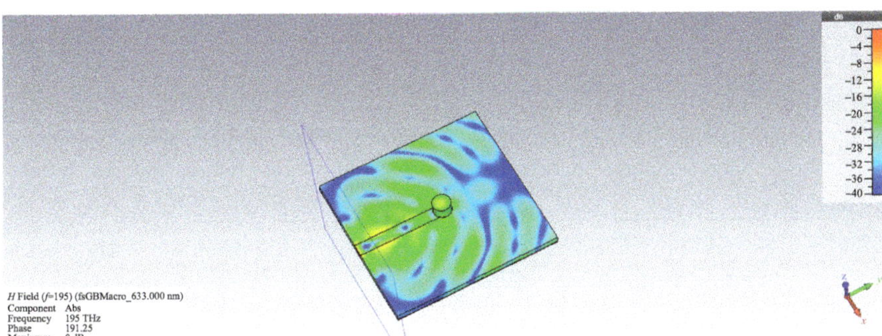
0
−4
−8
−12
−16
−20
−24
−28
−32
−36
−40
H Field (f=195) (fsGBMacro_633.000 nm)
Component Abs
Frequency 195 THz
Phase 191.25
Maximum 0 dB

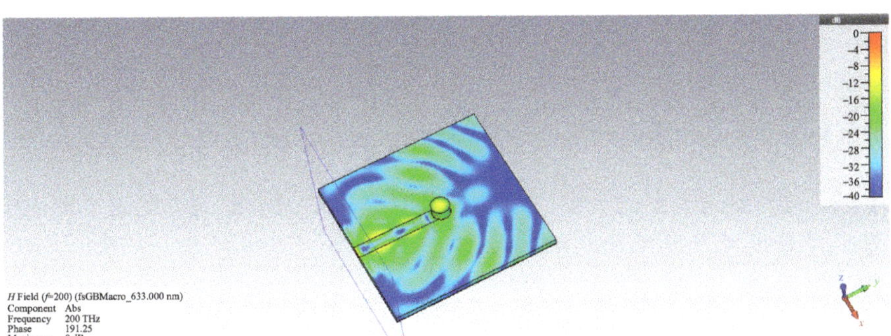
0
−4
−8
−12
−16
−20
−24
−28
−32
−36
−40
H Field (f=200) (fsGBMacro_633.000 nm)
Component Abs
Frequency 200 THz
Phase 191.25
Maximum 0 dB

Showing Super Directivity of DRA-radiated beam in the opposite direction of input applied (laser outputs):

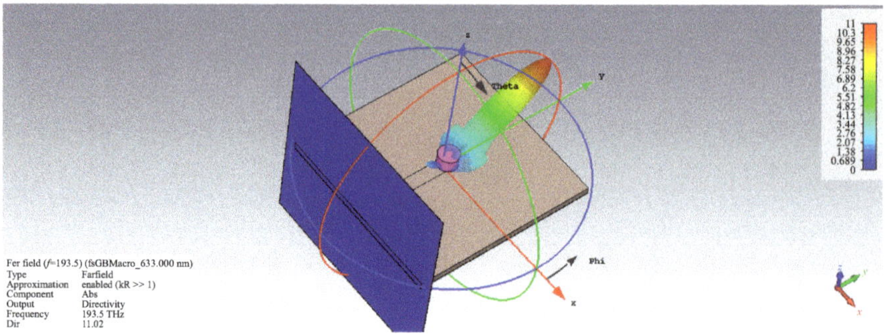

Far-field radiations with Gaussian Beam input excitation:

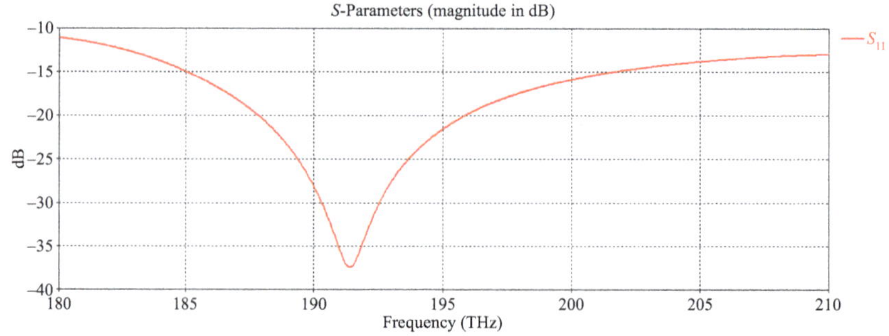

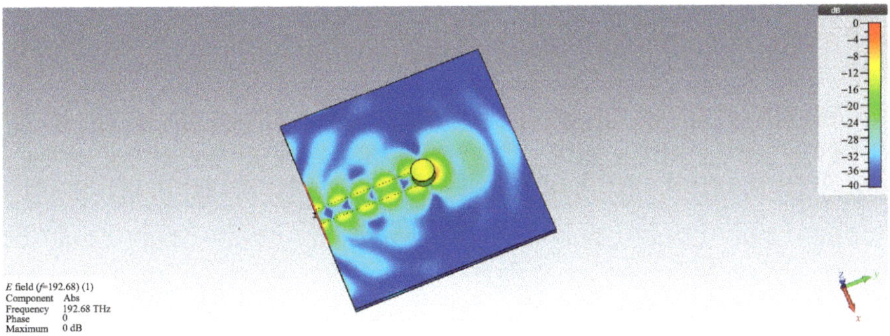
dB
0
−4
−8
−12
−16
−20
−24
−28
−32
−36
−40
E field (f=192.68) (1)
Component Abs
Frequency 192.68 THz
Phase 0
Maximum 0 dB

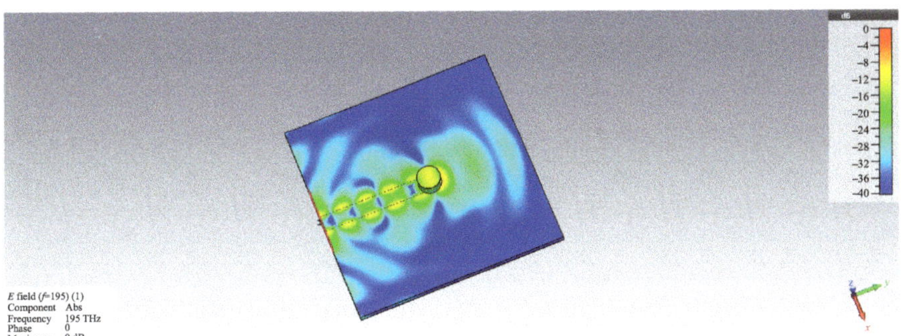
dB
0
−4
−8
−12
−16
−20
−24
−28
−32
−36
−40
E field (f=195) (1)
Component Abs
Frequency 195 THz
Phase 0
Maximum 0 dB

H Field:

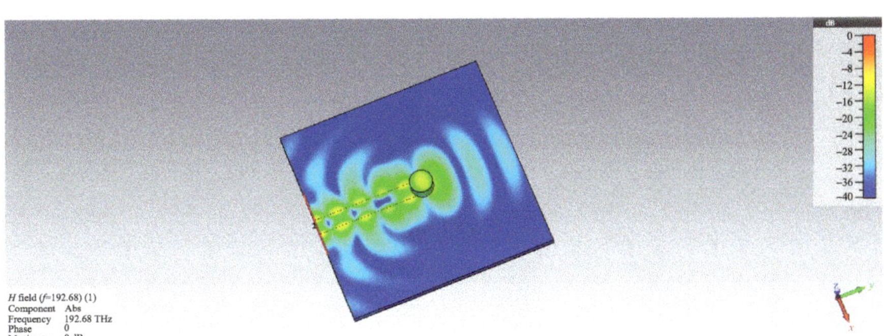
dB
0
−4
−8
−12
−16
−20
−24
−28
−32
−36
−40
H field (f=192.68) (1)
Component Abs
Frequency 192.68 THz
Phase 0
Maximum 0 dB

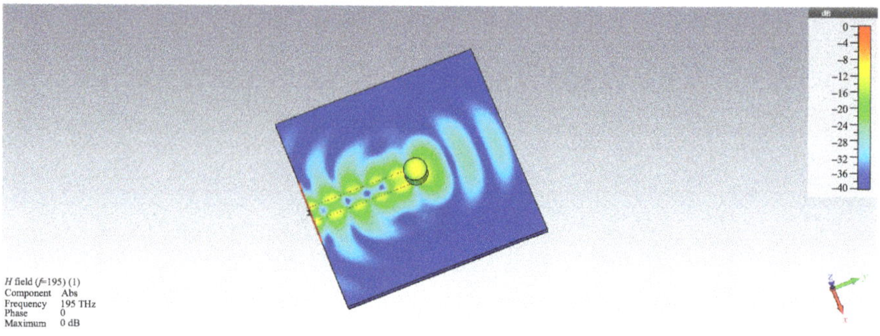

Radiation pattern with lumped port:

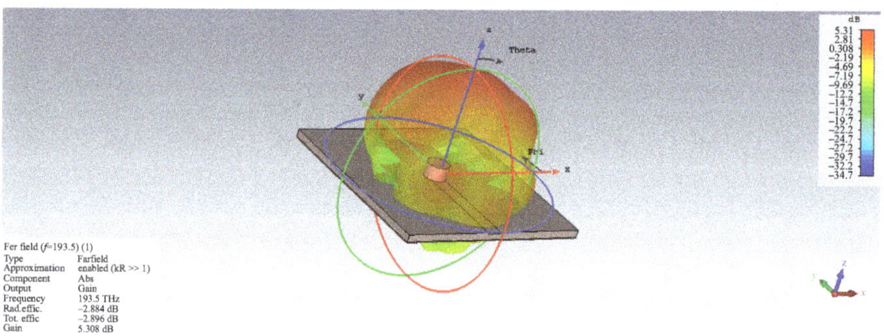

Far-field radiation with lumped-port input excitation:

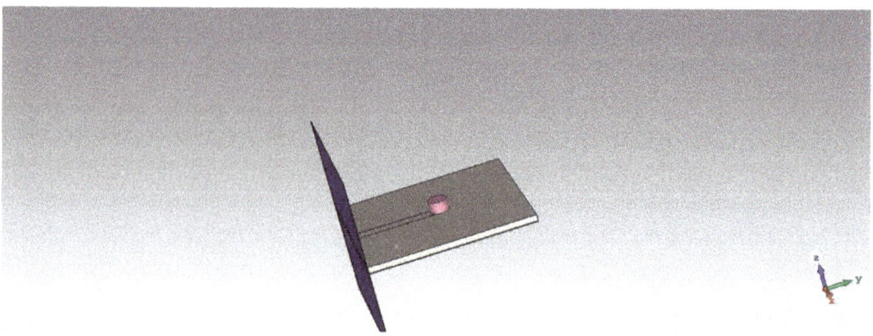

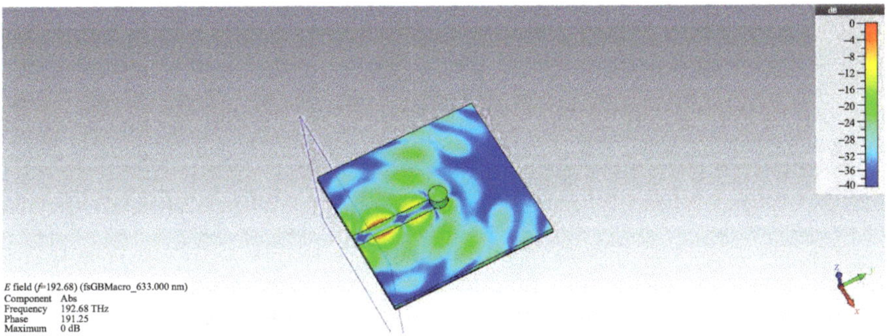
dB
0
-4
-8
-12
-16
-20
-24
-28
-32
-36
-40
E field (f=192.68) (fsGBMacro_633.000 nm)
Component Abs
Frequency 192.68 THz
Phase 191.25
Maximum 0 dB

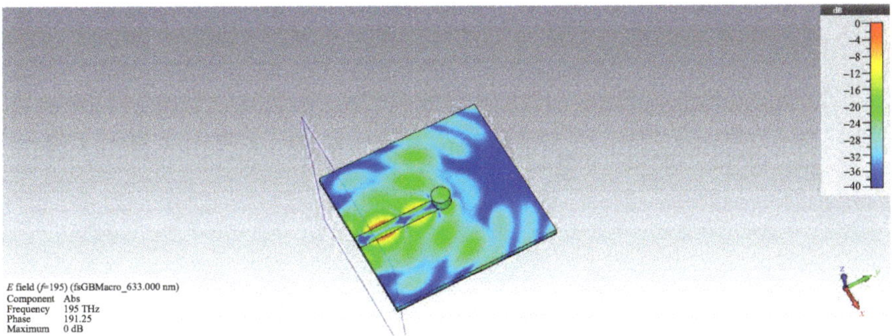
dB
0
-4
-8
-12
-16
-20
-24
-28
-32
-36
-40
E field (f=195) (fsGBMacro_633.000 nm)
Component Abs
Frequency 195 THz
Phase 191.25
Maximum 0 dB

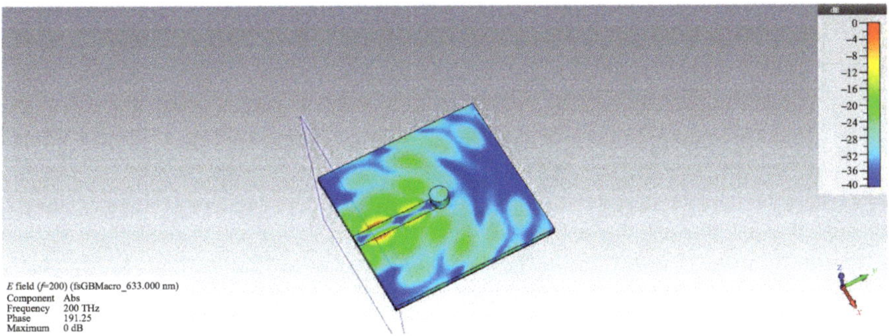
dB
0
-4
-8
-12
-16
-20
-24
-28
-32
-36
-40
E field (f=200) (fsGBMacro_633.000 nm)
Component Abs
Frequency 200 THz
Phase 191.25
Maximum 0 dB

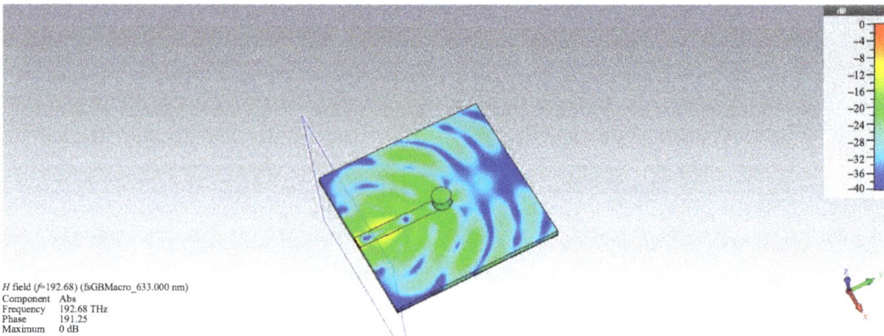
dB
0
−4
−8
−12
−16
−20
−24
−28
−32
−36
−40
H field (f=192.68) (fsGBMacro_633.000 nm)
Component Abs
Frequency 192.68 THz
Phase 191.25
Maximum 0 dB

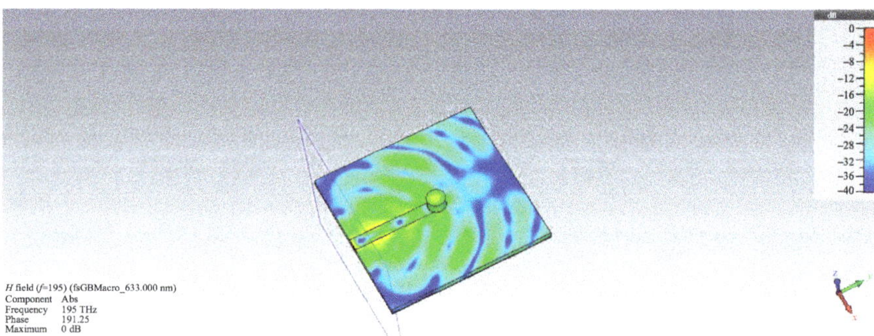
dB
0
−4
−8
−12
−16
−20
−24
−28
−32
−36
−40
H field (f=195) (fsGBMacro_633.000 nm)
Component Abs
Frequency 195 THz
Phase 191.25
Maximum 0 dB

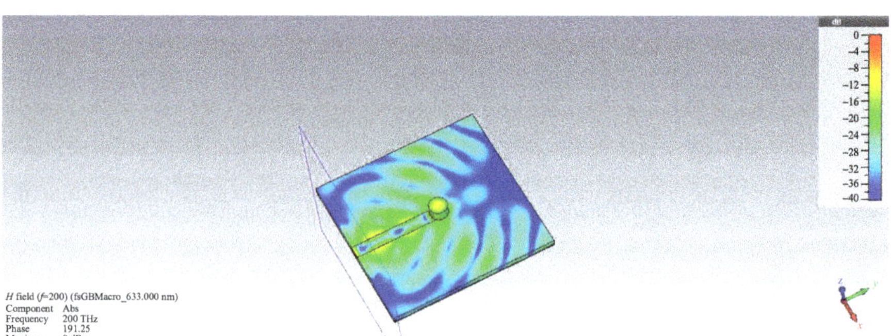
dB
0
−4
−8
−12
−16
−20
−24
−28
−32
−36
−40
H field (f=200) (fsGBMacro_633.000 nm)
Component Abs
Frequency 200 THz
Phase 191.25
Maximum 0 dB

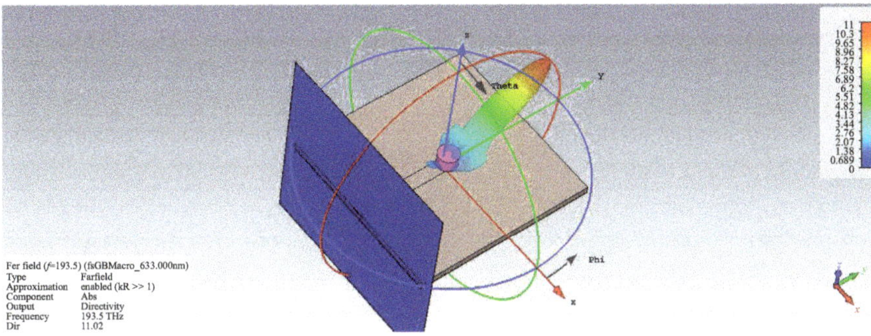

With waveguide port:

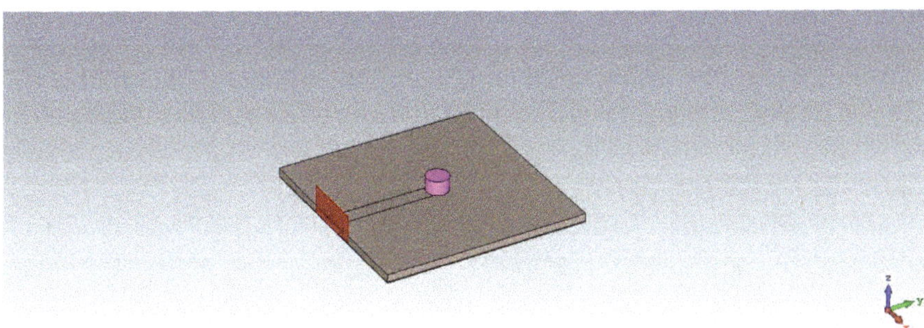

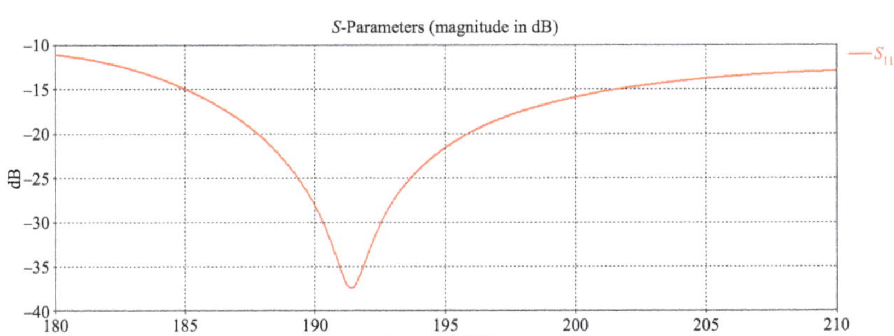

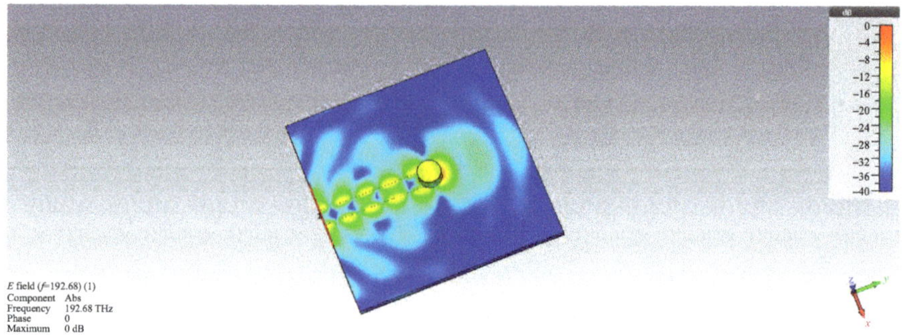
dB
0
-4
-8
-12
-16
-20
-24
-28
-32
-36
-40
E field (f=192.68) (1)
Component Abs
Frequency 192.68 THz
Phase 0
Maximum 0 dB

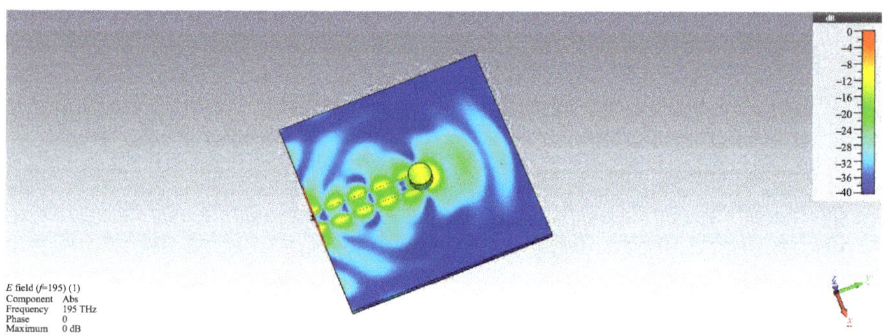
dB
0
-4
-8
-12
-16
-20
-24
-28
-32
-36
-40
E field (f=195) (1)
Component Abs
Frequency 195 THz
Phase 0
Maximum 0 dB

H Field:

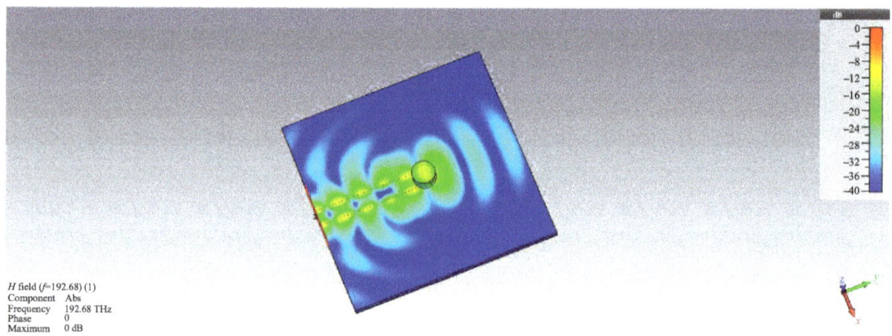
dB
0
-4
-8
-12
-16
-20
-24
-28
-32
-36
-40
H field (f=192.68) (1)
Component Abs
Frequency 192.68 THz
Phase 0
Maximum 0 dB

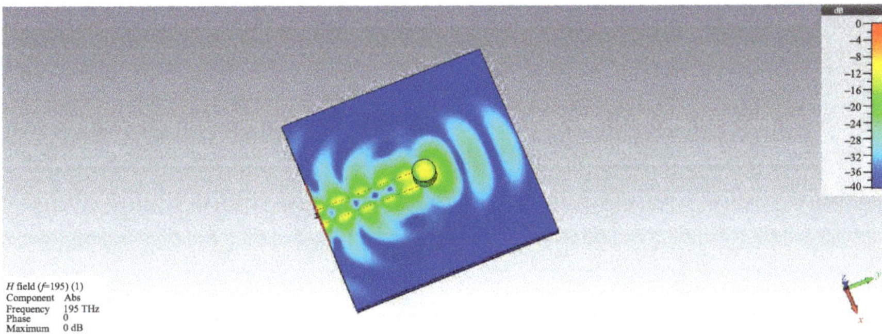
dB
0
-4
-8
-12
-16
-20
-24
-28
-32
-36
-40
H field (f=195) (1)
Component Abs
Frequency 195 THz
Phase 0
Maximum 0 dB

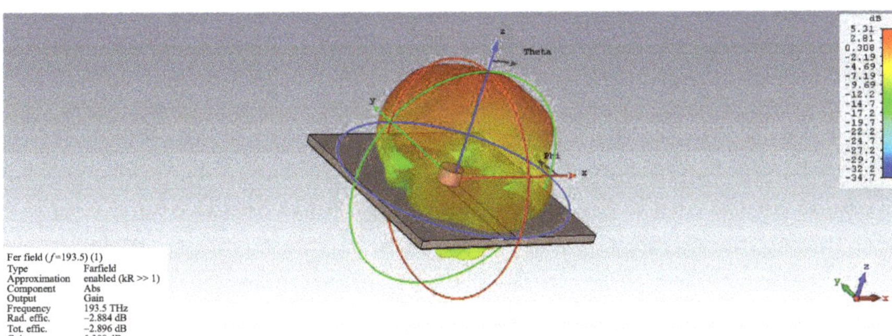
dB
5.31
2.81
0.308
-2.19
-4.69
-7.19
-9.69
-12.2
-14.7
-17.2
-19.7
-22.2
-24.7
-27.2
-29.7
-32.2
-34.7
Theta
Fer field (f=193.5) (1)
Type Farfield
Approximation enabled (kR >> 1)
Component Abs
Output Gain
Frequency 193.5 THz
Rad. effic. −2.884 dB
Tot. effic. −2.896 dB
Gain 5.308 dB

Appendix H
Silicon dielectric resonator antenna at 5-THz frequency

$$f_r = \frac{c}{2\pi r}\frac{6.324}{\sqrt{\varepsilon_r + 2}}\left[0.27 + 0.36\left(\frac{r}{2h}\right) + 0.02\left(\frac{r}{2h}\right)^2\right]$$

H.1 THz DRA fabrication process

Clean room for micromachining, spectroscopy, oscilloscope, spectrometer, dry and wet etching, angstrom engineering electron beam lithography, sputtering system, masking system, spin coater, bake plate, surface roughness tester, photo-resist chemical, microscope, spectrometer, signal generator, laser source, VNA and extension head (10 MHz to 500 THz). An SU-8 glue-type film or layer is used. It is used +1-μm precision in fabrication. A .GDS II design file is compatible with foundry/silicon CDRA on a silver surface and an SiO_2 substrate.

We intend to test this using terahertz TDS in reflection mode. Terahertz time-domain spectroscopy is used for testing. The terahertz time-domain spectroscopy system operates both in transmission and reflection modes. It will be a costly process. We can process on a 3″-wafer, and so, we may end up many samples for testing.

Use of complex Si binding process makes it hard. For terahertz, we use high-resistivity Si (intrinsic, float zone Si). Also, we use materials like TiO_2 for optical frequencies. By the sample size and integration (spot size 20 mm, 0.3–7.0 THz), not only by VNA, we can test in free-space mode at 5-THz frequency. The process may take about 4–6 weeks to fabricate once we finalize the design. To combine single crystal Si with SiO_2, it is a very complicated process. For it to work at terahertz, Si has to be intrinsic Si (no impurities), and so, cannot bond to SiO_2 by typical techniques (like anodic bonding) and needs a very thin cross-linking polymer (imagine it is like glue). This is the role of the thin SU-8. For terahertz, we use high-resistivity Si (intrinsic, float zone Si). The materials like TiO_2 are used at optical frequencies. The use of silver nano feed line is important. Both free-space testing capabilities (spot size 20 mm, 0.3–7.0 THz) and VNA to directly probe and test at 5 THz are important testing capabilities. Once we finalize the design, we can then get the **GDSII file**. To combine single-crystal Si with SiO_2, it is very complicated. For it to work at terahertz, the Si has to be intrinsic Si (no impurities), and so, cannot bond to SiO_2 by typical techniques (like anodic bonding) and needs a very thin cross-linking polymer (imagine it is like glue). This is the role of the thin SU-8 (Figure H.1 and Table H.1).

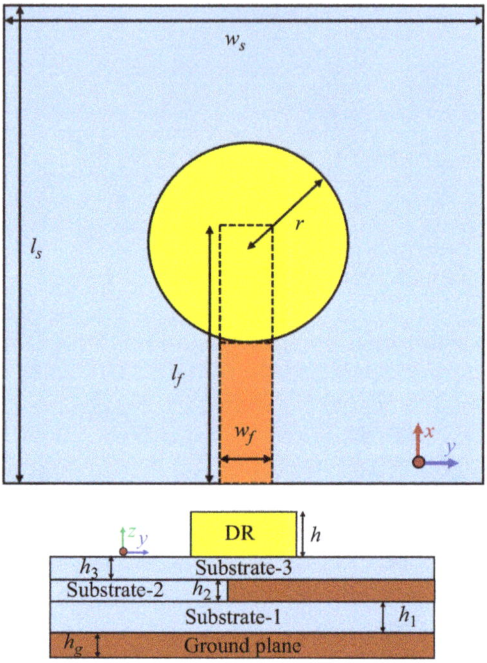

Figure H.1 The geometry of the proposed antenna

Table H.1 The dimensions of the proposed antenna geometry

Parameters	Dimensions (μm)	Materials name
Substrate width (w_s)	50	Silicon dioxide (SiO_2)
Substrate length (l_s)	50	Silicon dioxide (SiO_2)
Feed-line length (l_f)	26.5	Silver (Ag)
Feed-line width (w_f)	3	Silver (Ag)
Radius (r)	12	Silicon (Si)
Ground height (h_g)	0.035	Silicon (Si)
Substrate-1 height (h_1)	1.6	Silicon dioxide (SiO_2)
Substrate-2 height (h_2)	0.2	Silicon dioxide (SiO_2)
Substrate-3 height (h_3)	0.1	Silicon dioxide (SiO_2)
Dielectric resonator (h)	12	Silicon (Si)

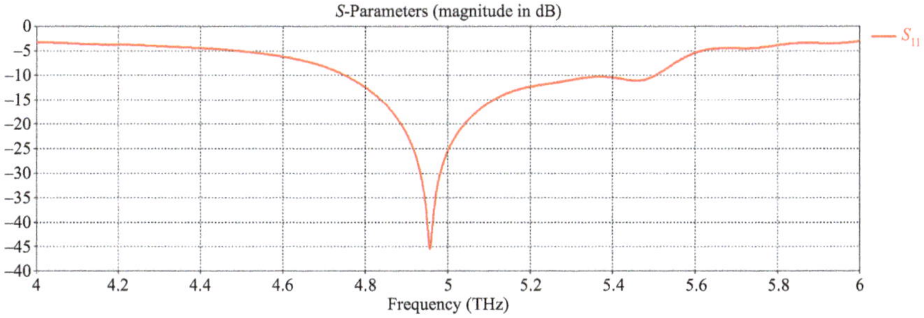

Simulated S_{11} parameter at 4.92 THz

Appendix I

DRA designing process

I.1 Design process of aperture coupled DRA

Dielectric resonator antenna (DRA) has low-loss and high-radiation efficiency. A DRA can excite multiple modes simultaneously using aspect-ratio between 0.5 and 2.5. They are most stable due to dielectric materials. DRA material is available in different dielectric constant values, i.e. 10–100. They can generate TE/TM/TEM modes. They are best candidates for getting multimode and in turn multiband antennas. These are basic requirements for any vehicular wireless communication system. An antenna designer can build an antenna of his choice of operating frequencies.

Let resonant frequency be 3.7 GHz, resonant mode be TE_{111} (fundamental mode), impedance 50 Ω and substrate used $FR4$, $\epsilon_r = 4.4$, height $= 0.8$ mm. λ_o is the frequency space wavelength, and λ_g is the guided wavelength.

Standard formulas used are

$$\begin{aligned} \lambda_o &= \frac{C}{f_o} \\ \lambda_g &= \frac{\lambda_o}{\sqrt{\epsilon_r \text{ effective}}} \end{aligned} \tag{I.1}$$

Microstrip line dimensions (width and length) can be computed as

$$\begin{aligned} B &= \frac{377\pi}{2\, z_o \sqrt{\varepsilon_r}}; \\ A &= \frac{20}{60}\sqrt{\frac{\varepsilon_r + 1}{2}} + \frac{\varepsilon_r - 1}{\varepsilon_r + 1}\left(0.23 + \frac{0.11}{\varepsilon_r}\right); \\ \frac{w}{d} &= \left\{\frac{8\, e^A}{e^{2A} - 2}\right\} \cdot \frac{2}{\pi}[B - 1 - \ln[2B - 1]] + \frac{\varepsilon_r - 1}{2\, \varepsilon_r}\left[\ln[B - 1] + 0.39 - \frac{0.61}{\varepsilon_r}\right]; \\ Z_o &= 50\ \Omega \end{aligned}$$

where w is the width of microstrip line, d is the thickness of FR-4 (0.8 mm).

Width obtained for microstrip line where the length of microstrip line characteristic $= \lambda_o$.

$\text{Stub} = \dfrac{\lambda_g}{8}$

$L_s = \text{slot} = 0.4\ \lambda_g$

$W_s = \text{slot} = 0.2\ L_s\ (0.2 \times 0.4\ \lambda_g)$

$\frac{W}{d} > 2$ (aspect-ratio mode dependent)

Ground plane = 4 times of length
Infinite ground plane = 4 times of width
Finite ground plane $6hhL$ (design DRA)
$6h + W$ (design DRA)

Let us now use 3.7 GHz as design frequency microstrip line.

$$B = \frac{377\pi}{2\,z_o\sqrt{\varepsilon_r}} = \frac{377 \times 3.14}{2 \times 50 \times \sqrt{4.4}} = 5.64$$

$$A = \frac{Z_o}{60}\sqrt{\frac{\varepsilon_r + 1}{2}} + \frac{\varepsilon_r - 1}{\varepsilon_r + 1}\left(0.23 + \frac{0.11}{\varepsilon_r}\right)$$

$$= \frac{50}{60}\sqrt{\frac{4.4 + 1}{2}} + \frac{4.4 - 1}{4.4 + 1}\left(0.23 + \frac{0.11}{4.4}\right) = 1.53$$

$$\frac{W}{d} = \left\{\frac{8e^A}{e^{2A} - 2}\right\} \cdot \frac{2}{\pi}\,[B - 1 - \ln[2B - 1]] + \frac{\varepsilon_r - 1}{2\,\varepsilon_r}\left[\ln(B - 1) + 0.39 - \frac{0.61}{\varepsilon_r}\right]$$

$$= \left\{\frac{8\,e^{1.53}}{e^{2(1.53)} - 2}\right\} \cdot \frac{2}{\pi}\,[5.64 - 1 - \ln[2(5.64) - 1]]$$

$$+ \frac{4.4 - 1}{2(4.4)}\left[\ln[5.64 - 1] + 0.39 - \frac{0.61}{4.4}\right]$$

$$= (1.218)[4.64 - 2.33] + (0.386)[1.786] \Rightarrow 3.5025$$

$$\frac{W}{d} = 3.5025,\ w = 3.5025 \times 0.8\ \text{mm}$$

$$w = 2.8\ \text{mm}$$

Stub:

$$f_r = 3.7\ \text{GHz},\ \lambda_o = 81\ \text{mm} = \frac{C}{f_r}$$

$$\lambda_o = \frac{3 \times 10^8 \times 10^3}{3.7 \times 10^9} = 81\ \text{mm}$$

$$\Rightarrow \lambda_g = \frac{\lambda_o}{\sqrt{\epsilon_r\ \text{effective}}} = \frac{81}{\sqrt{8.6}} = 27.62\ \text{mm}\left[\varepsilon_r\text{effective} = \frac{4.4 + 12.8}{2} = 8.6\right]$$

$$\Rightarrow \text{Length of stub} = \frac{\lambda_g}{8} = \frac{27.62}{8} = 3.4525\ \text{mm}$$

Slot:

$$L_s = 0.4\,\lambda_g = 0.4 \times 27.62 = 11.048 \text{ mm}$$
$$W_s = 0.2\,l_s = 0.2 \times 11.048 = 2.2096 \text{ mm}$$

DRA dimensions: (TE_{111} fundamental mode)
Formula used for rectangular DRA designing,

$$f_{mn\ell} = \frac{ck_{mn\ell}}{2\pi\sqrt{\mu_r \in_r}} = \frac{c}{2\pi\sqrt{\mu_r \in_r}}\sqrt{\left(\frac{m\pi}{a}\right)^2 + \left(\frac{n\pi}{b}\right)^2 + \left(\frac{\ell\pi}{d}\right)^2}.$$

$$a = 20 \text{ mm}$$
$$b = 20 \text{ mm}$$
$$d = 14 \text{ mm}$$

$$m = n = 1 = 1; \quad \frac{a}{b} = 0.5\text{–}2.5$$

$$\text{fix } d = 14 \text{ mm}; \quad \frac{b}{d} = 0.5\text{–}2.5$$

$$Q = \frac{f_r}{f_h - f_l} = \frac{f_r}{\text{bandwidth}}$$

Appendix J
DRA design case study

List of equipment used in terahertz dielectric resonator antenna (DRA) and microwave DRA design and testing are as follows:

A vector network analyzer comes with calibration kit, signal generator, spectrum analyzer, power meter, noise figure analyzer, cables and connectors, anechoic chamber, antenna test system, material test system, power source and power extenders.

Resonant frequency formulations used in rectangular DRA design (Figures J.1–J.14):

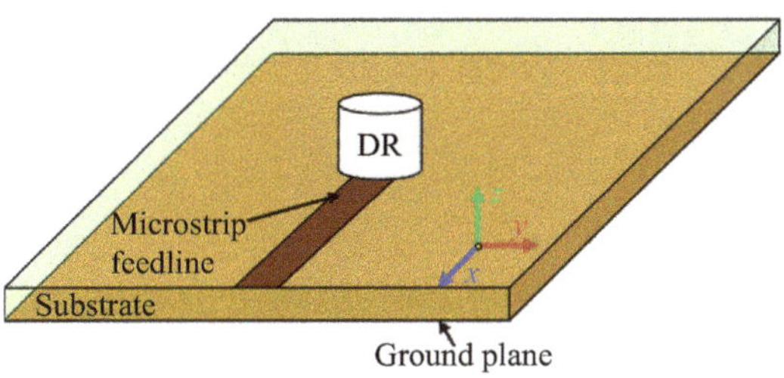

Figure J.1 Cylindrical DRA

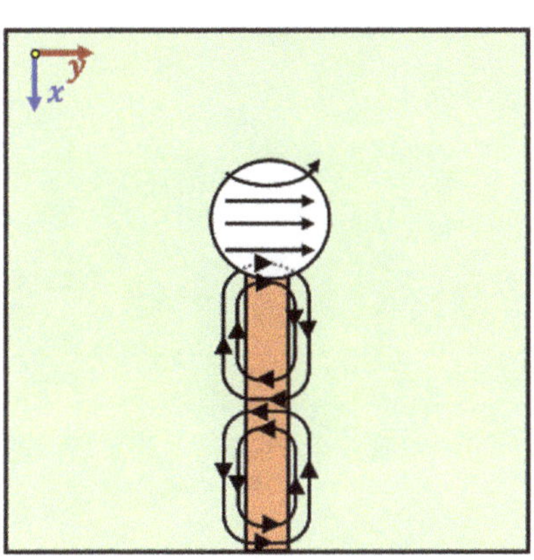

Figure J.2 CDRA with fields

$$\epsilon_r k_0^2 = k_x^2 + k_y^2 + k_z^2$$
$$k_0^2 = \omega_0^2 \mu_0 \epsilon_0$$

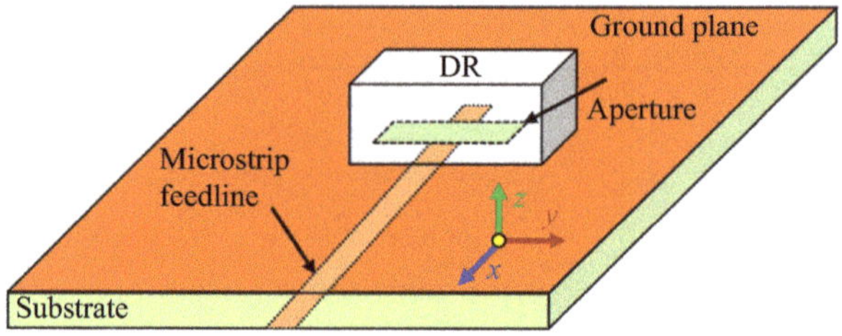

Figure J.3 Rectangular DRA with feed slot and stub components

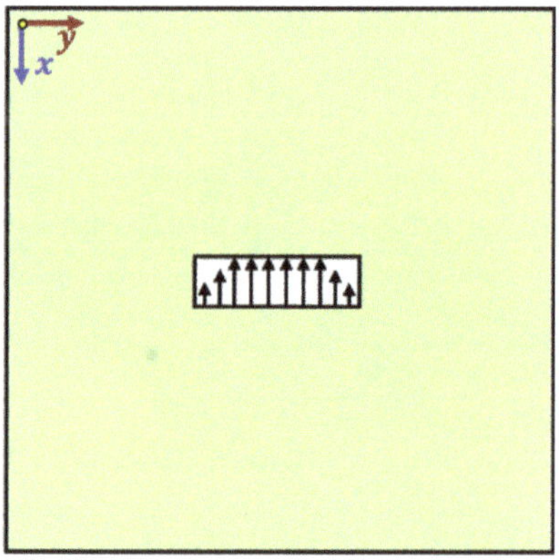

Figure J.4 DRA E-field dipole formation

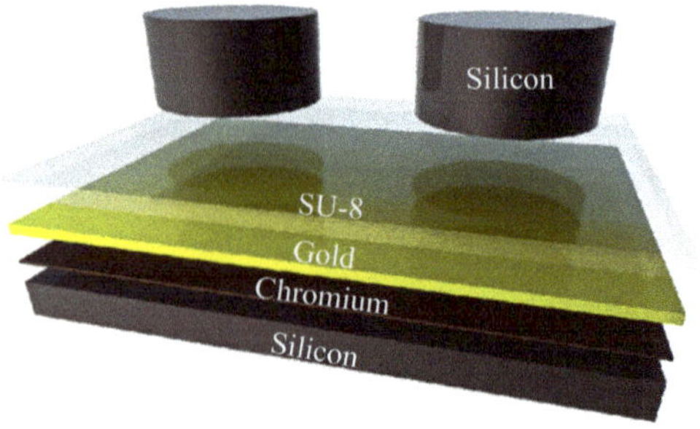

Figure J.5 Terahertz cylindrical DRA

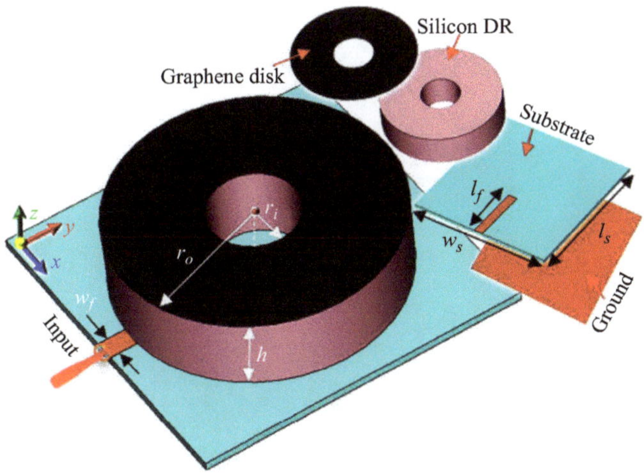

Figure J.6 Terahertz ring DRA

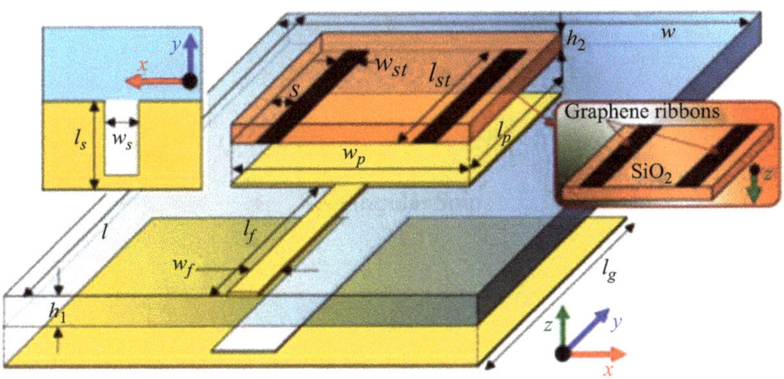

Figure J.7 Terahertz graphene DRA

Thus, the resonant frequency can be calculated and analyzed using k_0. For an isolated rectangular DR in free space, the resonant frequency can be determined as given in the following equation (Figures J.15–J.19):

$$(f)_r m,n,p = \frac{c}{2\pi\sqrt{\mu\epsilon}}\sqrt{\left[\left(\frac{m\pi}{a}\right)^2+\left(\frac{n\pi}{b}\right)^2+\left(\frac{p\pi}{h}\right)^2\right]}$$

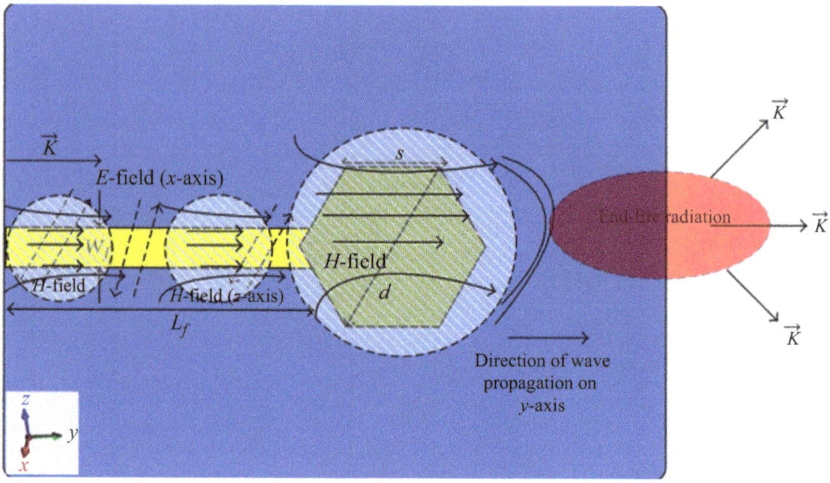

Figure J.8 Terahertz fields vectors

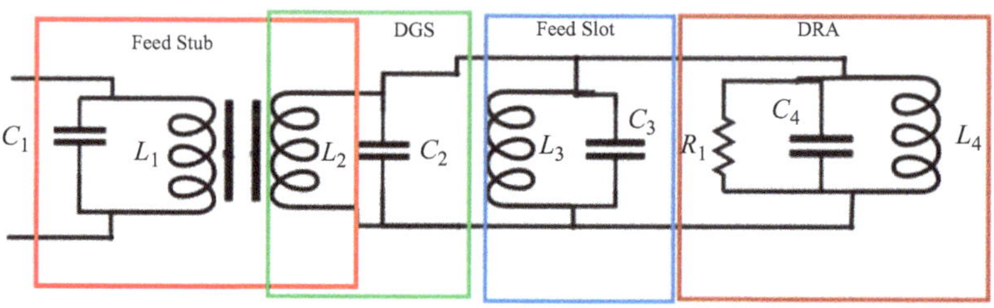

Figure J.9 Equivalent DRA circuit with R, L, C representation

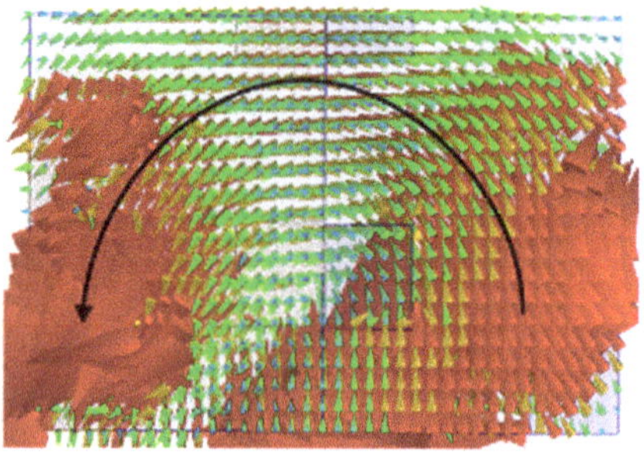

Figure J.10 DRA exciting resonant mode

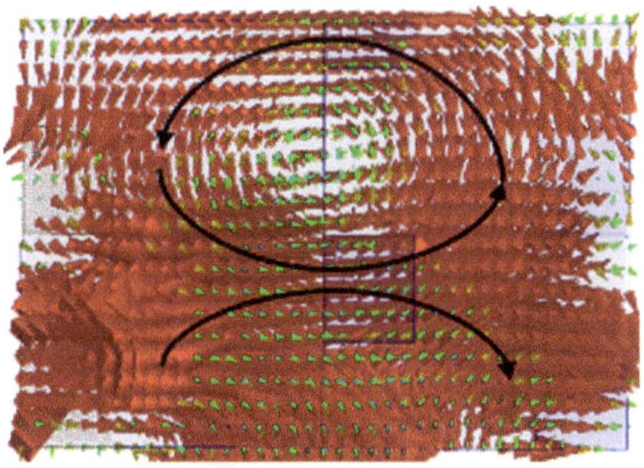

Figure J.11 DRA exciting higher order modes

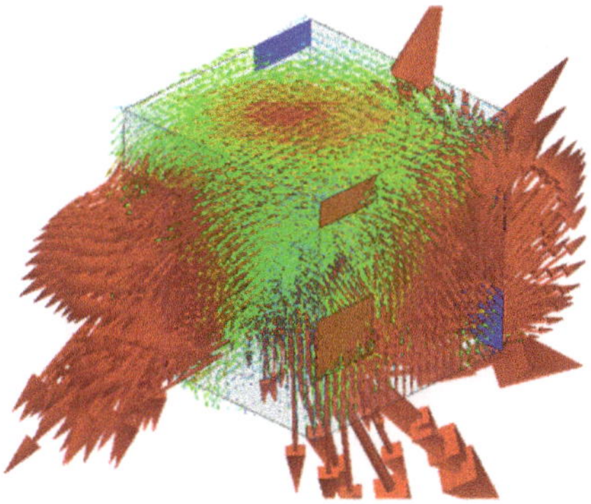

Figure J.12 Rectangular DRA fields

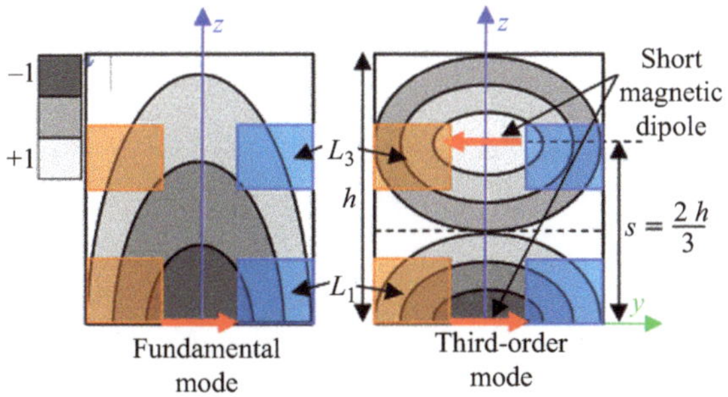

Figure J.13 Resonant modes field diagrams

Cylindrical DRA resonant frequency formulation:

$$f = \frac{6.324c}{2\pi d\sqrt{\varepsilon_r + 2}}\left[0.27 + 0.36\left(\frac{d}{2h}\right) + 0.002\left(\frac{d}{2h}\right)^2\right]$$

Figure J.14 Rectangular DRAs with metal strips for exciting higher order modes

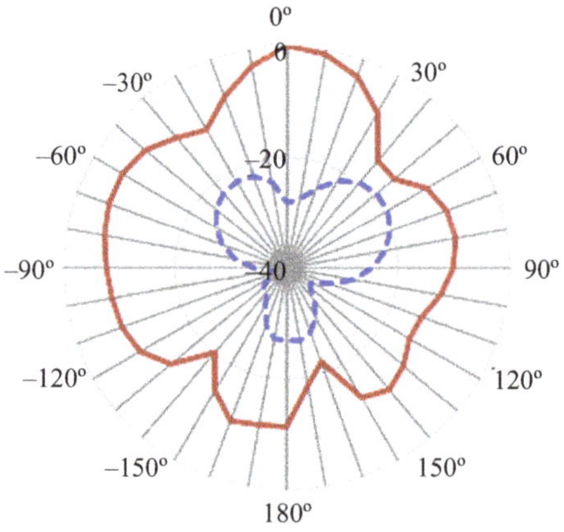

Figure J.15 Radiation pattern

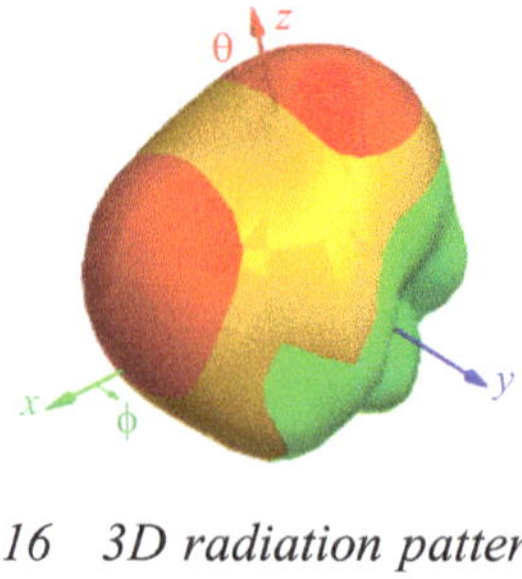

Figure J.16 3D radiation pattern of DRA

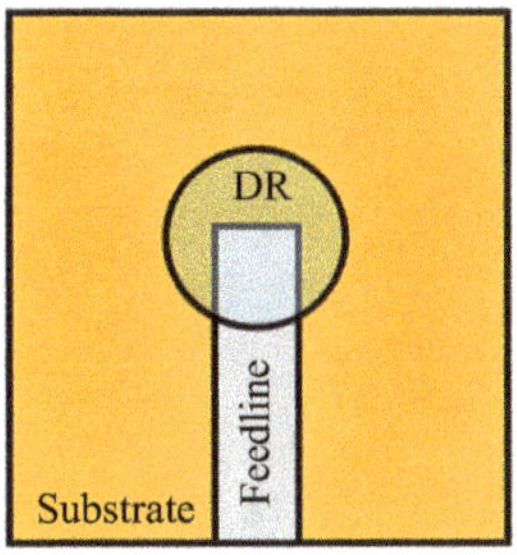

Figure J.17 Terahertz DRA with feed

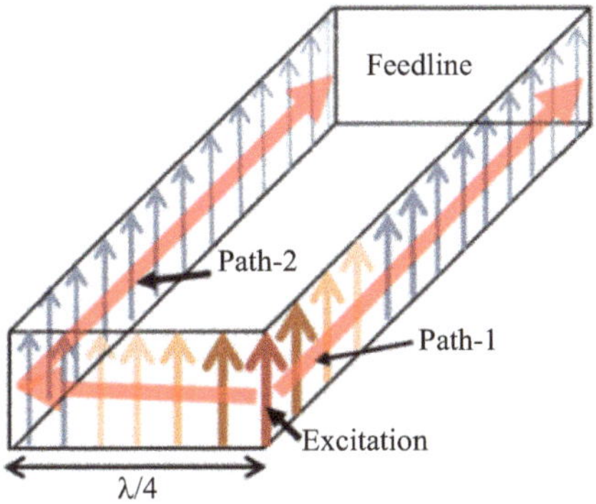

Figure J.18 Terahertz DRA feed

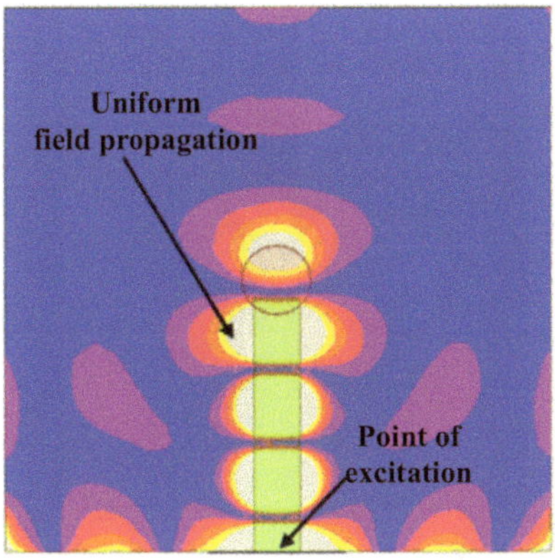

Figure J.19 Terahertz DRA fields

Appendix K

Vector network analyzer process for calibration

Used case 40-GHz VNA (Figure K.1).

1. Switch on VNA.
2. Click on calibration.
3. Click on calibration wizard.
4. Set frequency.
5. Click on next.
6. Select on port.
7. Starting frequency range and stop frequency.
8. View calibration/kit.
9. Calibration response Port 1/Port 2.
10. Set the calibration kit to 85052B (3.5 mm).
11. Apply the open, short and broadband at the load side.
12. First connect open then choose a male/female connector option. For an antenna as one-port select female option.
13. Second connect short to load and similarly choose male/female.
14. Third select broadband and select female for antenna as one port.
15. Now remove the broadband and connect the antenna for the measurement.
16. Check S parameters with the help of marker.
17. Data (for .csv) and graphs (.jpg).
18. For a two-port device, it will be a female option.
19. Repeat the previous points 4–6.
20. Rest all the same as was done for a single-port option.
21. Draw an equivalent circuit with CST software and compare results with MATLAB®.
22. In a post-processing window of CST, there is a tab with name network analysis and by using this tab, you can transform your desired model into lump components by mentioning the order of cascaded network along with specific frequency.
23. Just import the measured S parameters (S1p file) from the VNA into Design Studio canvas of CST software. Then create the lumped circuit model and optimize R, L, C values. S parameters are the same (minimizing the difference S_{11}).
24. The same can be done with ADS, Microwave Office and even MATLAB.

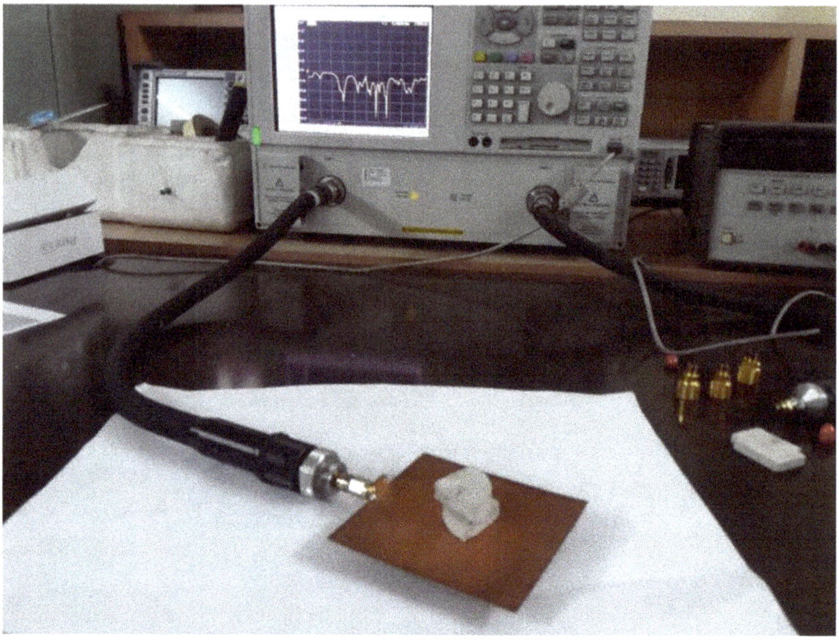

Figure K.1 VNA for DRA measurements

25. Mostly, the *RLC* approach will work for narrow band antennas.
26. You get the s-parameters, exported from your VNA.
27. On the Design Studio you ask for the item "Touchstone file block," which is under the class "Data Import." There you will point the path to your .s1p/.s2p file, which will show up as a box with one or two ports, depending on your measurement. It can then be connected to other circuit elements on your canvas—an *RLC* circuit model for return loss by MATLAB.
28. You can draw the S_{11} on the Smith chart from result in project tree and can determine the imaginary and real parts; then you can predict the circuit.
29. Drawing mode chart graph and also finding the coupling matrix of a coupled resonator filter using ADS.
30. Theoretically, you need to approximate the antenna in terms of *R*, *C* and *L*.
31. The input impedance of an antenna can be obtained from S_{11} since $S_{11} = (Z_L - Z_O)/(Z_L + Z_O)$.
32. For S_{11} magnitude and phase and given Z_O, one can calculate Z_L.
33. By changing the frequency, one can get Z_L as a function of frequency.
34. An equivalent circuit that is normally a parallel *R*, *L*, *C* **circuit shunted by the far field.**
35. **SPP**–nano-DRA–light consists of an electromagnetic wave that oscillates the electrons of metal.
36. Once electrons have deviated from their mean position, they oscillate back and forth.
37. When the frequency of incident wave matches with the free oscillation of metal, it generates local surface plasmon.

Glossary

DR	dielectric resonator
Q-Factor	quality factor
TE mode	transverse electric mode
TM mode	transverse magnetic mode
DRA	dielectric resonator antenna
HEM mode	hybrid electromagnetic mode
NDRA	nano dielectric resonator antenna
VSWR	voltage standing wave ratio
DWM	dielectric waveguide model
CP	circularly polarized
MIMO	multi-input–multi-output
AR	axial-ratio
FSS	frequency selective surfaces
EBG	electromagnetic band-gap
ECC	envelope correlation coefficient
TARC	total active reflection coefficient
DG	diversity gain
MEG	mean effective gain
CCL	channel capacity loss
DGS	defected ground structures
AMC	artificial magnetic conductor
SEM	scanning electron microscope
LHCP	left-hand circularly polarized
RHCP	right-hand circularly polarized
LP	linearly polarized

Index

www.ingramcontent.com/pod-product-compliance
Lightning Source LLC
LaVergne TN
LVHW051048210926
848486LV00001B/49

* 9 7 8 1 8 3 9 5 3 3 5 5 6 *